SUBUNIT ENZYMES

Biochemistry and Functions

ENZYMOLOGY

A Series of Textbooks and Monographs

Volume 1. Immobilized Enzymes, Antigens, Antibodies,
and Peptides: Preparation and Characterization,
edited by Howard H. Weetall

Volume 2. Subunit Enzymes:
Biochemistry and Functions,
edited by Kurt E. Ebner

In Preparation

Enzyme Structure and Function, *by S. Blackburn*
Enzyme Reactions and Enzyme Systems,
By Charles Walter

Additional Volumes in Preparation

SUBUNIT ENZYMES

Biochemistry and Functions

edited by Kurt E. Ebner

Department of Biochemistry
Oklahoma State University
Stillwater, Oklahoma

MARCEL DEKKER, INC.   New York

MARCEL DEKKER, INC.
270 Madison Avenue, New York, New York 10016

LIBRARY OF CONGRESS CATALOG CARD NUMBER: 74-29697

ISBN: 0-8247-6280-0

Current printing (last digit) :
10 9 8 7 6 5 4 3 2 1

PRINTED IN THE UNITED STATES OF AMERICA

PREFACE

The subunit character of a large number of proteins, including enzymes, has been well established during the past decade. It is also evident that the degree of association or interaction of subunits in proteins is highly variable. Examples occur from those wherein the interaction of subunits is so tight that the subunits act in concert to those wherein the association is so loose that it may be considered transitory.

The degree of subunit association can greatly influence the biological or catalytic activity of proteins and, in certain cases, changes in the degree of associate may be influenced by ligands such as substrates, products, inhibitors, and effector molecules. The activity of a catalytic site is dependent upon its unique three-dimensional structure, which in turn may be influenced by subunit interactions mediated by various ligands. The concept of induction of conformational changes via subunits interaction is also well established, and it is these changes that may depress or enhance catalytic activity.

An understanding of the nature and types of subunit interactions in proteins has led to a clearer understanding of metabolic regulation in a number of diverse systems and in particular a better understanding of the regulation of apparently unrelated pathways. The manner in which enzymes are regulated by subunit interactions is indeed varied and the consequences of such interactions are not totally predictable as evidenced by the examples presented in this text. An increasingly large number of enzymes are being studied with respect to subunit structure and the relationship of subunit interaction to enzymatic activity. A number of these enzymes are becoming well documented and examinations of their properties and modes of action can serve as guidelines for other enzymes containing subunits.

The purpose of this book is to examine in as much detail as possible examples of different types of subunit interactions that may exist in proteins and which can be considered representative of other less well-studied systems. The first chapter presents a theoretical basis and examines models for subunit interactions and their consequences in simpler systems. The other chapters give examples and details of enzymes of varying complexity. Glutamine synthetase is characterized by its interaction with a large number of ligands, many of which have independent but additive effects. In UDP-galactose-4-epimerase, the interactions are more subtle and less pronounced. The lactose synthetase system is representative of a case where two proteins are normally dissociated and where one protein acts as a modifier of the other protein. The steroid protein carrier protein is representative of a protein that acts as a unique carrier system, and acetyl CoA carboxylase and aspartate transcarbamylase are representative of proteins that have interactions and properties of an intermediate degree of complexity.

The proteins reviewed in this book are intended to be representative of a number of different types of subunit interactions which occur in proteins. The examples are not inclusive, but examinations of their properties should serve as guidelines for studying other systems.

CONTRIBUTORS TO THIS VOLUME

Irving P. Crawford, Microbiology Department, Scripps Clinic and Research Foundation, La Jolla, California

Robert A. Darrow, Charles F. Kettering Research Laboratory, Yellow Springs, Ohio

Mary E. Dempsey, Department of Biochemistry, University of Minnesota Medical School, Minneapolis, Minnesota

Kurt E. Ebner, Department of Biochemistry, Oklahoma State University, Stillwater, Oklahoma

Othmar Gabriel, Biochemical Research Laboratory, Harvard Medical School and the Massachusetts General Hospital, Boston, Massachusetts

A. Ginsburg, Laboratory of Biochemistry, National Heart and Lung Institute, National Institutes of Health, Bethesda, Maryland

Herman M. Kalckar, Biochemical Research Laboratory, Harvard Medical School and the Massachusetts General Hospital, Boston, Massachusetts

M. Daniel Lane, Department of Physiological Chemistry, The Johns Hopkins University, School of Medicine, Baltimore, Maryland

Alexander Levitzki, Department of Biophysics, The Weizmann Institute of Science, Rehovot, Israel

Steve C. Magee, Department of Biochemistry, Oklahoma State University, Stillwater, Oklahoma

Joel Moss, Department of Medicine, The Johns Hopkins University, School of Medicine, Baltimore, Maryland

S. Efthimios Polakis, Department of Physiological Chemistry, The Johns
    Hopkins University, School of Medicine, Baltimore, Maryland

E. R. Stadtman, Laboratory of Biochemistry, National Heart and Lung
    Institute, National Institutes of Health, Bethesda, Maryland

CONTENTS

Preface   iii

Contributors   v

1. SUBUNIT INTERACTIONS IN PROTEINS

Alexander Levitzki   1

    I.   Introduction   2
    II.   Basic Concepts of Allosteric Control   3
    III. The Structure of Multisubunit Proteins   4
    IV.   Cooperativity in Multisubunit Proteins — The Basic
          Concepts   8
    V.   Molecular Models for Cooperativity   13
    VI.   Protein Conformation — Equations of State   27
    VII. The Energetics of Subunit Interactions   28
    VIII. Association-Dissociation Phenomena   31
    IX.   Protein Design and Subunit Interactions   31
    X.   A Few Examples   34
    XI.   Concluding Remarks   39
          Acknowledgement   39
          References   39

2. GLUTAMINE SYNTHETASE OF ESCHERICHIA COLI:
    STRUCTURE AND REGULATION

A. Ginsburg and E. R. Stadtman   43

    I.   Introduction   44
    II.   Physical and Chemical Properties   45

    III.   Regulation by Enzyme Catalyzed Adenylylation and
           Deadenylylation   58

    IV.   Subunit Interactions   62

     V.   Concluding Remarks   78

         References   81

3.   UDP-GALACTOSE-4-EPIMERASE

   Othmar Gabriel, Herman M. Kalckar, and Robert A. Darrow   85

     I.   Introduction   86

    II.   General Molecular Properties of UDP-Galactose-4-
          Epimerase and Its Reaction Mechanism   87

    III.   UDP-Galactose-4-Epimerase from Yeast, Subunits,
          and Prosthetic Groups   90

    IV.   UDP-Galactose-4-Epimerase from E. coli   105

     V.   UDP-Galactose-4-Epimerase from Other Sources   120

    VI.   Overall View of Reactions Catalyzed by UDP-
          Galactose-4-Epimerase and Relations to Other
          Sugar Transformations   122

         Acknowledgements   128

         Addendum   128

         References   131

4.   LACTOSE SYNTHETASE: α-LACTALBUMIN AND β-(1 → 4)
   GALACTOSYLTRANSFERASE

   Kurt E. Ebner and Steve C. Magee   137

     I.   Introduction   138

    II.   Lactose Synthetase   139

    III.   Structural Similarities Between α-Lactalbumin
          and Lysozyme   142

    IV.   Evolutionary Relationship Between α-Lactalbumin
          and Lysozyme   148

     V. β-(1 → 4) Galactosyltransferase   149

    VI.   Interaction Between α-Lactalbumin and Galactosyl-
          transferase   167

VII.   Biological Significance   170

       Acknowledgements   173

       Addendum   173

       References   174

5.   ACETYL COENZYME A CARBOXYLASE

M. Daniel Lane, S. Efthimios Polakis, and Joel Moss   181

     I.   Introduction   182

     II.   The Acetyl CoA Carboxylation Reaction   183

     III. The Escherichia coli Acetyl CoA Carboxylase System 189

     IV.   Animal Acetyl CoA Carboxylases   193

     V.   Yeast and Plant Acetyl CoA Carboxylases   214

           Acknowledgement   215

           References   215

6.   TRYPTOPHAN SYNTHETASE

Irving P. Crawford   223

     I.   Introduction   224

     II.   Physical and Immunological Properties of the Enteric
           Bacterial Enzyme and Its Subunits   227

     III. Modification of Specific Amino Acid Residues of the
           Enteric Bacterial Enzyme   230

     IV.   Reaction Mechanism   234

     V.   Regulation of Enzyme Synthesis   235

     VI.   Evolutionary Relationships   241

     VII. Recent Developments   258

           References   259

7.   SQUALENE AND STEROL CARRIER PROTEINS

Mary E. Dempsey   267

     I.   Introduction   268

     II.   Discovery of Squalene and Sterol Carrier Protein   270

III.   Structural Characteristics of Squalene and Sterol Car-
          rier Protein   274

IV.   Functions of Squalene and Sterol Carrier Protein   282

V.   Ubiquitous Occurrence of Squalene and Sterol
          Carrier Protein   296

VI.   Regulatory Role of Squalene and Sterol Carrier Protein   297

VII.   Concluding Remarks   301

       Acknowledgements   302

       References   303

AUTHOR INDEX   307

SUBJECT INDEX   329

SUBUNIT ENZYMES

Biochemistry and Functions

Chapter 1

SUBUNIT INTERACTIONS IN PROTEINS

Alexander Levitzki

Department of Biophysics
The Weizmann Institute of Science
Rehovot, Israel

Department of Biological Chemistry
Institute of Life Science
The Hebrew University of Jerusalem
Jerusalem, Israel

I.	INTRODUCTION	2
II.	BASIC CONCEPTS OF ALLOSTERIC CONTROL	3
III.	THE STRUCTURE OF MULTISUBUNIT PROTEINS	4
	A. Other Types of Protein Assemblies	6
IV.	COOPERATIVITY IN MULTISUBUNIT PROTEINS — THE BASIC CONCEPTS	8
	A. The Hill Coefficient at 50% Saturation	10
	B. The Allosteric Dimer	10
V.	MOLECULAR MODELS FOR COOPERATIVITY	13
	A. The Monod-Wyman-Changeux Concerted Model	14
	B. The Koshland-Némethy-Filmer Sequential Model	19
	C. Diagnostic Tests Using Binding Curves	23
	D. Comparison of the Models	26
VI.	PROTEIN CONFORMATION — EQUATIONS OF STATE	27
VII.	THE ENERGETICS OF SUBUNIT INTERACTIONS	28
VIII.	ASSOCIATION-DISSOCIATION PHENOMENA	31
IX.	PROTEIN DESIGN AND SUBUNIT INTERACTIONS	31

X.   A FEW EXAMPLES   34
     A. Hemoglobin — The Mechanism of Cooperativity   34
     B. CTP Synthetase — The Role of an Allosteric Effector   36
     C. Rabbit Muscle Glyceraldehyde-Phosphate Dehydrogenase —
         Negative Cooperativity   37
     D. Yeast Glyceraldehyde-3-Phosphate Dehydrogenase   38

XI. CONCLUDING REMARKS   39

I. INTRODUCTION

The study of the mechanisms which control cellular processes has become of prime importance in recent years. The control mechanisms may be divided into three categories: (1) genetic control, that is the control of gene expression; (2) hormonal control of cell function; and (3) allosteric control of regulatory enzymes.

These three types of control are involved in the entire spectrum of life processes encountered in a living organism. Genetic control refers to the switching on or switching off of genes; hormonal control refers to the switching on or off of metabolic processes, usually via the second messenger cAMP; allosteric control refers to the regulation of enzyme activity via cooperative protein-ligand interactions and via effector ligands which either switch on or off individual enzymes. Allosteric control mechanisms may also apply to membranes in which properties such as permeability are activated by membrane-ligand interactions.

In virtually all the systems described above, the controlling event in the process involves the interaction of a ligand with a specific receptor, usually a macromolecule. This statement can be illustrated by the following examples:

1. Genetic Control — the expression of the lac operon depends on the interaction of the β-galactoside inducer with the lac repressor bound to a specific site on the DNA. Similarly, the attachment of histidine to the His-tRNA converts the latter to a potent repressor for the histidine operon. The small ligand histidine functions as a co-repressor which, upon covalent binding to its specific tRNA, increases the affinity of the latter toward DNA at the histidine region.

2. Hormonal Control — the interaction of adrenaline, glucagon, or a variety of other hormones with specific membrane receptors switches on adenylate cyclase which produces cAMP. This in turn initiates various other processes, such as glycogenolysis, depending on the target tissue.

3. Allosteric Control — the interaction of CTP with ATCase inhibits the activity of ATCase; similarly, the interaction of GTP with CTP synthetase activates CTP synthetase.

These three types of control mechanisms have one basic property in common — the interaction of a ligand with a specific receptor which triggers a certain biochemical event. It is generally accepted that conformational changes induced by these ligand-receptor interactions are transmitted specifically to the macromolecule or the subunit which is responsible for the biochemical function. This conformational change either switches on or switches off the biological activity executed by the target macromolecule. This is generally known as subunit interaction and is the basis for the wide spectrum of regulatory mechanisms found in vivo. It is the molecular nature of these interactions which will be discussed in some detail in this chapter. The regulatory processes involving subunit interactions are best understood in soluble regulatory enzymes and are less well understood in systems which involve membrane-bound receptors or in systems which involve protein-nucleic acid interactions. We will, therefore, limit this discussion to the analysis of subunit interactions in regulatory enzymes and other multisubunit proteins. The understanding of subunit interactions in regulatory proteins may become the key to the understanding of more complex regulatory phenomena.

II. BASIC CONCEPTS OF ALLOSTERIC CONTROL

In the late 1950s, a number of workers discovered that in bacteria metabolic pathways which lead to the synthesis of essential metabolites are subject to feedback (or end-product) inhibition [1-3]. It was established that, in many metabolic pathways, the terminal metabolite in the pathway functions as a specific inhibitor of the first enzyme in the pathway. Enzymological studies on a number of metabolic pathways revealed that the end product which is chemically distinct from the substrates of the initial enzyme in the pathway inhibits the enzyme's activity by binding to a site distinct from its active site. Since this feedback inhibitor is not isosteric with the substrate, the term allosteric effector was coined [4]. It was established that the allosteric effector interacts with a specific allosteric site on the enzyme which is topographically distinct from the active site. The binding of the allosteric effector to the allosteric site brings about the allosteric transition which consists of a specific conformational change at the active site (and other areas of the protein molecule), thus modulating its activity. It was very quickly realized that allosteric effectors are not necessarily inhibitors. They may also function as activators. Monod et al. [4], in their classical paper, give the example of phosphorylase b activation by 5'AMP. In this case, 5'AMP functions as a positive effector, switching on the phosphorylase reaction. It was, therefore, clear that allosteric effectors may be either negative effectors or positive effectors, depending on whether they inhibit or stimulate the reaction in question. The essence of allosteric effects involves the interaction between two classes of sites: the active site and the regulatory site (allosteric site). Site-site interactions are not limited to heterologous sites but may also occur between identical (homologous) sites. This can be illustrated by using hemoglobin as an example. The sigmoidal nature of the oxygen binding curve was noted long ago. This phenomenon was explained by stating that the oxygen binding sites in hemoglobin interact in a cooperative fashion such that the affinity of hemoglobin for oxygen increases as a function of hemoglobin saturation. Although, strictly speaking, hemoglobin is not an enzyme, the same phenomenon has been noted in many enzymes. In 1965 Monod et al. [5] stressed that many allosteric enzymes such as ATCase [6] and threonine deaminase [7] also possess this cooperative behavior toward the substrate molecule. Thus, ATCase binds aspartate cooperatively, and threonine deaminase [7] binds threonine cooperatively. Furthermore, in 1963 Monod et al. [4] noted that treatment of many allosteric enzymes with Hg^{2+} not only desensitizes those enzymes toward their respective allosteric effectors but also eliminates their substrate cooperative effect. In other words, the desensitized enzyme binds the substrate molecule in a noncooperative Michaelian fashion. This observation indicated that cooperative interactions between identical ligand sites (homologous sites) and the interactions between allosteric sites and active sites (heterologous sites) are linked functionally. In fact, both types of interactions are the two forms of a more general phenomenon: site-site interactions. When the interacting sites are identical, the interactions are termed homotropic (example: binding of oxygen to hemoglobin). When the interactions are among different types of sites, i.e., between active sites and regulatory sites, the interactions are termed heterotropic (example: the inhibition of ACTase by CTP). Many regulatory proteins display both homotropic and heterotropic interactions, as in the case of hemoglobin and ATCase.

Although allosteric proteins were originally defined as those proteins possessing regulatory sites distinct from their active sites, it is now generally accepted that every protein which exhibits site-site interactions is an allosteric protein, even when the only interacting sites are identical.

III. THE STRUCTURE OF MULTISUBUNIT PROTEINS

Regulatory enzymes are multisubunit structures. Structural analysis of regulatory enzymes has, in every single instance, revealed that the subunits interact with one another by way of specific noncovalent bonds. The subunits form well-defined geometrical structures, and their structure is intimately related to their mode of action. It will therefore be necessary for us to discuss the principles of design of oligomeric proteins in some detail.

The terminology used to describe the structure of multisubunit enzymes is as follows. A protein composed of a number of subunits is called an oligomeric protein, or an oligomer. The subunits building the oligomeric structure are referred to as protomers, monomers, or subunits. The subunits (protomers) are bound to one another at specific binding domains by noncovalent bonds. A subunit is composed of one polypeptide chain, although there are a limited number of cases in which each subunit is composed of more than one polypeptide chain. Some regulatory proteins are composed of identical subunits such as in glyceraldehyde-3-phosphate dehydrogenase, or from nonidentical subunits as in the case of hemoglobin. In the latter case the α- and β-subunits have identical functions. In other oligomeric proteins, such as ATCase and tryptophan synthetase, the two types of subunits are different and also have different functions.

Most oligomeric proteins are composed of a small number of subunits which form closed oligomeric structures that rarely contain more than 12 subunits. In certain cases, such as multienzyme complexes or spherical viruses, the number of protomers is much larger. In their classical paper, Monod, Wyman and Changeux [5] observed that the specificity of subunit recognition is so great that monomers of an oligomeric protein will associate exclusively with their identical partners even at high dilution and in the presence of other proteins. This principle has been verified by detailed renaturation studies on numerous oligomeric enzymes [8]. The existence of strong and specific noncovalent subunit interactions which form these geometrically defined aggregates indicates that subunits interact at specific binding domains. According to Monod, Wyman and Changeux [5], two modes of subunit interaction are possible:

(a) ISOLOGOUS   (b) HETEROLOGOUS
    ASSOCIATION   ASSOCIATION

FIG. 1. Modes of Subunit Association

1. Isologous Association — the binding domain is made of two identical binding sets (Fig. 1). As shown in Figure 1, there are two identical binding sets, each consisting of an "ab" contact. These binding domains are related to each other by a twofold rotational axis of symmetry.

2. Heterologous Association — the binding domain is made of two different binding sets (Fig. 1). In Figure 1, there are two binding sets, but they differ since one is a "bc" contact and the other is an "a" contact. The subunits are arranged in a "head to tail" arrangement.

The majority of proteins composed of subunits are either dimers or tetramers. The isologous mode of subunit interaction is the prevalent mode of aggregation in known protein dimers and tetramers. However, some heterologous tetramers such as tryptophanase [9] and pyruvate carboxylase [10] do seem to have cyclic symmetry (C_4). Trimers, pentamers, and hexamers, in which the mode of subunit aggregation is heterologous and which therefore possess cyclic symmetry, are also known [11].

A. Other Types of Protein Assemblies

The construction of oligomeric structures with more than four subunits requires the use of both heterologous and isologous subunit interactions. Thus, for example, an oligomer of point symmetry 2:3, such as the dodecameric glutamine synthetase [12], has three twofold axes and four threefold axes of rotational symmetry. In this case, the oligomer possesses one isologous binding set and two heterologous binding sets. Point group symmetry 432 refers to 24 subunits arranged in cubic symmetry, and the highest point group symmetry 532 relates to 60 subunits arranged in icosahedral symmetry, as is seen in the protein shells (capsids) of spherical viruses. Higher protein assemblies which involve closed geodesic dome-like arrangements are possible when one allows the subunits to arrange themselves in such a way that their environment is not always identical. This has been observed in large spherical viruses and has been termed quasi equivalence [13]. The complex assemblies are beyond the scope of this article and will not be discussed further. It should, however, be noted that some multienzyme complexes, such as pyruvate decarboxylase from bacteria and mammals, are arranged in the cubic point group symmetry [14].

As is shown in Figure 1, an isologous association of two subunits has a twofold axis of rotational symmetry in the binding domain. The isologous dimer can further associate to form an isologous tetrahedral structure (Fig. 2), in which each subunit is attached to three other subunits via

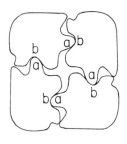 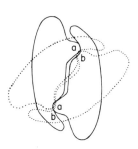

(a) HETEROLOGOUS TETRAMER   (b) ISOLOGOUS TETRAMER

FIG. 2. Heterologous and Isologous Tetramer Assemblies

three types of isologous binding domains. The tetrahedral structure thus
formed possesses three twofold axes of rotational symmetry. Since the
subunits are chiral themselves, no more elements of symmetry exist.
This is why ping-pong balls are not good models for protein subunits.
Indeed, all of the tetrameric enzymes composed of identical subunits
which have been thus far analyzed by x-ray crystallography possess what
is known as 2:2:2 point group symmetry, which means the existence of
three twofold axes of rotational symmetry, as described above.

In an isologous dimer, the binding sets composing the binding domain
are saturated within the dimer. The binding domains within a heterologous
dimer are not internally saturated and will therefore lead to continued aggregation, forming polydisperse "open" structures (Fig. 1). Closed
structures can be formed once the subunits form rings. This is exempli-
fied by the heterologous tetramer shown in Figure 2. In this heterologous
tetramer no element of symmetry exists within the subunit in binding
domains. The symmetry element in closed heterologous structures is a
cyclic rotational symmetry, $360°/n$, where n is the number of subunits
composing the closed heterologous structure; a heterologous tetramer
possesses a fourfold axis of rotational symmetry [4]. Thus, such subunit
associations will lead to ring structures in which the smallest closed heterologous structure is a trimer. Dimers would not be stable in a heterologous assembly. It follows therefore that, when stable dimers are
encountered as native species or as the dissociation products of a larger
protein assembly, they are isologous in nature.

In the following sections it will become apparent that a detailed knowledge of the enzyme's quaternary structure is essential for the rigorous
analysis of its interaction with ligands. The number of interacting

surfaces between subunits is determined by their three-dimensional organization in space, and any attempt to understand protein-ligand interactions
involves the explicit knowledge of their geometrical arrangement. A more
detailed discussion of the structure and symmetry of oligomeric enzymes
can be found elsewhere [11, 15].

IV. COOPERATIVITY IN MULTISUBUNIT PROTEINS —
THE BASIC CONCEPTS

The binding of oxygen to hemoglobin has intrigued scientists since 1904
when the sigmoidal nature of hemoglobin's oxygen binding curves was first
noted by Bohr [16]. During the following 70 years, hemoglobin has played
a key role in the different phases of understanding of allosteric phenomena.

In 1910 and 1913 Hill [17, 18] treated the oxygen binding as a single step
phenomenon. He suggested that hemoglobin is aggregates of hemoglobin
Hb_n, where n is a specific but, at that time, unknown integer. According
to Hill each Hb molecule possesses one oxygen binding site, namely one
iron moiety. The binding of a family of oxygen is therefore described by:

$$Hb_n + nO_2 \rightleftharpoons Hb_n(O_2)$$ (1)

$$K = \frac{[Hb_n(O_2)n]}{[Hb_n][O_2]^n}$$ (2)

Hill suggested plotting $\log [Hb_n(O_2)_n]/[Hb_n]$ versus $\log [O_2]$; this is
known as the Hill plot. The ratio $[Hb_n(O_2)_n]/[Hb_n]_{total}$ is the fraction
of hemoglobin oxygen binding sites occupied by oxygen. This fraction is
designated as $\overline{Y}$. It follows that the fraction of free hemoglobin sites is
$1-\overline{Y}$. By rewriting Eq. (2) one obtains:

$$K = \frac{\overline{Y}}{(1-\overline{Y})[O_2]^n}$$ (3)

and the Hill equation is given by the expression:

$$\log \frac{\overline{Y}}{1-\overline{Y}} = \log K + n \log [O_2].$$ (4)

According to Hill, a plot of $\log \frac{\overline{Y}}{1-\overline{Y}}$ versus the partial pressure of oxygen
should yield a straight line with a slope n, a term known as the Hill coef-
ficient, n_H. For oxygen binding to hemoglobin, Hill obtained the value 2.8
for n_H.

According to the scheme suggested by Hill, the Hill coefficient should originally have been a whole number. Hill explained his finding by stating: "In point of fact n does not turn out to be a whole number but this is due simply to the fact that aggregation is not into one particular type of molecule but rather into a whole series of different molecules so equation (1) is a rough mathematical expression for the sum of several similar quantities with n equal to 1, 2, 3, 4, and possibly higher integers."

In 1925 Adair [19] established that hemoglobin is a molecule containing four equivalent binding sites for oxygen and that the hemoglobin molecule does not dissociate in the absence of oxygen. Adair noted that the cooperative binding of oxygen was not accurately described by the Hill scheme, since Hill did not take into account the intermediate species $Hb(O_2)$, $Hb(O_2)_2$ and $Hb(O_2)_3$, but treated the oxygen binding as a single step reaction [Eq. (1)].

Adair therefore wrote the binding steps for oxygen as a series of equilibria:

$$Hb + O_2 \xrightleftharpoons{K_1} Hb(O_2) \tag{5}$$

$$Hb(O_2) + O_2 \xrightleftharpoons{K_2} Hb(O_2)_2 \tag{6}$$

$$Hb(O_2)_2 + O_2 \xrightleftharpoons{K_3} Hb(O_2)_3 \tag{7}$$

$$Hb(O_2)_3 + O_2 \xrightleftharpoons{K_4} Hb(O_2)_4 \tag{8}$$

Each step is characterized by an equilibrium association constant K_1, K_2, K_3 and K_4, respectively. The Adair equation is given by Eq. (9):

$$\overline{Y} = \frac{K_1[O_2] + 2K_1K_2[O_2]^2 + 3K_1K_2K_3[O_2]^3 + 4K_1K_2K_3K_4[O_2]^4}{4(1 + K_1[O_2] + K_1K_2[O_2]^2 + K_1K_2K_3[O_2]^3 + K_1K_2K_3K_4[O_2]^4)} \tag{9}$$

The mono-, di-, and tri-bound hemoglobins are all included in this equation. The Hill equation, based on Hill's simplified treatment predicted a straight-line relationship over all the substrate concentration range with a Hill coefficient (slope) of 4.0. As is well known, the Hill plot is straight only in the neighborhood of $\overline{Y} = 1/2$ and the n value is 2.8 and not 4.0. In view of Adair's finding that hemoglobin does not undergo aggregation or dissociation, Hill's original explanation for the coefficient's being 2.8 is not acceptable.

It is now obvious that the fact that the Hill plot is only linear within a limited ligand concentration range and that the Hill coefficient is frequently a fractional number is due in both cases to the fact that intermediate species exist and should be taken into account in any analysis of a binding process. These are intermediate species in terms of the degree of saturation, not in terms of aggregation. The Hill equation [Eq. (4)] is still applied as a very useful diagnostic tool for estimating the degree of cooperativity, in the analysis of protein-ligand interactions. The Hill coefficient is also a useful guide in estimating the minimal number of interacting binding sites. The Adair approach is a general one and does not assume any molecular mechanism, since the equilibria considered [Eq. (5) - (8)] are written on purely thermodynamic grounds. One should therefore stress that an Adair type of analysis is <u>always</u> valid and therefore extremely useful in the analysis of ligand binding to proteins. As will become apparent later, the molecular models developed to explain allosteric phenomena differ in the interpretation of the Adair binding constants.

A. The Hill Coefficient At 50% Saturation

From this discussion until now, it may appear that the Hill coefficient is a vague concept and should be used only as an auxiliary tool. This conclusion is, however, incorrect since the Hill coefficient can be expressed in terms of the intrinsic binding constants of the Adair equation. To demonstrate the possibility of obtaining an analytical expression for the Hill coefficient, I chose to analyze the binding of ligand to a protein dimer. The analytical expressions in the dimer case are useful since they combine simplicity on the one hand, and demonstrate the fundamental characteristics of cooperativity on the other. The Hill coefficient is usually measured at 50% saturation (n_H), a region easily accessible experimentally.

The correlations developed between the Hill coefficient and the Adair parameters enable one also to understand more fully the correct meaning of the Hill coefficient.

B. The Allosteric Dimer

Consider a dimer E interacting with the ligand S according to Eq. (10) and (11):

$$E + S \rightleftharpoons ES \tag{10}$$

$$ES + S \rightleftharpoons ES_2 \tag{11}$$

The concentrations of ES and ES_2 at equilibrium are given by Eq. (12) and (13):

$$[ES] = 2 K_1[E][S] \tag{12}$$

$$[ES_2] = 1/2 K_2[ES][S] = K_1 K_2[E][S]^2 \tag{13}$$

where K_1 and K_2 are the intrinsic (statistically corrected*) association constants. The saturation function will be given by Eq. (14) where N_S is the average number of ligand molecules bound per protein dimer. $\overline{Y}$ is the fractional saturation and is always N_S divided by the total number of binding sites (two in this case).

$$N_S = \frac{[ES] + 2[ES_2]}{[E] + [ES] + [ES_2]} \tag{14}$$

or

$$N_S = \frac{2K_1(S) + 2K_1K_2(S)^2}{1 + 2K_1(S) + K_1K_2(S)^2} \tag{15}$$

so that

$$\overline{Y} = \frac{K_1(S) + K_1K_2(S)^2}{1 + 2K_1(S) + K_1K_2(S)^2} \tag{16}$$

and

$$\frac{\overline{Y}}{1 - \overline{Y}} = \frac{K_1(S) + K_1K_2(S)^2}{1 + K_1(S)} \tag{17}$$

The Hill coefficient or Hill slope is therefore given by (18):

$$\frac{d \log \frac{\overline{Y}}{1 - \overline{Y}}}{d \log (S)} = n \ (\text{Hill slope}) \tag{18}$$

It is therefore clear that the Hill coefficient is not constant and is a function of [S]. This is, indeed, the case as has been observed in Hill plots for numerous proteins. In general, the Hill coefficient is measured at 50% saturation (where $\overline{Y} = 1/2$). It is therefore useful to obtain the explicit formula for the Hill coefficient at $\overline{Y} = 1/2$. One must first solve the value of $S_{0.5}$ (ligand concentration at 50% saturation)** in terms of K_1 and K_2 using Eq. (16) or (17). The value of $S_{0.5}$ is given by Eq. (19):

---

*It is always advisable to use intrinsic (statistically corrected) binding constants since they are directly related to the intrinsic affinity of the protein binding site toward the ligand. Also, molecular mechanisms deal actually with changes in <u>intrinsic</u> affinity.

**$S_{0.5}$ is a very useful term suggested by Koshland et al. [24]; it designates the ligand concentration required to achieve 50% saturation ($\overline{Y} = 1/2$). In cooperative cases $S_{0.5}$ is a complex function of the consecutive binding constants. In a noncooperative case $S_{0.5}$ equals the single dissociation constant or the Michaelis constant when enzyme kinetics is performed.

$$S_{0.5} = \frac{1}{\sqrt{K_1 K_2}} \tag{19}$$

After taking the derivative (18) according to Eq. (17) and inserting $S_{0.5}$ according to Eq. (19) into (18), one obtains the formula for the Hill coefficient at $\overline{y} = 1/2$.

$$n_H = \frac{2}{1 + \sqrt{\dfrac{K_1}{K_2}}} \tag{20}$$

It should now be clear that three situations are possible:

When $K_1 = K_2$, $n_H = 1$, noncooperativity (Michaelian binding). (21)

When $K_1 < K_2$, $n_H > 1$, positive cooperativity. (22)

When $K_1 > K_2$, $n_H < 1$, negative cooperativity. (23)

It is interesting to note that, using the Adair approach with no molecular assumptions, three types of ligand binding curves are immediately predicted (Fig. 3). Positive cooperativity ($n_H = 2$) results in a sigmoidal curve. Noncooperativity ($n_H = 1$) results in the classical Michaelian hyperbola. Negative cooperativity ($n_H = 1/2$) results in a flattened hyperbola.

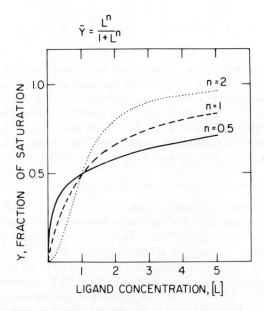

FIG. 3.   Positive, Negative, and Noncooperativity

V. MOLECULAR MODELS FOR COOPERATIVITY

In the development of molecular models to explain cooperativity, hemoglobin once again played an important role. The first attempts to describe
oxygen binding to hemoglobin in terms of site-site interaction between
oxygen binding sites was made by Pauling in 1935 [20]. Pauling used a
restricted Adair scheme by which he explained the cooperativity of oxygen
binding by a progressive increase in the affinity of hemoglobin for oxygen.
He postulated that this progressive increase in affinity for oxygen is brought
about by the fact that the heme groups interact with each other. By assuming that direct heme-heme interaction occurred, Pauling was able to mathe-
matically account for the cooperative oxygen binding.

It took many years to demonstrate that actually Pauling's model was
incorrect, since the heme groups are too far from each other to interact
directly [21]. Also, by using various spectroscopic techniques, it was
found that the structure of the protein moiety is changed, while the heme
groups themselves remain unaffected by ligation of oxygen [22]. Since
primary heme-heme interactions were excluded, it appeared that the
cooperative nature of oxygen binding results from interactions between the
protein subunits, thereby indirectly affecting the heme groups and their
affinity for oxygen. Attention was therefore focused on the nature of these
subunit interactions in hemoglobin.

Parallel to the developments in understanding hemoglobin, Monod et al.
[4, 5] and Gerhart and Pardee [6, 23] showed that cooperative binding of
a ligand is a general feature of the regulatory enzymes which occupy key
points in metabolic pathways. The idea that allosteric activation or inhibition and cooperative binding of ligands are two common properties of regulatory enzymes was formulated elegantly by Monod, Jacob and Changeux
in 1963 [4]. Subsequently, attempts were initiated to understand regulatory
processes in molecular terms. These attempts resulted in the development
of two molecular models which were proposed in the mid-1960s. Both the
"concerted model," suggested by Monod, Wyman, and Changeux [5], and
the "sequential model," suggested by Koshland, Némethy, and Filmer [24],
consider the coupling between ligand binding and conformational (tertiary
and quaternary) changes which occur in the protein molecule. The cooperative nature of ligand binding is, in both models, a function of the degree
of coupling between ligand binding and subunit interactions. To assess the
applicability of either of these models to any experimental system, certain
characteristics of the system must first be established. One must have a
clear understanding of the enzyme's quaternary structure, including the
number of subunits involved and their spatial arrangements. In addition,
one should determine the type of cooperativity in ligand binding and establish the fundamental parameters, such as the Hill coefficient and the Adair
constants. In the general discussion of the two models which follows, it

will become apparent that the two models differ in their molecular interpretation of the Hill coefficient and the Adair parameters.

A. The Monod-Wyman-Changeux Concerted Model

According to the Monod-Wyman-Changeux model [5], any allosteric enzyme exhibits the following properties:

1. The enzyme exists in two conformations, T and R, which are in equilibrium. The T conformation has the lower affinity for the ligand.
2. The change in conformation from T to R or vice versa occurs with conservation of symmetry. Thus, all subunits in the R-state or in the T-state have the same conformation and exhibit identical intrinsic affinities toward the ligand S (Fig. 4).

1. The Allosteric Dimer as Analyzed by the MWC Model

We will now consider two cases: the case of exclusive binding and the case of nonexclusive binding to a protein dimer.

a. <u>Exclusive Binding.</u> In this case the binding of ligand occurs to the R-state only.

$$T \underset{}{\overset{L}{\rightleftharpoons}} R \underset{}{\overset{K_R}{\rightleftharpoons}} RS \underset{}{\overset{K_R}{\rightleftharpoons}} RS_2 \qquad\qquad (24)$$

L is the allosteric equilibrium constant $(L = [T_0]/[R_0])$, and K_R is the intrinsic ligand affinity to the R-state. In this case, if [R] is expressed in terms of E, it will easily be seen that:

$$[E] = [T] + [R] \qquad\qquad (25)$$

$$[E] = [R]L + [R] \qquad\qquad (26)$$

$$[R] = \frac{[E]}{1+L} \qquad\qquad (27)$$

Thus, $[RS] = 2K_R \dfrac{1}{1+L} [E][S]$

$\qquad [RS_2] = K_R^2 \dfrac{1}{1+L} [E][S]^2.$

From (12) and (13) it is also clear that at the same time:

$\qquad [RS] = 2K_1[R][S]$

$$[RS_2] = K_1 K_2 [R][S]^2 .$$

It follows therefore that:

$$K_1 = \frac{K_R}{1+L} \text{ and } K_2 = K_R .$$

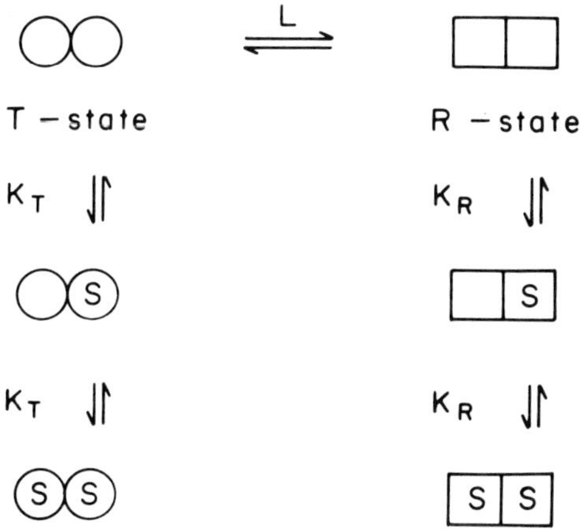
L
T – state
R – state
K_T
K_R
S
S
K_T
K_R
S S
S S

$$K_1 = \frac{K_T L + K_R}{1+L} \text{, intrinsic binding constant}$$

$$K_2 = \frac{K_T^2 L + K_R^2}{K_T L + K_R} \text{, intrinsic binding constant}$$

$$n_H = \frac{2}{1 + \sqrt{\frac{K_1}{K_2}}}$$

$$n_H = \frac{2}{1 + \sqrt{\frac{(K_T L + K_R)^2}{(1+L)(K_T^2 L + K_R^2)}}}$$

$$1 \leq n_H \leq 2$$

FIG. 4. The MWC Model for an Allosteric Dimer

Since the Hill coefficient is given by (20) one obtains (28):

$$n_H = \frac{2}{1 + \sqrt{\dfrac{1}{1+L}}} \tag{28}$$

From this relationship it follows that $1 \leq n_H \leq 2$. The MWC model cannot account for negative cooperativity because n_H cannot assume values less than 1.0.

b. <u>Nonexclusive Binding.</u> In the nonexclusive case, both the T-form and the R-form bind the ligand S with intrinsic binding constants K_T and K_R, respectively (Fig. 4). In this case one can write:

$$[ES] = [TS] + [RS] = 2K_T[T][S] + 2K_R[R][S]$$

$$[ES] = 2 \left(\frac{K_T L}{1+L} + \frac{K_R}{1+L} \right) [E][S] = 2 \frac{K_T L + K_R}{1+L} [E][S]$$

and

$$[ES_2] = [TS_2] + [RS_2] = \frac{K_T^2 L + K_R^2}{1+L} [E][S]^2$$

since $[ES] = 2K_1[E][S]$

and

$$[ES_2] = K_1 K_2 [E][S]^2 .$$

One obtains the following expressions for K_1 and K_2:

$$K_1 = \frac{K_T L + K_R}{1+L}$$

$$K_2 = \frac{K_T^2 L + K_R^2}{K_T L + K_R}$$

From (20) it is clear that the expression for the Hill coefficient will be given by (29):

$$n_H = \frac{2}{1 + \sqrt{\dfrac{(K_T L + K_R)^2}{(1+L)(K_T^2 L + K_R^2)}}} \tag{29}$$

Again one can see that when $L = 0$, $n_H = 1$, and when $n_H > 0$, $n_H > 1$. As in the case of exclusive binding, one can see that $1 \le n_H \le 2$ and again negative cooperativity cannot be accommodated by the allosteric model of Monod et al.

c. The General Case. Let us consider an enzyme with n subunits which occurs in two conformations T and R, respectively. The binding to the two states occurs with intrinsic affinity constants K_T and K_R, respectively:

$$T \underset{L}{\overset{L}{\rightleftharpoons}} R$$

$$T + S \underset{K_T}{\rightleftharpoons} TS \qquad R + S \underset{K_R}{\rightleftharpoons} RS$$

$$TS_{n-1} + S \underset{K_T}{\rightleftharpoons} TS_n \qquad RS_{n-1} \underset{K_R}{\rightleftharpoons} RS_n$$

The saturation function for this case will be

$$\overline{Y} = \frac{K_T L[S](1 + K_T[S])^{n-1} + K_R[S](1 + K_R[S])^{n-1}}{L(1 + K_T[S])^n + (1 + K_R[S])^n} \tag{30}$$

The saturation curves derived from Eq. (30) are either positively cooperative or noncooperative. One can analyze the Hill slope,

$$\frac{d \log \dfrac{\overline{Y}}{1 + \overline{Y}}}{d \log [S]} = n \text{ (Hill slope)}$$

at $\overline{Y} = 0.5$ and investigate its behavior as a function of K_R / K_T and of L. As in the case of a dimer, the Hill coefficient can attain values only between 1.0 and n:

$$1 \le n_H \le n \tag{31}$$

Negative cooperativity cannot be accounted for by the general model.

2. Allosteric Inhibition and Allosteric Activation in the MWC Model

The Monod-Wyman-Changeux model offers an elegant molecular explanation for allosteric inhibition and allosteric activation.

The allosteric inhibitor I has a higher affinity for the T-state, whereas the allosteric activator A has a higher affinity for the R-state. Therefore, in the presence of the allosteric inhibitor a larger fraction of the enzyme is pulled to the T-state, thereby increasing the apparent value of L. The apparent affinity of the protein toward the substrate S is thus reduced, and the response to S becomes more cooperative. Similarly, in the presence of the allosteric activator the apparent affinity of the system toward the substrate is increased, and the curve becomes less cooperative toward the substrate (Fig. 5). Upon saturation of the system with the activator, the system may become Michaelian toward the substrate if the R-state binds the activator exclusively.

3. The Advantages and Limitations of the MWC Model

The model assumes that an oligomeric enzyme is always symmetric. It accounts for cooperative binding by the fact that an equilibrium exists between the R- and T-states which have different affinities for the ligand concerned. The model can therefore analyze a binding curve in terms of only three parameters: L, K_R, and K_T [Eq. (30)]. The beauty of the model is that it uses these same three parameters for an allosteric dimer

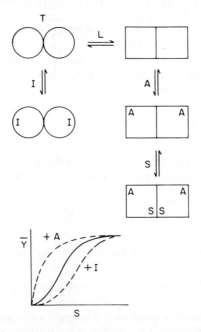

FIG. 5. The Effect of an Allosteric Inhibitor or an Activator on the Ligand Binding Curve

and also for an n-mer, n being any number. Indeed, all of the positively cooperative binding curves reported to date in the literature can be easily explained by a MWC molecular model. However, this fact, as we will shortly see, is not sufficient to establish the validity of the model.

The limitations of the MWC model stem from the very basic assumptions of the model which are the sources of its beauty, namely, the conservation of symmetry throughout the binding process and the assumption of a preexisting equilibrium between an R-state and a T-state in the absence of added ligand. For this model to be acceptable, one must do more than merely fit the data to the model. There must also be independent proof of (1) the existence of a T and R conformations in the absence of ligand and (2) the conservation of symmetry throughout the binding process.

Point 1 is difficult to prove since L values which are used to fit the binding data of the best-studied allosteric system (ATCase, hemoglobin, and yeast GPDH) are quite high (Table 1). This means that only a small fraction of the protein is in the R-state in the absence of ligand. It is, therefore, impossible to detect the existence of the two states by physico-chemical techniques, since one of them (the T state) is always in a large excess. Point 2 is even more difficult to prove, since no experimental approach less sophisticated than x-ray crystallography can be used to detect molecular symmetry. It has often been claimed, especially by advocates of the MWC model, that the state function, namely the function which correlates the conformation of the oligomeric protein as a function of added ligand, gives the clue. This, however, is a dangerous over-simplification, as will be demonstrated below, after discussing the Koshland-Némethy-Filmer sequential model.

B. The Koshland-Némethy-Filmer Sequential Model

An alternative model for allosteric transitions was proposed in 1966 by Koshland, Némethy, and Filmer [24]. The basic assumptions of this model are:

TABLE 1

L Values for Some Allosteric Proteins

	L	Reference
Hemoglobin	9054	5
ATCase	4	25, 26
Yeast GPDH	9	57-59

1. The oligomeric protein exists in one conformation prior to ligand binding.
2. Upon ligand binding a conformational change is induced in the subunit which binds the ligand.
3. The conformational change is not restricted to the filled subunit but may be transmitted to neighboring vacant subunits via the subunit interaction domains.

As with the MWC model, one may most easily understand the KNF sequential model by looking at the case of an allosteric dimer.

1. The Allosteric Dimer as Analyzed by the KNF Model

In Figure 6 the scheme of ligand binding to a dimer according to the KNF sequential model is given. Definitions of the different parameters are also given in the figure. Kt_{AB} is the equilibrium constant of the square conformation (Conformation B) over the round one (Conformation A). The unliganded state is always defined as the standard state of the protein, and therefore K_{AA} is defined as 1.0. K_{AB} is the subunit interaction constant between a circle (A) conformation and a square (B) conformation divided by K_{AA}. K_{BB} is also defined as the subunit interaction between two B conformations divided by K_{AA}.

It is easily seen from Figure 6 how the Adair equation is derived for the KNF model and how the expression for the Hill coefficient n_H at the midpoint is arrived at:

$$n_H = \frac{2}{1 + \sqrt{\dfrac{K^2_{AB}}{K_{BB}}}} \qquad (32)$$

It can be seen from Eq. (32) that the ratio between K^2_{AB} and K_{BB} will determine whether n_H will be larger, smaller, or equal to 1.0. It is interesting that when $K^2_{AB} = K_{BB}$ then $n_H = 1.0$; this means that not every conformational change coupled to ligand binding in a multisubunit enzyme must result in cooperativity. On the contrary, conformational changes which obey the relation $K^2_{AB} = K_{BB}$ will result in noncooperative saturation curves. Indeed, in many multisubunit enzymes the ligand binding curves are noncooperative although conformational changes do occur. Thus, for example, in the tetrameric enzyme lactate dehydrogenase the conformational changes which accompany NAD binding are noncooperative but can be observed directly by x-ray crystallographic studies [27]. It is immediately seen from Eq. (32) that negative cooperativity will occur

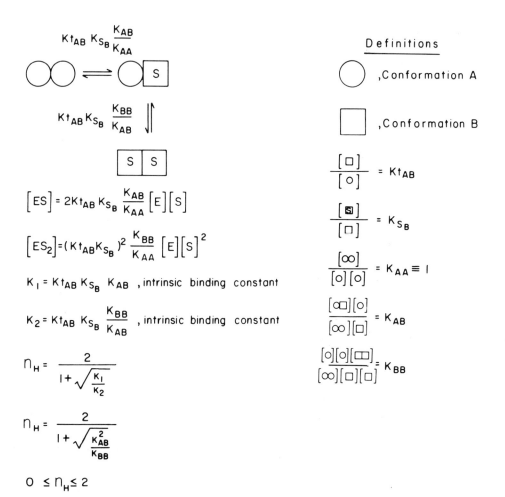

FIG. 6.   The KNF Sequential Model for an Allosteric Dimer

when $K^2_{AB} > K_{BB}$. Thus, the sequential model can easily account for negative cooperativity, whereas the MWC concerted model cannot. It is also seen from Eq. (32) that positive cooperativity will occur when $K_{BB} > K^2_{AB}$. Thus the KNF sequential model can account for all types of cooperativity predicted by the Adair approach.

The Michaelian case in the MWC means either (1) $K_R = K_T$ or (2) $L = 0$ and all the enzyme is in the R-form, or (3) $L \to \infty$ and all the enzyme is in the T-form (see Fig. 4). In either case, no conformational changes will accompany the Michaelian ligand binding process. The existence of negative cooperativity and the occurrence of conformational changes in Michaelian noncooperative binding process indicate that the KNF sequential model has a broader application to biological systems than the MWC concerted model.

2. The General KNF Approach

Similar to the derivation of equations for the allosteric dimer, one may derive the equations for any general case. As in the dimer case, the first step is the formulation of the explicit assumption; the second is the writing down of all the species; the third is the writing of the mathematical expression for the binding curve.

3. Allosteric Activation and Allosteric Inhibition in the KNF Model

Let us consider the general case, where the protein can bind the ligand S, the allosteric activator A, and the allosteric inhibitor I, and treat it according to the KNF sequential model (Fig. 7). It can be seen from Figure 7 that a number of assumptions have already been made implicitly. The first is that the oval conformation is the final conformation whenever the activator A is bound. It can easily be visualized that a subunit with A can attain different conformations, depending on whether S is bound to it or not. The second assumption is that the enzyme does not bind A and I at the same time; in the more general case one should consider species containing both A and I bound to them.

The scheme in Figure 7 shows that many terms will appear in the Adair equation, each one specifying the specific combination of conformations. There is no difficulty in fitting binding curves by adjusting many parameters, as in the case discussed here. The major problem is to find ways of proving the existence of the different conformations or the absence of some conformations. This becomes a formidable task, even in the case of a dimer. One must find experimental methods of determining the different accessible conformations of the protein subunit by direct physicochemical and chemical methods. As a guideline, therefore, one should fit binding curves with the maximal number of conformations that can be accounted for experimentally. As in the case of the concerted model, one usually finds it difficult to prove the existence of more than three conformations of an enzyme subunit, and therefore one should eliminate some of the conformations shown in Figure 7, also for the sequential case. A simplified version of the scheme is shown in Figure 8.

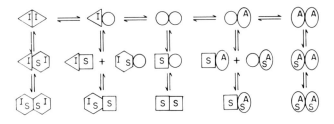

$\bigcirc$ Conformation of naked subunit

$\boxed{S}$ Conformation of subunit with only S bound

$\genfrac{(}{)}{0pt}{}{A}{S}$ Conformation of subunit with A bound or A and S bound

$\triangleleft^{I}$ Conformation of subunit with I bound

$\langle^{I}_{S}\rangle$ Conformation of subunit with I and S bound

FIG. 7. The Conformations of a Dimer in the Presence of Substrate, Activator, and Inhibitor According to the KNF Model

C. Diagnostic Tests Using Binding Curves

It has already been mentioned that the most rigorous way of fitting ligand binding curves is by using the Adair equation. This is true because the Adair scheme considers the binding process without assuming any molecular model. The two models discussed above allow one to interpret the Adair constants in terms of molecular parameters. The question, therefore, is whether some features of the binding curves themselves may lead to a better understanding of the properties of the protein under study.

The fraction of enzyme sites saturated by ligand ($\overline{Y}$) as a function of the free ligand concentration ($[S]$) can be plotted in a number of ways as is shown in Figure 9; the $\overline{Y}$ versus $[S]$ plot, the double reciprocal plot according to Lineweaver and Burk [28], the Scatchard plot [29], and the Hill plot [18]. $\overline{Y}$ is determined either by a direct binding measurement or by taking the ratio $\overline{Y} = v/V_{max}$ where v is the rate of an enzyme reaction. In this case, the fraction of sites saturated is given by the velocity v at any ligand concentration $[S]$ divided by the maximal velocity V_{max} at infinite substrate concentrations. The accurate determination of V_{max} or the maximal saturation in a binding curve can be attained by extrapolations

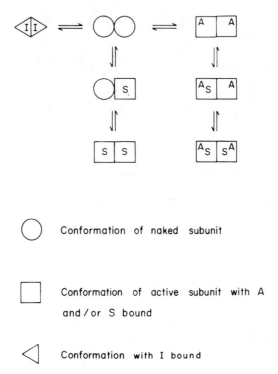

FIG. 8. The Binding of Substrate, Activator, and Inhibitor According to a Simplified KNF Model

in either the double reciprocal plot or the Scatchard plot. Whenever positive cooperativity occurs, the deviations from the characteristic Michaelian binding will be immediately recognizable in all four plots. In the case of "negative cooperativity," the $\overline{Y}$ versus [S] plot seems at first glance normal, since it appears hyperbolic as in the Michaelian case. This is probably the reason why such cases were overlooked until quite recently [30]. When the same data are plotted in the double reciprocal plot, Scatchard plot, or the Hill plot (Fig. 9), interesting features are revealed: concavity downward in the double reciprocal plot, concavity upward in the Scatchard plot, and a Hill slope smaller than 1.0. If one looks carefully at the $\overline{Y}$ versus [S] plot it can be seen that, with negative cooperativity curve, the hyperbola is more flattened than in the normal Michaelian case — it is steeper than a Michaelian curve at the first part of the curve, but approaches the limiting value of $\overline{Y} = 1$ at a slower rate than a Michaelian curve. It is interesting that negative cooperativity and its

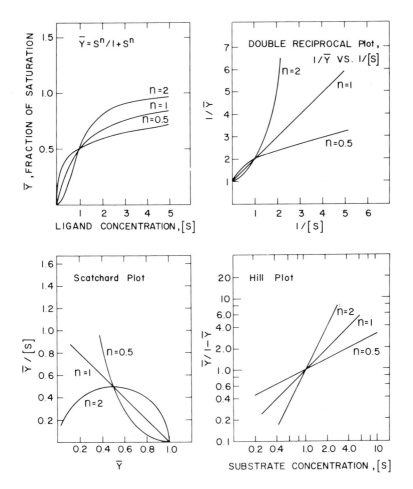

FIG. 9. Diagnostic Plots for Binding Data

diagnosis were not recognized as a general phenomenon in multisubunit enzymes until quite recently [30]. Since a mixture of proteins possessing different affinities toward the ligand will also demonstrate saturation curves with all the characteristics of negatively cooperative binding, it is therefore important to establish the homogeneity and nativity of the protein studied [30].

Whenever negative cooperativity can be established, the molecular model which should be adopted for this protein will be the sequential one [30]. Of course, positive cooperativity or noncooperativity can be explained by both models in most cases. When both models are possible, further experiments determine which is valid.

In conclusion, binding curves are essential for the determination of the number of binding sites in a protein molecule and are needed to establish the type of site-site interactions. In cases of positively cooperative binding curves, both models are equally acceptable; thus we are left without a clear understanding of the molecular events which occur upon ligand binding. In the case of negatively cooperative and mixed positive-negative cooperativity, one is immediately forced to exclude the concerted MWC model [30-32].

General mathematical fitting procedures for the Adair equation are now available. These methods are instrumental in obtaining the Adair constants for any type of binding curve.

D. Comparison of the Models

The MWC model is based on the assumption that symmetry is an essential feature of the protein and must be preserved even during protein function. Thus, whenever a protein oligomer changes its structure during a binding process, all the subunits must change their conformation simultaneously. The MWC model assumes a preexisting conformational equilibrium between two or more states in which the different conformations differ in their affinity toward the ligand. Therefore, the added ligand stabilizes the conformation to which it has higher affinity. The KNF model assumes that the protein is indeed symmetric in design but that this symmetry is not necessarily preserved during the process of ligand binding. The conformational distortions occurring in the protein subunit which binds the ligand are induced and may or may not propagate to vacant subunits through the subunit interaction domains. Thus, a protein oligomer can acquire a broad spectrum of conformations, depending on the strength of subunit interactions and their direction in space. The existence of intermediate states in which the different protomers within an oligomer can acquire different conformations is a major difference between the two models. The sequential model is much more general than the concerted model since it accounts for all types of cooperativities, including negative cooperativity, whereas the concerted model does not allow negative cooperativity. Furthermore, the limiting case of the sequential model for positively cooperative systems is phenomonologically similar to the concerted model. It is generally believed that the advantage of the MWC model is its simplicity. This is, indeed, true, but it is now apparent that the generalized approach of Koshland is probably the more realistic one.

VI. PROTEIN CONFORMATION - EQUATIONS OF STATE

The binding of a ligand by an oligomeric protein may be associated with conformational changes in the binding as well as in the unbound subunits. Because of subunit interactions, the structural changes accompanying lig- and binding may propagate to vacant subunits. This is, in fact, the basis of all cooperative phenomena. It is, therefore important to analyze not only the ligand binding curves but also to directly analyze the state function $\bar{R}$ which directly determines the fraction of enzyme subunits which are in the bound conformation, whether they have ligand bound to them or not. Let us consider again the allosteric dimer. In the MWC concerted model (Fig. 4), the equation of state $\bar{R}$ will be given by (33):

$$\bar{R} = \frac{\boxed{}\,\boxed{} + \boxed{S}\,\boxed{} + \boxed{S}\,\boxed{S}}{\bigcirc\!\bigcirc + \boxed{}\,\boxed{} + \boxed{S}\,\boxed{} + \boxed{S}\,\boxed{S}} \tag{33}$$

where a square represents the unbound conformation and a circle the bound conformation. The corresponding saturation function is given by (34):

$$\bar{Y} = \frac{\boxed{S}\,\boxed{} + 2\,\boxed{S}\,\boxed{S}}{2\,\bigcirc\!\bigcirc + 2\,\boxed{}\,\boxed{} + 2\,\boxed{S}\,\boxed{} + 2\,\boxed{S}\,\boxed{S}} \tag{34}$$

It is immediately apparent from these two equations that $\bar{R} > \bar{Y}$ throughout the titration curve and if one plots $\bar{R}$ versus $\bar{Y}$, one will obtain a concave upward curve (curve b in Fig. 10). When the simple sequential model is used (Fig. 6), the saturation function $\bar{Y}$ and the state function $\bar{R}$ will be:

$$\bar{Y} = \frac{\bigcirc\!\boxed{S} + 2\,\boxed{S}\,\boxed{S}}{2\,\bigcirc\!\bigcirc + 2\,\bigcirc\!\boxed{S} + 2\,\boxed{S}\,\boxed{S}} \tag{35}$$

$$\bar{R} = \frac{\bigcirc\!\boxed{S} + 2\,\boxed{S}\,\boxed{S}}{2\,\bigcirc\!\bigcirc + 2\,\bigcirc\!\boxed{S} + 2\,\boxed{S}\,\boxed{S}} \tag{36}$$

namely, $\overline{R} = \overline{Y}$ at any point (curve A in Fig. 10). When the molecule undergoes conformational changes according to a more general sequential mech-
anism (Fig. 8), $\overline{R}$ will always be higher than $\overline{Y}$. Therefore, whenever $\overline{R} =$
$\overline{Y}$ the only applicable model is the sequential one. On the other hand, whenever $\overline{R} > \overline{Y}$, both models are possibly valid and it is not true that the
molecule must obey the MWC concerted model.

VII. THE ENERGETICS OF SUBUNIT INTERACTIONS

A very important question related to the whole subject of subunit interactions remains: What is the order of magnitude of free energies involved in
regulatory phenomena? When one speaks about changes of conformation,
how many calories are involved and what types of molecular force are
likely to play a role in these energy interchanges?

It was pointed out by J. Wyman [33] that the Hill plot approaches a
limiting slope of 1.0 both at low and high ligand concentrations. In the
region of very low ligand occupancy ($S \rightarrow 0$) the only significant binding
process is the binding to the first site described by the appropriate association constant K_1. The only significant binding process at very high ligand
concentration ($S \rightarrow \infty$) is the binding to the last available site. This process

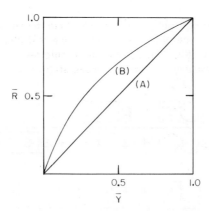

FIG. 10. The Dependence of the State Function on the Saturation Function.
$\overline{R}$ is the fraction of enzyme subunits in the bound conformation (square) as
a function of the fractional ligand saturation ($\overline{Y}$). (A) Simple sequential.
(B) MWC or general sequential.

is described by the association constant K_n. To find the value of K_1 one continues the Hill plot with a slope of unity ($n_H = 1$) from the low ligand concentration end (low left in Fig. 11). At the point where $\log \overline{Y}/1-\overline{Y} = 0$ one obtains the value of $\log (1/K_1)$ on the log S axis. Similarly, $\log (1/K_n)$ is obtained from the Hill plot possessing a slope of unity constructed from the high ligand concentration end (high right in Fig. 11). It is clear that the difference between K_1 and K_n determines the total free energy of interaction among the protein subunits. As is shown in Figure 11, K_1 and K_n can be determined easily. The free energy changes may be evaluated by using equations derived in the following manner:

$$\Delta G_1 = RT\ln K_1 \tag{37}$$

$$\Delta G_n = RT\ln K_n \tag{38}$$

therefore

$$\Delta(\Delta G) = RT\ln \frac{K_n}{K_1} \tag{39}$$

If h is defined as the distance between the two limiting $n = 1$ plots and X the distance between the two values: $\log (1/K_n)$ and $\log (1/K_1)$ on the log S axis (Fig. 11), then:

$$X^2 = h^2 + h^2 \tag{40}$$

or

$$X = \sqrt{2}\, h \tag{41}$$

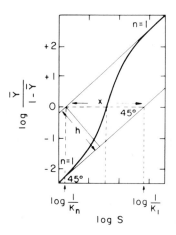

FIG. 11. The Subunit Interaction Energy in an Enzyme with n Subunits

since

$$X = \log \frac{1}{K_1} - \ln \frac{1}{K_n} \tag{42}$$

namely

$$X = \log \frac{K_n}{K_1} \tag{43}$$

one obtains

$$\sqrt{2}\, h = \log \frac{K_n}{K_1} \tag{44}$$

Using (39) one obtains:

$$\Delta(\Delta G) = 2.3RT \sqrt{2}\, h. \tag{45}$$

In the case of hemoglobin $\Delta(\Delta G)$ equals 3,000 calories per mole tetramer (3 kcal/mol). Since hemoglobin is a tetrahedral molecule, six subunit interaction domains share this quantity of 3 kcal/mol. It has been shown [34] that most of the interaction energy occurs at the $\beta_1 - \beta_2$, $\beta_1 - \alpha_2$ $\beta_2 - \alpha_1$, $\alpha_1 - \beta_1$ and $\alpha_2 - \beta_2$ interfaces with no contribution from the $\alpha_1 - \alpha_2$ binding domain. If one assumes an equal distribution of this interaction energy, then the interaction energy of a subunit interaction domain is around 600 to 700 calories per mole (0.6 to 0.7 kcal/mol). This amount of free energy is rather low. It is therefore obvious that significant cooperativities are easily accessible energetically. This calculation is not actually surprising since we know that the dependence of free energy on ligand concentration is a logarithmic one. The low energies involved in subunit interactions do not necessarily imply hydrophobic interactions as has sometimes been stated. In hemoglobin, for example, it is explicitly known that electrostatic salt bridges play a key role in the subunit interactions of the protein [35, 36].

Since the quantity RT at $25°$ is 600 calories, one would expect a sharp difference of cooperativity parameters (such as the Hill coefficient) on temperature. Surprisingly, the temperature effects on subunit interactions have been investigated in detail on only one protein [37]. The subunit interaction energy is a positive quantity in a positively cooperative system, as shown in Figure 11. In negative cooperativity the interaction energy is negative.

An example: Haber and Koshland [34] prepared the hemoglobin derivative $[\alpha Fe(III)\beta]_2$ where the two α chains are in the cyanomet form. In this molecule they could selectively investigate the $\beta - \beta$ subunit interactions. They found that the Hill slope was 1.3 and the interaction energy 700 kcal/mol. This value of the $\beta - \beta$ interaction energy confirmed earlier estimates of Wyman [33]. See also the discussion above.

VIII.   ASSOCIATION-DISSOCIATION PHENOMENA

We have thus far learned that cooperativity results from subunit interactions and their alteration upon ligand binding. Self-association or dissociation of a protein molecule as a result of ligand binding also yields cooperative binding curves [11, 38-41]. This is not surprising since association or dissociation of protomers are essentially also conformational changes. Cooperativity in ligand binding to proteins which undergo aggregation (or dissociation) coupled to ligand binding differs markedly from that of nonaggregating systems. The distinct characteristics [41] of such systems are: (1) high cooperativities (Hill coefficient) can occur at low ligand concentrations, and (2) the cooperativity decreases as a function of the total protein concentration in the system. It is interesting to note that protein dissociation or aggregation coupled to ligand binding is observed in all cases in which a high degree of cooperativity was found [40, 41]. It was suggested [41] that whenever high degrees of cooperativities had to be achieved, nature chose the maximal conformational changes which could occur upon ligand binding, namely dissociation or association.

IX.   PROTEIN DESIGN AND SUBUNIT INTERACTIONS

One of the serious disadvantages in the concerted model is that it does not take into account the precise geometrical arrangement of the subunits in the protein assembly. It has already been pointed out in Section III that protein subunits can be arranged in a number of ways in space. In protein tetramers, for example, one finds that the tetrahedral arrangement is quite common (Fig. 2). In this arrangement every subunit makes three different isologous chemical contacts with the neighboring subunits. It is therefore likely that the protein subunit interactions across each contact will be different. This has already been demonstrated in hemoglobin (see above).

In tetrahedral molecules one might expect that the subunit interactions will differ from one another. For a tetramer, the possible interactions are pp, qq and rr (Fig. 12). This fact must be taken into account when one follows the molecular events which accompany ligand binding. For example, when a ligand binds to a protein tetramer, three di-bound species are possible (Fig. 12). It can be seen that when the tetramer has two ligands bound, three geometrical isomers are possible. Each of the di-bound species will have a certain stability depending on the subunit interaction energy, as is indicated in Figure 12. In a large number of protein tetramers, it has been found that one set of binding domains (say, pp) contributes much more to the behavior of the molecule than the other two sets (qq and rr). This behavior may lead to half-of-the-sites reactivity

where the binding of a ligand to two active sites switches off completely the
two remaining sites (Fig. 13). The binding of the first ligand to the
a priori symmetric tetramer switches off the subunit across the pp binding

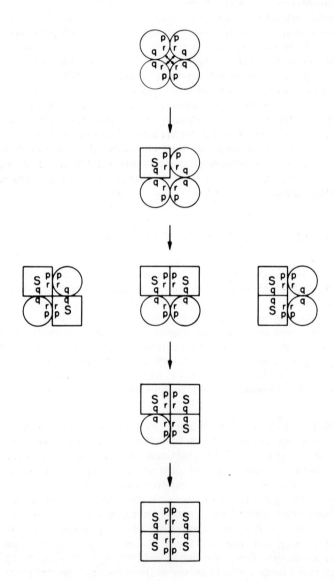

FIG. 12. Binding of a Ligand to a Tetrahedral Protein

domain and affects the subunits across the rr and the qq binding domains less strongly. The second ligand molecule to enter again switches off the subunit across the pp binding domain. The end result of the entire process is the binding of ligand to two of the four binding sites, producing an asymmetric di-ligand tetramer (Figure 13). This phenomenon has been observed in many enzymes [42].

It has been suggested [42] that a protein tetramer which exhibits half-of-the-sites reactivity is in fact a dimer of dimers because the subunit interactions across one set of binding domains is much stronger than the others. Studies on the assembly pathway of LDH isozymes [43, 44] from their unfolded subunits and the dissociation pattern of the native LDH tetramers [45] in dilute solutions and in denaturants also favor the dimer of dimers hypothesis. In those studies it was shown that LDH dimers are important intermediates in both processes.

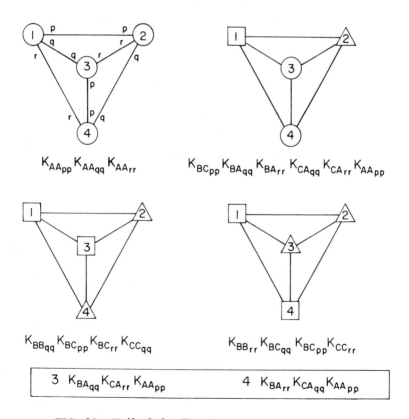

FIG. 13. Half-of-the-Sites Reactivity in a Tetramer

In protein assemblies where more binding domains exist and where both isologous and heterologous domains appear, the situation is more complex. The detailed theoretical aspects of such systems was recently discussed [46].

X. A FEW EXAMPLES

So far we have analyzed theoretical molecular models which can explain cooperative and allosteric inhibition or activation. Most of all, we have demonstrated that binding curves alone can serve only as the first diagnostic step in the analysis of an allosteric protein. To understand more fully the detailed behavior of an allosteric molecule, one must apply other methods to determine the structural changes occurring upon ligand binding by a protein. Only a combination of binding data and the data derived from other experimental techniques can unravel the molecular events occurring in an allosteric protein. In a number of specific cases, this knowledge has been achieved. Through these examples we will now try to illustrate which tools are used to investigate the behavior of a protein.

A. Hemoglobin — The Mechanism of Cooperativity

Hemoglobin played a key role in understanding allosteric phenomena. Almost every possible tool accessible to the experimentalist has been used to analyze this protein. Three basic molecular phenomena have to be explained in hemoglobin: (1) the cooperative binding of oxygen ($n_H = 2.8$), (2) the alkaline Bohr effect, which is the tendency of hemoglobin to release four protons upon oxygenation:

$$Hb \ + \ 4O_2 \ \rightleftharpoons \ Hb(O_2)_4 \ + \ 4H^+ \tag{46}$$

and (3) the allosteric stabilization of hemoglobin in the deoxy conformation by the binding of DPG (2, 3 diphosphoglyceric acid), the allosteric effector.

The first attempts to clarify the conformational changes and the molecular events occurring upon oxygenation of hemoglobin were done carrying out detailed analysis of saturation curves. This approach proved unsuccessful because both the MWC model and the KNF model were equally compatible with the experimental data [5, 24]. Ogawa and McConnell [47] later examined the dependence of the state function $\overline{R}$ on $\overline{Y}$. They attached a spin label to cysteine 93 of the β-subunit and followed its EPR spectral changes as a function of hemoglobin oxygenation. Using this technique they found a linear dependence of $\overline{R}$ on $\overline{Y}$ which suggested that oxygen binding follows the simple sequential scheme. Schulman et al. [22]

demonstrated that the heme itself does not undergo any conformational changes upon oxygenation. This observation persuaded the workers in the field that one should look for conformational changes in the protein molecule only, as Ogawa and McConnell had done. Measurements of oxygen binding [34] to hybrid molecules $\alpha[Fe(III)CN]_2\beta_2$ and $\alpha_2\beta[Fe(III)CN]_2$ revealed that the cooperativity in oxygen binding results mainly from $\beta_1-\beta_2$, $\alpha_1-\beta_2$ and $\alpha_2-\beta_1$ interactions. It was found that the $\alpha_1-\alpha_2$ interactions do not contribute to the cooperative oxygen binding. This evidence led to the conclusion that the functional hemoglobin molecule is a tetrahedron.

On the basis of detailed crystallographic studies of hemoglobin and its derivatives from various animals, Perutz [35, 36] proposed an elegant theory which describes the sequence of events which occur upon hemoglobin oxygenation. He suggested that oxygen binds first to the two α subunits, transforming the high-spin Fe(II) atoms to low-spin Fe(II) atoms. This transformation shrinks the Fe(II) shell by 0.8 Å and "squeezes" the atom into the heme plane. This movement pulls the whole heme plane toward a helical section at the other side of the heme plane, which in turn bumps into another helical section of the molecule. This triggering event is followed by a series of molecular movements leading to the breakage of two salt bridges ($-COO^-$ . . . ^+H_3N-bridges) and the release of two Bohr protons. The breakage of these two salt bridges causes the two β subunits to move apart from each other by 7 Å. This, then, allows the two empty heme groups to move to a conformational position more accessible to oxygen. Now these two hemes can bind the third and the fourth oxygen molecule with increased affinity. The binding of the third and fourth oxygen molecule occurs without significant conformational changes taking place because the hemoglobin molecule has already "clicked" from the deoxy state to the oxy state upon binding of the first two oxygen molecules. Theoretically, part of the binding energy of oxygen to the first two sites was used to "click" the molecule into the complete oxy conformation. Therefore, the binding process of the two remaining oxygen molecules can occur more easily — that is, with increased affinity. Perutz's theory also explains the effect of 2,3 DPG on oxygen binding. One molecule of DPG binds across the twofold symmetry axis between the two β chains in deoxyhemoglobin. When the β chains move away from each other by 7 Å upon oxygenation, DPG falls out of the now much bigger cleft. This is the molecular explanation for the well-established fact that DPG stabilizes the deoxy conformation and therefore shifts the oxygen binding curves toward higher oxygen pressure. It is interesting to note that the allosteric effector, 2,3 DPG, in the hemoglobin system binds at the subunit interaction domain and is bound <u>between</u> two subunits. The only other known case for a binding site shared between two polypeptide chains is the antibody molecule.

The Perutz mechanism is essentially a sequential mechanism because the molecular events described follow the binding of the first one or two oxygen molecules. Although there is a great deal of experimental data which support Perutz's theory, some important experiments are not in complete agreement with it. It is, however, the theory that is most compatible with the accumulated experimental data and seems to be the best mechanistic explanation for the behavior of hemoglobin.

B. CTP Synthetase — The Role of an Allosteric Effector

CTP synthetase catalyzes the following reactions

$$UTP + ATP + NH_3 \xrightarrow{Mg^{2+}} CTP + ADP + Pi \tag{47}$$

$$UTP + ATP + Glutamine \xrightarrow[GTP]{Mg^{2+}} CTP + ADP + Pi + Glutamate \tag{48}$$

The enzyme is a dimer with a molecular weight of 108,000 [40, 42] and is composed of two identical subunits. Each subunit possesses one site for each of the substrates ATP, UTP, NH_3 and L-glutamine and the allosteric activator GTP as determined by direct binding studies [37, 40]. The enzyme dimer dimerizes to form a tetramer in the presence of either ATP or UTP. The effects of ATP and UTP are synergistic and their binding is strongly positively cooperative. It has also been established that the functional form of the enzyme is the tetrameric form which does not dissociate even at 10^{-9} M concentrations. It was found that both the reactions [Eq. (47) and (48)] involve the formation of two common covalent UTP intermediates as depicted in Figure 14. When glutamine is used as the nitrogen donor, GTP functions as the switch-on allosteric effector. Through detailed chemical studies [37, 48] it has been found that the GTP effector activates the glutamine reaction by facilitating the conversion of

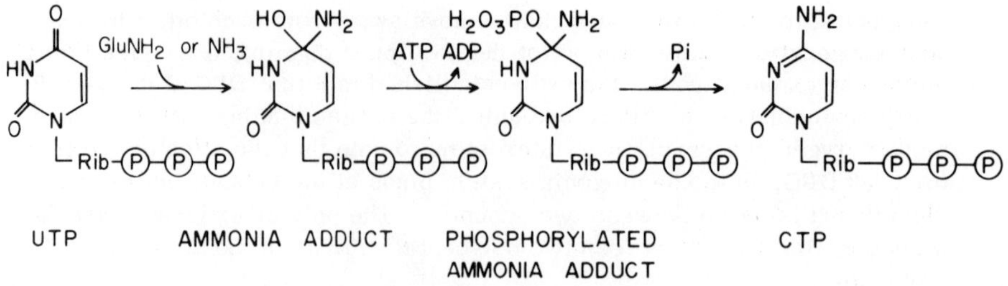

FIG. 14. The CTP Synthetase Reaction

the enzyme-glutamine Michaelis complex to a glutamyl-enzyme thioester intermediate and enzyme bound unprotonated ammonia. In this way ammonia is provided for the reaction as shown in Figure 14. Detailed kinetic analyses of the GTP effect and affinity labeling experiments have revealed that the glutamylation step is indeed the target of the GTP activator. It has been established that ATP and UTP, which function as substrates in the amination step, also function as allosteric activators of the preceding glutamylation step. Only in the presence of all ligands — ATP, UTP, and GTP — is glutamine cleaved at the maximal rate. The effect of GTP is manifested by both the sixfold decrease of K_M toward glutamine and a twelvefold increase in the k_{cat} of the glutamine dependent reaction. GTP has no effect whatsoever on the probably nonphysiological ammonia reaction. The GTP effects on both K_M and k_{cat} are due to a single molecular event: the facilitation of the glutamylation reaction by glutamine. These detailed chemical and kinetic analyses have revealed the precise molecular events which are induced in the enzyme active site by an allosteric effector. Although many enzymes are activated or inhibited by allosteric effectors, CTP synthetase was the first instance in which the mechanism of such phenomena was analyzed.

The binding data, as well as the chemical reactivity and changes in fluorescence [37, 40] of the enzyme subunit as a function of added ligands, all imply that the enzyme subunit can attain a number of conformations. Every ligand induces its share of conformational change [37, 40]. In conclusion, it has been demonstrated for CTP synthetase that, although the naked unliganded molecule is symmetric, this enzyme is a flexible structure which undergoes dynamic changes of conformation. By affinity labeling techniques it was found that the tetrameric form of CTP synthetase possesses the property of half-of-the-sites reactivity toward glutamine. This finding indicates that the enzyme possesses one strong pair of subunit interactions and two weaker pairs of subunit interactions in the protein tetramer. As mentioned previously, this seems to be a general phenomenon of enzymes displaying half-of-the-sites reactivity.

C. Rabbit Muscle Glyceraldehyde-Phosphate Dehydrogenase — Negative Cooperativity

Rabbit muscle GPDH was found to bind the coenzyme NAD^+ in a very strongly negatively cooperative fashion [49]. The intrinsic binding constants differ by more than an order of magnitude ($K_1 \sim 10^{10}$ M, $K_2 \sim 10^8$ M, $K_3 \sim 5 \times 10^5$ M, $K_4 \sim 5 \times 10^4$ M). It was established that the enzyme is a homogeneous molecule both by sequencing the protein [50] and by electrofocusing experiments [Levitzki and Mowbray, unpublished]. The binding of NAD^+ therefore probably reflects either an a priori asymmetry in the tetrameric apoenzyme or genuine negative cooperativity. By

studying properties such as viscosity, SH reactivity, and differential sedimentation [49, 51], a number of workers have reported that the enzyme
undergoes sequential conformational changes as a function of NAD binding.
This evidence tends to support the hypothesis that the enzyme is symmetric
a priori and that NAD^+ binding is needed negatively cooperative because
the conformational changes are induced by ligand binding. The enzyme also
exhibits the property of half-of-the-sites reactivity toward both acylating
agents [52] and alkylating agents [53]. Thus, the enzyme can be easily
acylated by two acyl groups, but the introduction of two additional acyl
groups is extremely difficult [52]. In addition, it has been found that
alkylating two of four active site SH groups irreversibly inhibits the enzyme.
This half-of-the-sites reactivity can be due either to an a priori asymmetric assembly of the protein tetramer or to an induced conformational change
where binding of ligands to half of the active sites switches off the other
half. Bernhard and coworkers [52, 54] claim that the mammalian holo-
GPDH is an a priori asymmetric tetramer. Recent studies in our own
laboratory [53, 55] indicate that the half-of-the-sites reactivity of both the
apoenzyme and the holoenzyme fit the sequential model better than the
a priori asymmetric assembly. It was claimed that a GPDH molecule to
which substituents were introduced into its active site may indeed be asymmetric, depending on the nature of the substituents. Low resolution x-ray
crystallography of SH derivatized holo-GPDH indeed suggests the absence of
the 2:2:2 symmetry among the NAD binding sites [56]. However more refined
x-ray data is required to establish the details of the molecular symmetry.

D. Yeast Glyceraldehyde-3-Phosphate Dehydrogenase

This enzyme is totally different from the muscle GPDH. One group of
experimenters [57-59] shows positively cooperative binding curves toward
NAD^+ although investigations in another laboratory [60, 61] have suggested
mixed positive-negative cooperativity. Elegant T-jump experiments and
rapid mixing experiments by Kirschner and coworkers [57-59] were
interpreted in terms of the MWC scheme. It was shown that NADH binds
with equal affinity to both the T- and the R-state of the enzyme tetramer,
whereas NAD^+ has a preferential affinity toward the R-state. The data of
Kirschner, therefore, are in full agreement with a simple MWC scheme.

It has recently been found [62] that yeast GPDH also possesses the
property of half-of-the-sites reactivity toward active site alkylating
reagents. This result indicates either an a priori asymmetry in the
molecule or induced conformational changes in the protein tetramer
brought about by blocking half of the enzyme sites. These observations
cannot be rationalized on the basis of a MWC scheme. More experiments
are needed to explain the two sets of results which support the two different
models of cooperativity.

XI. CONCLUDING REMARKS

A mechanistic understanding of the behavior of multisubunit regulatory enzymes can be revealed by using molecular models which tie together the architecture of their multisubunit structure with their dynamic function. Simple models such as the Hill scheme, the simple concerted model, and the simple sequential model initiated the breakthrough in the field, which took place in the mid-1960s. Later it became apparent that more sophisticated models were required to account for a variety of molecular phenomena in multisubunit proteins in general and in regulatory enzymes in particular. In the case of hemoglobin, x-ray crystallography, protein chemistry, EPR, NMR, optical methods, and many other techniques have all led to a high level of understanding of the hemoglobin molecule. As in other cases, too, some of which have been discussed here, a concerted effort involving many techniques is required to achieve maximal understanding of the molecular basis for enzyme regulation.

In addition to the investigation of regulatory enzymes, many of these techniques are also applicable to the study of more complex regulatory systems. Using these methods, significant progress has recently been made toward an understanding of regulatory systems such as acetylcholine receptor-drug interactions, membrane (Na^+, K^+, Mg^{2+}) ATPase, and hormone-adenyl cyclase interactions.

ACKNOWLEDGEMENT

The author acknowledges Dr. Michael L. Steer for his extremely useful suggestions and for critically reading the manuscript.

REFERENCES

1. A. Novick and L. Szilard, in Dynamics of Growth Processes, Princeton University Press, Princeton, N. J. , 1954, p. 21.

2. H. E. Umbarger, Science, 123, 848 (1956).

3. R. A. Yates and A. B. Pardee, J. Biol. Chem. , 221, 757 (1956).

4. J. Monod, J.-P. , Changeux, and F. Jacob, J. Mol. Biol. , 6, 306 (1963).

5. J. Monod, J. Wyman, and J.-P. Changeux, J. Mol. Biol. , 12, 88 (1965).

6. J. C. Gerhart and A. B. Pardee, Cold Spr. Harb. Symp. Quant. Biol., 28, 491 (1963).

7. J.-P. Changeux, Brookhaven Symp. Biol., 17, 232 (1964).

8. R. A. Cook and D. E. Koshland, Jr., Proc. Natl. Acad. Sci., 64, 247 (1969).

9. Y. Morino and E. E. Snell, J. Biol. Chem., 242, 5591 (1967).

10. R. C. Valentine, N. G. Wrigley, M. C. Serutton, J. J. Irias, and M. F. Utter, Biochemistry, 5, 3111 (1966).

11. I. M. Klotz, N. R. Langerman, and D. W. Darnall, Ann. Rev. Biochem., 25, (1970).

12. R. C. Valentine, B. M. Shapiro, and E. R. Stadtman, Biochemistry, 7, 2143 (1968).

13. D. L. D. Caspar and A. Klug, Cold Spr. Harb. Symp. Quant. Biol., 27, 1 (1962).

14. D. J. Derosier, R. M. Oliver, and L. J. Reed, Proc. Natl. Acad. Sci., 68, 1135 (1971).

15. B. W. Matthews and S. A. Bernhard, Ann. Rev. Biophys. Bioeng., 2, 257 (1973).

16. C. Bohr, K. Hasselbach, and A. Krogh, Skand. Arch. Physiol., 16, 402 (1904).

17. A. V. Hill, J. Physiol. (London), 40, IV-VIII (1910).

18. A. V. Hill, Biochem. J., 7, 471 (1913).

19. G. S. Adair, J. Biol. Chem., 63, 529 (1925).

20. L. Pauling, Proc. Natl. Acad. Sci., 21, 186 (1935).

21. M. F. Perutz, H. Muirhead, J. M. Cox, and L. C. G. Goaman, Nature, 219, 131 (1968).

22. R. G. Schulman, S. Ogawa, K. Wütrich, T. Yamane, J. Peisach, and W. E. Blumberg, Science, 165, 251 (1969).

23. J. C. Gerhart and A. B. Pardee, Fed. Proc., 23, 727 (1964).

24. D. E. Koshland, G. Némethy, and D. Filmer, Biochemistry, 5, 365 (1966).

25. J.-P. Changeux, J. C. Gerhart, and H. K. Schachman, Biochemistry, 7, 531 (1968).

26. J.-P. Changeux and M. M. Rubin, Biochemistry, 7, 553 (1968).

27. M. G. Rossmann, M. J. Adams, M. Buehner, G. C. Ford, M. I. Hockert, P. I. Lentz, Jr., A. McPherson, Jr., R. W. Schevitz, and I. E. Smiley, Cold Spr. Harb. Symp. Quant. Biol., 36, 179 (1971).

28. H. Lineweaver and D. Burk, J. Am. Chem. Soc., 56, 658 (1934).

29. G. Scatchard, Ann. N. Y. Acad. Sci., 51, 660 (1949).

30. A. Levitzki and D. E. Koshland, Proc. Natl. Acad. Sci., 62, 1121 (1969)

31. J. Teipel and D. E. Koshland, Biochemistry, 8, 4656 (1969).

32. R. A. Cook, Biochemistry, 11, 3792 (1972).

33. J. Wyman, J. Am. Chem. Soc., 89, 2202 (1967).

34. J. E. Haber and D. E. Koshland, Jr., J. Biol. Chem., 246, 7790 (1971).
35. M. F. Perutz, Cold Spr. Hrbr. Symp. Quant. Biol., 36 (1971).
36. M. F. Perutz, Nature, 228, 726 (1970).
37. A. Levitzki and D. E. Koshland, Jr., Biochemistry, 11, 241 (1972).
38. L. W. Nichol, W. T. H. Jackson, and O. J. Winzor, Biochemistry, 6, 2449 (1967).
39. M. H. Klapper and I. M. Klotz, Biochemistry, 7, 223 (1968).
40. A. Levitzki and D. E. Koshland, Jr., Biochemistry, 11, 247 (1972).
41. A. Levitzki and J. Schlessinger, Biochemistry, 13, 5214 (1974).
42. A. Levitzki, W. B. Stallcup, and D. E. Koshland, Jr., Biochemistry, 10, 337 (1971).
43. A. Levitzki, FEBS Lett., 24, 301 (1972).
44. A. Levitzki and H. Tenenbaum, Israel J. Chem., 12, 327 (1974).
45. I. C. Cho and H. Swaisgood, Biochemistry, 12, 1572 (1973).
46. A. Cornish-Bowden and D. E. Koshland, Jr., J. Biol. Chem., 245, 6241 (1970).
47. S. Ogawa and H. McConnell, Proc. Natl. Acad. Sci., 58, 19 (1967).
48. A. Levitzki and D. E. Koshland, Jr., Biochemistry, 10, 3365 (1971).
49. A. Conway and D. E. Koshland, Jr., Biochemistry, 7, 4011 (1968).
50. J. E. Harris and R. N. Perham, Nature, 219, 1025 (1968).
51. J. K. F. Noel and V. N. Schumaker, J. Mol. Biol., 68, 523 (1972).
52. R. A. MacQuarrie and S. A. Bernhard, Biochemistry, 10, 2456 (1971).
53. A. Levitzki, J. Mol. Biol., 90, 451 (1975).
54. S. A. Bernhard and R. A. MacQuarrie, J. Mol. Biol., 74, 73 (1973).
55. J. Schlessinger and A. Levitzki, J. Mol Biol., 82, 547 (1974).
56. H. C. Watson, E. Duée, and W. D. Mercer, Nature New Biology, 240, 130 (1972).
57. K. Kirschner, M. Eigen, R. Bittman, and B. Voigt, Proc. Natl. Acad. Sci., 56, 1661 (1968).
58. K. Kirschner, E. Gallego, I. Schuster, and D. Goodall, J. Mol. Biol., 58, 29 (1971).
59. G. V. Ellenrieder, K. Kirschner, and I. Schuster, Eur. J. Biochem., 26, 220 (1972).
60. R. A. Cook and D. E. Koshland, Jr., Biochemistry, 9, 3337 (1970).
61. S. C. Mockrin and D. E. Koshland, Jr., in preparation.
62. W. B. Stallcup and D. E. Koshland, Jr., Biochem. Biophys. Res. Comm., 49, 1108 (1972).

Chapter 2

GLUTAMINE SYNTHETASE OF ESCHERICHIA COLI:
STRUCTURE AND REGULATION

A. Ginsburg
E. R. Stadtman

Laboratory of Biochemistry
National Heart and Lung Institute
National Institutes of Health
Bethesda, Maryland

I. INTRODUCTION   44

II. PHYSICAL AND CHEMICAL PROPERTIES   45
    A. Molecular Structure   45
    B. Divalent Cation Activation   48
    C. Catalysis of γ-Glutamyl Transfer   52
    D. Stoichiometry in Binding Effectors and Substrates   54
    E. Antigenic Determinants   57

III. REGULATION BY ENZYME CATALYZED ADENYLYLATION
     AND DEADENYLYLATION   58

IV. SUBUNIT INTERACTIONS   62
    A. In Biosynthetic Catalysis   63
    B. In Binding Effectors   70
    C. In Stability   72
    D. Formation of Hybrid Molecules from Subunits   75

V. CONCLUDING REMARKS   78

I. INTRODUCTION

Glutamine synthetase catalyzes the synthesis of L-glutamine in the reaction:

$$\text{L-glutamate} + \text{ATP} + \text{NH}_4^+ \xrightarrow{\text{Me}^{2+}} \text{L-glutamine} + \text{ADP} + \text{P}_i \qquad (1)$$

The catalytic activity [1] and the metal ion (Me^{2+}) requirement [2] depend on the form of the enzyme from E. coli. Reaction (1) favors the biosynthesis of L-glutamine, with a free energy change of about -5.2 kcal mol^{-1} at pH 7 and $37°C$ [3, 4]. Glutamine is an important intermediate in the assimilation of ammonia by E. coli. The amide nitrogen of glutamine is utilized in the biosynthesis of AMP, CTP, L-tryptophan, L-histidine, glucosamine 6-phosphate, and carbamyl phosphate [5], and also of L-glutamate in a TPNH-dependent reaction catalyzed by glutamate synthase [6, 7]. A coupling of the reactions catalyzed by glutamine synthetase, glutamate synthase, and various transaminases additionally provides a pathway for ATP-dependent synthesis of most amino acids [7]. Consequently, glutamine synthetase has a central role in the nitrogen metabolism of this microorganism [8].

The complex regulation of glutamine synthetase activity in E. coli has been reviewed previously [8-14]. Control of glutamine synthetase activity in this microorganism involves the following regulatory mechanisms: (1) repression of enzyme synthesis by high concentrations of ammonia salts [10, 15]; (2) feedback inhibition by end products of glutamine metabolism [8, 16, 17]; (3) environmental availability and type of divalent cation present [18-22]; (4) adenylylation and deadenylylation of glutamine synthetase modulated by a cascade system consisting of several metabolite-regulated enzymes and a small regulatory protein [13, 14, 23, 24]. The covalent modification of glutamine synthetase by adenylylation [25] or deadenylylation dramatically affects the catalytic potential and divalent cation specificity of the enzyme (1, 2) and, to a lesser extent, the susceptibility of glutamine synthetase to feedback inhibition in biosynthetic catalysis [2, 20]. The latter two regulatory mechanisms (3 and 4) will be included in the present discussion. Subunit interactions that arise from both the dodecameric structure of glutamine synthetase and the introduction of 5'-adenylyl groups will be considered also. The direct and indirect structural alterations induced by the novel, regulatory mechanism of adenylylation provide an example of single site modification affecting both intra- and intersubunit interactions of an oligomeric protein.

II. PHYSICAL AND CHEMICAL PROPERTIES

A. Molecular Structure

Some physical and chemical properties of E. coli glutamine synthetase, together with references to particular studies, are summarized in Table 1. These studies have been reviewed elsewhere [26].

The native enzyme has 12 identical subunits which are molecularly arranged in two face-to-face hexagonal rings (Fig. 1). Figure 1a shows a picture of a 75% adenylylated enzyme preparation [200 µg $GS_{\bar{5}}$/ml in 0.01 M imidazole buffer (pH 7.0) with 0.001 M $MnCl_2$ after 20 min fixation with glutaraldehyde (0.5%, pH 7.3)]. The line on the photograph is equal to 500 Å. In Figure 1b, the binding of the subunits to make a closed ring is shown to depend on sites A and B which participate in heterologous associations. These sites are separated on any given unit by 120° so that the resultant structure is a hexagon. The sites C and D were assigned to isologous associations between the two rings; a second hexagon inverted on top of the first has all C sites above associating with D sites below, and similarly D sites bind to C sites. The center indicates a sixfold axis perpendicular to the page, and the arrows indicate twofold axes in a plane parallel to the page and halfway between the two hexagonal layers.

In hydrodynamic studies, the enzyme behaves as a spherical particle of 600,000 mol wt. The dodecamer is very stable in the presence of the activating divalent cations, Mn^{2+} or Mg^{2+}; certain monovalent cations also stabilize the quaternary structure. A specific tyrosyl residue of each of the 12 subunits can be adenylylated enzymatically to form a stable 5'-adenylyl-O-tyrosyl derivative [35]. Amino acids of the subunit polypeptide chain have been partially sequenced in the regions of the adenylylation site and of the amino and carboxy terminals [34, 36]. The carboxy terminal region of each subunit in the native structure apparently is buried in such a way as to make it quite resistant to digestion by carboxypeptidase A [26]. The intact dodecameric structure of glutamine synthetase is required for activity expression [19, 31, 37] and for the covalent attachment of 5'-adenylyl groups catalyzed by the adenylyltransferase [37, 38]. The specific tyrosyl residue of each subunit involved in adenylylation was found by Cimino et al. [39] not to be the most reactive tyrosyl group in the native structure toward chemical modification; a selective nitration of the enzyme with tetranitromethane or acetylation of the enzyme with N-acetyl-imidazole modified tyrosyl residues other than those at the adenylylation sites.

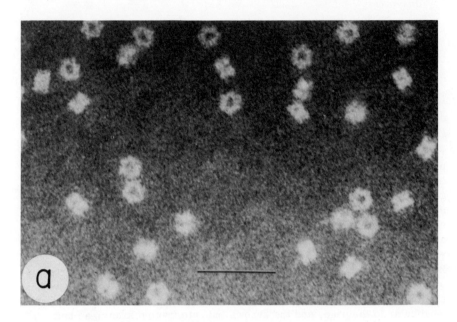

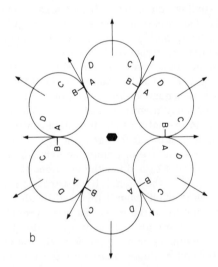

FIG. 1. Electron micrograph of glutamine synthetase from E. coli (a)
and the molecular geometry (b) proposed by Valentine et al. [32] for the
association of twelve identical subunits showing sixfold dihedral symmetry.
(Courtesy of the American Chemical Society.)

TABLE 1

E. Coli Glutamine Synthetase

		Ref.
Native Protein at pH 7		
Molecular weight	600,000	19
Apparent specific volume	~ 0.707 ml/g	19
Sedimentation coefficient ($s_{20,w}^{\circ}$)	20.3S	19
Isionic pH	4.9	27
α-Helical structures	$\sim 36\%$	28
Specific absorption at 2900 Å ($A_{1\ cm}^{0.1\%}$)	0.385	29
Reactive sulhydryl groups	–	30
Stability in presence of Mn^{2+}	+	15,19,31
Removal of Mn^{2+} or Mg^{2+}	Inactivation	18
Appearance in electron microscopy	Dodecamer: double hexagon	32
Hexagonal ring cross section	129Å	33
Subunit		
Molecular weight	50,000	15,32
Molecular dimensions	45 x 45 x 53Å	32
Sulfhydryl groups	5 (no disulfides)	30
Tyrosyl residues	15	25
Tryptophanyl residues	3	28
Amino terminal	Serine	34
Carboxy terminal	Valine	26,34
Adenylylation site	R-Asp-Asn-Leu-Tyr-Asp-R' $\overset{\mid}{AMP}$	35,36
(where R = H_2N-Ile-His-Pro-Gly-Glu-Ala-Met-Lys-; R' = Leu-Pro-Pro-Glu-Gly-Glu-Ala-Lys-COO$^-$)		36

Purified preparations of adenylylated and unadenylylated glutamine synthetase forms have the same amino acid composition but exhibit an ultraviolet difference spectrum that is the spectrum of 5'AMP [25]. Covalently bound 5'-adenylyl groups do not contribute to or perturb the absorbance of the protein at 2900Å (Table 1), while contributing linearly to the absorbance at 2600Å in an amount equivalent to the same concentration of free AMP [29, 40]. Thus, an average state of adenylylation ($\bar{n}$) of a purified enzyme preparation can be estimated from:

$$\bar{n} = 15.0 \; \frac{(A_{2600\text{Å}} - 2.92A_{3400\text{Å}})}{(A_{2900\text{Å}} - 1.89A_{3400\text{Å}})} - 13.3 \tag{2}$$

where $0 \le \bar{n} \le 12$, and the absorbancies at 2600Å and at 2900Å are corcorrected for any light scattering produced by the protein solution by subtracting the appropriate factors times the apparent absorbance measured at 3400Å [40]. The specific absorption coefficient at 2800Å can be calculated from the following empirical formula [40].

$$A^{0.1\%}_{2800\text{Å}, \; 1 \; cm} = 0.733 + 0.05 \, (\bar{n}/12) \tag{3}$$

The molecular geometry proposed by Valentine et al. [32] for the association of 12 identical subunits into a molecule having sixfold dihedral symmetry is shown in Figure 1b. The long dimension of each subunit (Table 1) is in the radial direction of the hexagonal ring. The monodisperse double hexagons observed in electron microscopy (Fig. 1a) have a molecular symmetry of 622, with the hexagonal rings joined face-to-face so that top and bottom surfaces are identical. The double hexagon structure was also observed [32] to be the basic unit in metal ion-induced tubular aggregation, which is initiated by the removal and readdition of Mn^{2+}, Mg^{2+}, or Co^{2+} to the enzyme in dilute buffer solution. More recently, Zn^{2+} addition in the presence of $MgCl_2$ has been shown by Miller et al. [41] also to promote paracrystalline aggregation of glutamine synthetase. The fact that dihedral symmetry is maintained in the repeating monomer units [32] means that metal ion-induced aggregation occurs at the molecular level. Interactions within the double hexagon ring structure occasionally may expose binding sites on a face which then will promote intermolecular associations in linear polymer formation.

B. Divalent Cation Activation

Glutamine synthetase is inactivated by the removal of Mn^{2+} or Mg^{2+} (Table 1). Consequently, fluctuations in the environmental free concentrations of specific divalent cations in the cell could be important. The inactivation of glutamine synthetase by the removal of Mn^{2+} causes a conformational change (relaxation) in the protein structure that leads to an exposure of sulfhydryl groups [30], to an increased susceptibility to dissociation [31], and (without a change in molecular weight) to an ultraviolet spectral perturbation (Fig. 2) and a decrease in sedimentation rate of $\Delta s_{20,w}^{\circ} = -0.6S$ [19].

Relaxation, however, does not produce changes in the secondary structure that can be detected by optical rotatory dispersion or circular dichroism measurements [28]. A tightening of the metal ion-free enzyme by the addition of Mn^{2+}, Mg^{2+}, or Ca^{2+} reactivates glutamine synthetase [18] and reverses the conformational changes produced by relaxation [19, 30]. As shown in Figure 2, the ultraviolet difference spectra produced by adding Zn^{2+}, Co^{2+}, or Cd^{2+} to the metal ion-free enzyme are much smaller in magnitude than that induced by the addition of Mn^{2+}, Mg^{2+}, or Ca^{2+} to relaxed enzyme. Certainly, Zn^{2+} [41], Co^{2+}, or Cd^{2+} [21] stabilize a distinctly different conformation of glutamine synthetase than does Mn^{2+}, Mg^{2+}, or Ca^{2+} [19, 28, 43].

Associated with the tightening or activation process is the binding of one metal ion (Mn^{2+}, Mg^{2+}, or Ca^{2+}) per enzyme subunit [28, 42, 43]. The tightening process produced by Mn^{2+} is first order with respect to unreacted enzyme in rate measurements of ultraviolet absorbency changes (Figs. 2 and 3) and has an Arrhenius activation energy of 21 kcal (mol subunit-Mn)$^{-1}$ [28].

The kinetics of the interaction of glutamine synthetase with activating and stabilizing divalent cations have some interesting features [18, 19, 28, 43, 44]. The binding of either Mn^{2+}, Mg^{2+}, or Ca^{2+} to an enzyme subunit of the dodecamer causes the release of two protons: one proton is displaced from the enzyme in a rapid reaction while the other is released in a slow first-order process. The half-time of the slow proton release (Fig. 3) is similar to that measured for activation [18] or for the reburial of aromatic residues in ultraviolet difference spectral measurements [19, 28]. The interaction of Mn^{2+}, Mg^{2+}, or Ca^{2+} with metal ion-free glutamine synthetase thus appears to occur sequentially as a rapid bimolecular reaction followed by a slow first order reaction, involving an induced conformational change in the enzyme structure. Certainly, this is an oversimplification since calorimetric [44] and fluorescence [18, 45] measurements have detected additional steps in the tightening process. Figure 3 shows that longer half-times were observed at $37\,^{\circ}C$ in calorimetric measurements in which a small endothermic contribution was not resolved from the slow proton release in the apparently first-order slow thermal process [44]. In fluorescence measurements with a hydrophobic probe [18], the addition of divalent cations to the metal ion-free enzyme causes a rapid increase in fluorescence. In recent measurements utilizing the intrinsic fluorescence of tryptophanyl residues [45], the addition of Mn^{2+} or Mg^{2+} to the relaxed enzyme produces a rapid fluorescence decrease in the millisecond time

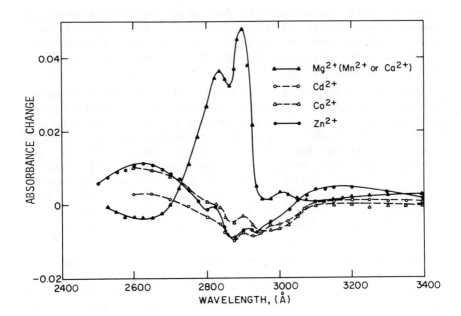

FIG. 2. Ultraviolet difference spectra at 25° C for unadenylylated gluta-
mine synthetase with Mg^{2+} (Mn^{2+} or Ca^{2+}) or Zn^{2+} (2.0 mg $GS_{\bar{1}}$/ml) or
with Co^{2+} or Cd^{2+} (1.9 mg $Gs_{\bar{1}}$/ml) vs the same concentration of metal
ion-free enzyme. Solutions of enzyme were in 0.01 M imidazole − 0.1 M
KCl buffer at pH 7. The Mg^{2+} curve illustrates difference spectra produced 25 min after the addition of 1 mM $MgCl_2$, 1 mM $MnCl_2$, or 1 mM
$CaCl_2$ to the metal ion-free enzyme in the sample compartment. A stoichiometric amount of Zn^{2+} per subunit (0.04 mM $ZnSO_4$) produced the
final absorption change recorded after 3 min; a doubling of the $ZnSo_4$ concentration had little effect on this difference spectrum. Difference spectra produced by 0.1 mM $CoCl_2$ ($\triangle$-----$\triangle$) or 0.1 mM $CdCl_2$ (o-----o) to
relaxed enzyme are shown also. The ultraviolet difference spectral data
were compiled from results of Shapiro and Ginsburg [19], Segal and
Stadtman [21], R. E. Miller et al. [41], and Hunt and Ginsburg [28, 43].

range. A comparison of the kinetics of the ultraviolet spectral change at
2900Å with those of the intrinsic fluorescence change of tryptophanyl
residues in the tightening process suggests that different parameters are
being measured.

The calorimetric experiments of Hunt et al. [44] showed that little net
heat is associated with the interaction of Mn^{2+} (or Mg^{2+}) with glutamine
synthetase. For the binding of each of 12 equivalent Mn^{2+} to unadenylylated

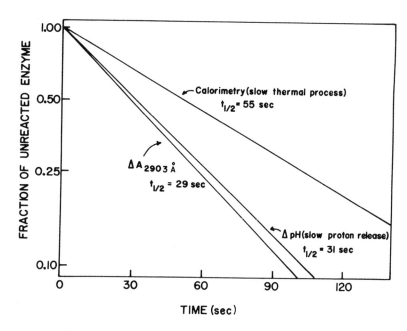

FIG. 3. Idealized first order rate plots for the tightening process. These rate plots were constructed from an average of half-time values observed at 37° C and pH 7.2 in measurements of the slow thermal process [44], of the slow proton release by pH changes in dilute buffers [28], or of the ultraviolet absorbance change at 2903 Å [28] for the binding of 12 molar equiv of Mn^{2+} (or Mg^{2+}) to unadenylylated glutamine synthetase. (From Ref. 14, p. 19, courtesy of Academic Press.)

glutamine synthetase [K_{eq} = $K'/(H^+)^2$] at 37° C and pH 7.2, the following thermodynamic parameters were indicated: $\Delta G'$ = -8.9 kcal mol^{-1} (standard state for hydrogen ions at activity of $10^{-7 \cdot 2}$ M), $\Delta H \simeq$ +3 kcal mol^{-1}, and $\Delta S' \simeq$ +38 cal deg^{-1} (mol of subunit-Mn)$^{-1}$. A similar large entropy increase was observed for the binding of Mg^{2+} to the enzyme under the same conditions. As was established for other protein-metal ion interactions [46], a large increase in entropy might result from a charge neutralization of solvated carboxylate ions upon complex formation with an attendant liberation of water. If so, the activating metal ion binding site of the subunit will be relatively less polar after chelation with Mn^{2+} or with Mg^{2+}. Displacement of the two protons from the enzyme subunit by its reaction with the activating divalent cation at neutral pH normally is related to the ionization of nitrogenous groups, but other ionizable groups may be involved. These amino acid residues may form a part of the metal ion binding cluster; alternatively, one or both groups may be structurally perturbed in the tightening process so that there is an induced ionization.

C. Catalysis of γ-Glutamyl Transfer

In addition to the biosynthetic reaction [Eq. (1)], glutamine synthetase also actively catalyzes the transfer of the glutamyl group from glutamine to hydroxylamine [15]:

$$\text{L-glutamine} + \text{NH}_2\text{OH} \xrightarrow[\text{ADP, arsenate}]{\text{Me}^{2+}} \gamma\text{-glutamylhydroxamate} + \text{NH}_4^+ \quad (4)$$

In contrast to the biosynthetic reaction (see Sec. IV, A below), heterologous subunit interactions within hybrid glutamine synthetase molecules (i.e., molecules containing both adenylylated and unadenylylated subunits) are not evident in reaction (4). In fact, Mn^{2+} supports the γ-glutamyl transferase activity of both types of subunits [2, 10-12, 47]. With a mixed buffer of imidazole, 2-methyl imidazole, and 2,4 dimethylimidazole there is an isoactivity point at pH 7.15, at which pH Mn^{2+} gives equivalent activation of glutamine synthetase forms that are adenylylated to different extents [48]. Although each subunit of the dodecamer appears to express γ-glutamyl transfer activity independent of the other subunits, adenylylation of the subunit profoundly influences the catalysis by that subunit as follows: the pH optimum is changed from pH 8.0 to pH 6.9 [12, 48], the affinity of the subunit for Mn^{2+}, glutamine, and ADP is decreased [42, 43, 49], and the sensitivity toward feedback inhibitors is altered [25], as is also the nucleotide and divalent cation specificity [8, 43]. Certain of these properties have provided useful assay methods [43, 48] for determining the average extent of adenylylation ($\bar{n}$) of glutamine synthetase forms in either a pure or impure state. The assay methods obey the following linear expression, which has been shown to be valid for many enzyme preparations of extreme and intermediate adenylylation states [48]:

$$\bar{n} = 12 - 12(b)/(a) \quad (5)$$

where $\bar{n}$ may vary from 0 to 12 mol 5'-adenylyl groups per mole of active enzyme; (a) is a measure of the total transferase activity of both adenylylated and unadenylylated subunits; (b) is a measure of the transferase activity of only unadenylylated subunits. In one method, the value of (a) is determined by measuring transferase activity in the presence of 0.3 mM Mn^{2+} at pH 7.15 in the mixed imidazole buffer, and the value of (b) is determined by measuring activity in the presence of 0.3 mM Mn^{2+} plus 60 mM Mg^{2+} at pH 7.15 [48]. Alternatively, the total transferase activity, (a), can be

measured at pH 7.57 in 3,3'-dimethylglutarate-triethanolamine buffer
with 0.3 mM Mn^{2+} plus 0.05 mM ADP, while the transferase activity of
only unadenylylated subunits in (b) is measured under the same conditions
with the ADP concentration decreased to 0.005 mM [43]. This latter
method is valid for purified preparations of glutamine synthetase, and it is
based on the observation [43] that adenylylation of the enzyme increases
K'_m for ADP, from $0.03 \mu M$ to $85 \mu M$ in the Mn^{2+}-dependent γ-glutamyl transfer
assay. The γ-glutamyl transferase activity of unadenylylated, but not of
adenylylated, glutamine synthetase can be supported by Mg^{2+}, Ca^{2+},
Zn^{2+}, or Co^{2+} [43]. Cd^{2+}, like Mn^{2+}, supports the γ-glutamyl transferase activity of either type of subunit [43].

   Binding and kinetic data for the Mn^{2+}-supported γ-glutamyl transfer
reaction catalyzed by the unadenylylated form of glutamine synthetase (GS)
have been correlated recently [43]. Each of the twelve subunits of the
enzyme catalyzes the reaction (Sec. II, D below), and there is an absolute
requirement for two Mn^{2+} binding sites of the subunit being occupied for
activity expression. The binding of Mn^{2+} to these two subunit sites is
random (i.e., independent). Glutamine, by causing a net uptake of protons
in binding to the enzyme, reduces from three to about one the number of
protons released upon the binding of Mn^{2+} to these two subunit sites. A
mechanism for forming an active Mn_2-enzyme-ADP complex (EM_2A) which
is consistent with the observations is:

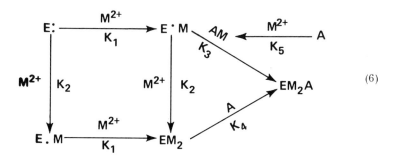

(6)

where E: = free GS subunit, M^{2+} = free Mn^{2+}, E·M or E.M = GS-Mn
subunit with the second or first Mn^{2+} binding site free, respectively,
EM_2 = GS-Mn_2 subunit with both Mn^{2+} sites occupied, A = ADP, AM =
ADP-Mn, $K_1 = 1.89 \times 10^6 \ M^{-1}$ (not affected by glutamine or arsenate),
$K_2 = 5.9 \times 10^5 \ M^{-1}$ (increased $\sim$ thirtyfold by glutamine), $K_3 = 2.86 \times 10^7$
M^{-1} (increased $\sim$ one hundredfold by arsenate), $K_5 = 7.6 \times 10^3 M^{-1}$, and
$K_4 = (K_3 \cdot K_5)/K_2 = 3.7 \times 10^5 \ M^{-1}$.

In assays with 1 μM ADP, EM_2A is formed mainly by the lower route. In this case, [AM] ≤ 10^{-8} M, and activity expression is roughly proportional to the probability of two Mn^{2+} being bound to the same subunit [43]. With 50μM ADP in assays, [AM] > 10^{-8} M and active EM_2A complex formation is mainly by the upper path; (i.e., AM addition to the second Mn^{2+} binding site). Under these conditions, calculated ratios of active EM_2A complex to total enzyme subunit concentration at various free Mn^{2+} concentrations gave a good fit of activity data [43].

With ADP, inorganic phosphate, and arsenate, there are ligand-induced fluorescent changes of the unadenylylated enzyme in the presence of Mn^{2+} or Mg^{2+} [50]. A random addition of ADP and phosphate (or arsenate) and a reciprocal enhancement of binding (particularly with the Mn^{2+}-enzyme) were observed [50]. In the presence of Mn^{2+} and glutamine, arsenate markedly increases the affinity of the unadenylylated enzyme subunits for ADP [43] [K_3 of the scheme (6)]. This may be due to an arsenate-induced conformational change in the enzyme or to formation of an arsenate· Mn· ADP complex at the enzyme surface. There is, however, no evidence that such a ternary complex exists in solution. Nevertheless, if Mg^{2+} is substituted for Mn^{2+} in the assay of the unadenylylated enzyme, K_m' for ADP is increased from 3 x 10^{-8} M to 2 x 10^{-4} M ADP [43]. Adenylylation also substantially lowers the affinity of glutamine synthetase for ADP [43, 49, 50]. It has been proposed [51] that the inactivity of the unadenylylated enzyme in Mn^{2+}-supported biosynthetic catalysis (reaction 1) arises, in fact, from the high affinity of this enzyme form for ADP in the presence of phosphate and Mn^{2+}. Orthophosphate is a potent inhibitor of reaction (1), as might be anticipated from the above considerations.

Ba^{2+} by itself does not support reaction (4), nor is Ba^{2+} effective in tightening or reactivating the divalent cation-free enzyme [18, 19]. Nevertheless, Ba^{2+} (5 mM) has the capacity to activate 1.5-fold the Mn^{2+}-supported activity of the unadenylylated enzyme at pH 7.0 [43]. The activating effect of Ba^{2+} is pH-dependent. A similar activation by 200 mM KCl (with NaCl slightly less effective) is observed, although the activating effect of Ba^{2+} at pH < 7.6 is considerably more specific than that of KCl or of a simple ionic strength effect. Ba^{2+} appears to bind to different metal ion binding sites of glutamine synthetase than those high affinity sites binding Mn^{2+} and, in so doing, enhances catalytic activity and antagonizes the binding of Mn^{2+} by the enzyme [43].

D. Stoichiometry in Binding Effectors and Substrates

The divalent cation Mn^{2+} binds to glutamine synthetase in sets of 12 which are separable on the basis of affinity [42]. The affinity of the enzyme for the first 12 molar equivalents Mn^{2+} bound, which are involved in the tightening or reactivation process (Sec. II, B), is decreased by the adenylylation of glutamine synthetase [42]. Each subunit appears to have apparently equivalent, noninteracting, Mn^{2+}-binding sites of three affinity types: at pH 7.2 in the presence of M/10 KCl, $K_1' = 2 \times 10^5 - 10^6 \ M^{-1}$ [28, 42]; $K_2' = 2 \times 10^4 \ M^{-1}$ [28, 42]; $K_3' = 5 \times 10^2 \ M^{-1}$ [42]. Under the same conditions, the enzyme also has 12 high affinity sites for Mg^{2+} with $K' = 2 \times 10^4 \ M^{-1}$ [28], for Co^{2+} with $K' = 5 \times 10^4 \ M^{-1}$ [28], for Ca^{2+} with $K' = 5 \times 10^4 \ M^{-1}$ [43], or for Zn^{2+} with $K' \simeq 10^7 \ M^{-1}$ [43]. The enzyme, however, is capable of binding simultaneously 12 equivalents of each Zn^{2+} and Mn^{2+}, with the Mn^{2+} addition producing the usual ultraviolet difference spectrum illustrated in Figure 2 [43]. Another indication that the highest affinity sites for Mn^{2+} and Zn^{2+} are different is that the binding of one equivalent of Zn^{2+} per subunit causes the displacement of only one proton equivalent [28]. It may be recalled that much smaller ultraviolet spectral perturbations are produced by Zn^{2+} [41], Cd^{2+}, or Co^{2+} [21] than by Mg^{2+}, Mn^{2+}, or Ca^{2+} (Fig. 2).

Each of the 12 subunits of glutamine synthetase has a potential catalytic site, the activation of which depends on the divalent cation present (Sec. II, C and IV, A). In turn, the divalent cation specificity is dictated by the absence or presence of a covalently bound 5'-adenylyl group to the enzyme subunit [Eq. (2)]. Both kinetic and equilibrium binding studies support this view. The reliability of expression (5) of Sec. II, C for determining the average state of adenylylation by γ-glutamyl transferase activity measurements suggests that each subunit of the dodecamer catalyzes this reaction in the absence of added Mg^{2+}. Furthermore, a kinetic analysis [43] by the method of continuous variation developed by Job [52, 53] shows that there is a 1 : 1 complex formation between glutamine synthetase subunits and ADP (Fig. 4) in activity expression. The kinetic analysis of Figure 4 is possible because ADP is a nonconsumable substrate in the γ-glutamyl transfer reaction (4) and because the stability constant of the unadenylylated enzyme subunits for ADP-Mn is sufficiently high (Sec. II, C above). Thus, the activation of glutamine synthetase by ADP is proportional to binding. In direct binding measurements with unadenylylated enzyme preparations, either ATP-Mn or ADP-Mn was bound to the extent of one per subunit, with half-saturation occurring at 5×10^{-6} M ATP-Mn [54] or 3×10^{-6} M ADP-Mn [43].

The feedback inhibitors AMP and L-tryptophan were each found to bind to the extent of 12 equivalents per mole glutamine synthetase in equilibrium binding studies [55]. The divalent cation (Mn^{2+} or Mg^{2+}) present or the state of adenylylation of the enzyme did not influence the binding of either inhibitor. The binding sites of the enzyme for AMP and L-tryptophan were shown to be separate by binding and calorimetric studies [55, 56]. The calorimetric results presented in Table 2 show that the sum of the heats (Q) measured for the individual effectors was equal within experimental error to that measured for a saturating mixture of these inhibitors [56].

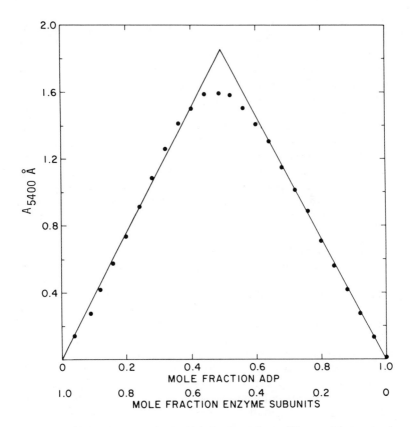

FIG. 4. Stoichiometry of ADP binding to glutamine synthetase in the γ-glutamyl transfer reaction. The analysis is by the procedure of Job [52, 53] in which the mole fraction of ADP and mole fraction of enzyme subunits were varied reciprocally from 0 to 1.0. The γ-glutamyl transferase assay mixtures at pH 7.25 contained 30 mM L-glutamine, 20 mM NH$_2$OH, 2.0 mM tris-AsO$_4$, 0.1 M KCl, 1.0 mM MnCl$_2$, and 0 to 0.3 μM unadenylylated enzyme subunits. After incubating for 30 min at 37°C, the reactions were stopped with FeCl$_3$ reagent [29] and absorbencies at 5400 Å determined. (From Ref. 43, courtesy of Academic Press.)

Since glutamine synthetase is a dodecamer, the results strongly suggest that each subunit has separate allosteric sites for binding AMP and L-tryptophan. Separate sites for most, if not all, of the other feedback inhibitors have been suggested also from kinetic evidence [17], the effects of inhibitors on the inactivation of glutamine synthetase by mercurials [30], and binding studies [55]. There are, however, interactions

TABLE 2

Binding and Calorimetric Studies on the Interaction of
Glutamine Synthetase with Inhibitors [a]

Experiment (inhibitors present at saturating concentrations)	$[S]_{0.5}$ [b] $(M \times 10^4)$	Q [c] (mcal/mg enzyme)
I. GS + L-tryptophan	9.1	-0.147
II. GS + AMP	1.25	-0.040
III. GS + L-tryptophan + AMP	—	-0.192

[a] Binding and calorimetric data are from Ref. 55 and Ref. 56, respectively. The calorimetric experiments were performed at 25°C, using a buffer of 0.02 M imidazole-chloride-0.1 M KCl-0.001 M $MnCl_2$ at pH 7.07 and 17-33 mg of unadenylylated glutamine synthetase (GS) at 4-11 mg/ml concentration.

[b] $[S]_{0.5}$ is the concentration of inhibitor required for half-saturation of the enzyme, with saturation occurring when 12 mol of each inhibitor are bound per mole of glutamine synthetase.

[c] Q is the measured heat of binding after correction for a small endothermic heat of dilution of the enzyme (GS) in a batch-type microcalorimeter.

between inhibitor sites and between substrate and inhibitor sites [8, 17, 30, 55].

E. Antigenic Determinants

Tronick et al. [57] have induced antibodies in the rabbit to purified unadenylylated and adenylylated forms of glutamine synthetase from E. coli. Complement fixation studies with purified γ-globulin fractions indicated that adenylylated and unadenylylated enzyme forms share a number of antigenic determinants; therefore, an immunological separation of these forms was not possible. However, a striking immunological homology was found in the cross-reactivity of antibodies prepared against the E. coli enzyme with glutamine synthetases from other microorganisms. The antibodies to E. coli glutamine synthetase cross-reacted in immunodiffusion assays with glutamine synthetases from five other organisms for which there is evidence of an adenylylation system of regulation in vivo. In contrast, there was no cross-reaction of these antibodies either with the glutamine synthetase from the gram-positive organism Bacillus subtilis [57, 58] or with the mammalian glutamine synthetase from sheep brain, which are two enzymes that are not adenylylated [9, 37]. This immunological specificity occurs despite a pronounced similarity in the gross physical structures of the E. coli and B. subtilis glutamine synthetases [58]. Since the ability to covalently modify the enzyme by enzymatically catalyzed adenylylation correlated well with the immunological cross-reactivity [57], the antigenic determinants perhaps relate to the specificity of the adenylylation-deadenylylation enzyme system. The sites on glutamine synthetase for the modifying enzyme system, which would be comprised of the sites of adenylylation together with the specific conformation of neighboring groups, could be the major antigenic determinants.

III. REGULATION BY ENZYME CATALYZED ADENYLYLATION AND DEADENYLYLATION

Adenylylation of glutamine synthetase markedly affects various catalytic parameters of the enzyme [8]. Adenylylation and deadenylylation are controlled by a metabolite-regulated cascade enzyme system. Some of the properties of the adenylylation-deadenylylation enzyme system will be summarized here.

Both adenylylation [Eq. (7)] and deadenylylation [Eq. (8)] reactions are catalyzed by adenylyltransferase (ATase) in the presence of Mg^{2+} [14, 23, 59, 60].

$$12 \text{ ATP} + \text{GS} \xrightarrow[\text{Me}^{2+}]{\text{ATase}} \text{GS} \cdot (\text{AMP})_{12} + 12 \text{ PP}_i \tag{7}$$

$$\text{GS} \cdot (\text{AMP})_{12} + 12 \text{ P}_i \xrightarrow[\text{Me}^{2+}]{\text{ATase}} \text{GS} + 12 \text{ ADP} \tag{8}$$

In reaction (7), unadenylylated glutamine synthetase (GS) can be adenylylated to the extent of 12 moles of 5'-adenylyl groups incorporated per mole enzyme [2], with a corresponding release of pyrophosphate [61]. Mantel and Holzer [62] have shown that reaction (7) is reversible and that the adenylyl-O-tyrosyl bond in glutamine synthetase is apparently energy rich. A value of $\Delta G_{obs}^{o} = -1.0$ kcal/mol is calculated [63] from extrapolated equilibrium measurements [62, 64] at pH 7.0 and $pMg^{2+} = 2$.

In reaction (8), deadenylylation of adenylylated glutamine synthetase (GS·AMP) was shown by Anderson and Stadtman [65] to occur by a phosphorolytic cleavage of the adenylyl-O-tyrosyl bond. This reaction is exergonic by about 1.0 kcal/mol also but is apparently irreversible.

Since both adenylylation and deadenylylation reactions are catalyzed by the same enzyme, coupling between these reactions must be prevented by an appropriate control of each function. Otherwise, a futile cycle will exist in which glutamine synthetase fluctuates between adenylylated and unadenylylated forms, and ATP will undergo phosphorolysis to ADP and pyrophosphate, as shown by the sum of reactions (7) and (8) in Eq. (9).

$$12 \text{ ATP} + 12 \text{ P}_i \longrightarrow 12 \text{ ADP} + 12 \text{ PP}_i$$

$$(\Delta G^o_{obs} = -2.0 \text{ kcal/mol})$$

(9)

An aimless coupling of reactions (7) and (8) is prevented by an elaborate regulatory system involving metabolite control of the adenylyltransferase (Table 3 and an interaction of adenylyltransferase with a small regulatory protein (P_{II}), which exists in two interconvertible forms (Fig. 5).

The scheme of Figure 5 and the properties of the adenylyltransferase outlined in Table 3 summarize our present knowledge of the cellular control of the adenylylation and deadenylylation reactions. At least four discrete proteins and a cascade-type of enzyme regulation are involved [66]. The adenylyltransferase of 130,000 $\pm$ 15,000 mol wt [38, 67, 68, 69] and the P_{II} regulatory protein of about 46,000 mol wt [23, 60, 70, 71] are easily separated by gel filtration [59, 70]. The uridylylation of the small regulatory protein is catalyzed by an enzyme (of about 160,000 mol wt) that has been recently separated from the adenylyltransferase by column chromatography [24, 66]. The uridylyltransferase (UTase) is activated by Mg^{2+} or Mn^{2+}, α-ketoglutarate, and ATP [23] in catalyzing the covalent attachment of UMP to P_{IIA} to form P_{IID} [23, 24]. The uridylylated regulatory protein (P_{IID}) is the form of P_{II} that stimulates ATase-catalyzed deadenylylation. Glutamine and inorganic phosphate are potent inhibitors of the uridylylation reaction [23]. The unmodified form of the regulatory protein (P_{IIA}) can be regenerated from P_{IID} by the action of a Mn^{2+}-dependent uridylyl removing (UR) enzyme that catalyzes the hydrolytic cleavage of UMP from P_{IID} [24]. The unmodified regulatory protein (P_{IIA}) interacts with ATase to stimulate the adenylylation reaction. Since the UR-enzyme activity has not as yet been separated from UTase [24, 71], it is possible that one protein is responsible for both activities involved in the interconversion of the regulatory protein ($P_{IIA} \rightleftharpoons P_{IID}$). The preliminary studies of Mangum et al. [24] indicated that Mg^{2+} cannot replace Mn^{2+} in the deuridylylation reaction; however, more recent studies of Adler et al. [66] indicate that Mg^{2+} will support deuridylylation if ATP and α-ketoglutarate are included in the reaction mixture. Specific metabolite controls (more of which may be discovered later) of the P_{IIA}-P_{IID} interconversion (Fig. 5) prevent a futile cycle of UTP hydrolysis to UMP and PP_i, which is the net reaction involved.

TABLE 3

Some Catalytic Properties of Adenylyltransferase

Activity	Requirements	Positive Effectors	Negative Effectors	Ref.
Adenylylation (Optimum pH = 8)	Mg^{2+} or Mn^{2+}, ATP, GS	Glutamine, P_{IIA}	α-ketoglutarate, UTP, P_i	1, 2, 12, 23, 24, 37, 67, 72
Deadenylylation (Optimum pH = 7.2)	Mg^{2+} or Mn^{2+} P_i, GS·(AMP)	P_{IID} with ATP & α-ketoglutarate	Glutamine	23, 24, 60, 70

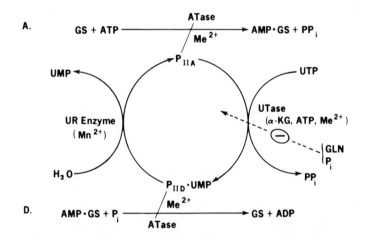

FIG. 5. Scheme of the regulatory protein ($P_{IIA} \rightleftharpoons P_{IID}$) interconversion from the studies of Brown et al. [23] and Mangum et al. [24]. The unmodified form of P_{II} (P_{IIA}) reacts with the adenylyltransferase (ATase) to stimulate adenylylation (A) of glutamine synthetase. The uridylylated form of P_{II} (P_{IID}) interacts with ATase to activate deadenylylation (D) of adenylylated glutamine synthetase. The uridylyltransferase (UTase) is activated by α-ketoglutarate, ATP, and Mg^{2+} or Mn^{2+}, and is markedly inhibited by L-glutamine or orthophosphate. A uridylyl removing (UR) enzyme is specifically activated by Mn^{2+} in hydrolyzing UMP from P_{IID}. (From Ref. 14, p. 27, courtesy of Academic Press.)

As indicated in Table 3, the adenylyltransferase has an absolute requirement for either Mg^{2+} or Mn^{2+} in the adenylylation or deadenylylation reactions. Deadenylylation is activated by P_{IID} (Fig. 6), ATP, and α-ketoglutarate. Adenylylation is stimulated by P_{IIA} (Fig. 6) and/or glutamine, with the divalent cation present also having a specific role [72]. The ratio of activities: (ATase + P_{IIA})/ATase for P_{IIA} stimulation of adenylylation is about 5 in a Mn^{2+}-supported assay without glutamine and about 2 in a Mg^{2+}-supported assay with or without saturating glutamine present. The Mg^{2+}-supported (P_{IIA}-stimulated) activity is extremely responsive toward activation by glutamine (twentyfold) whereas with Mn^{2+}, this activity is almost unchanged by varying glutamine concentrations [14]. Without glutamine, ATase + P_{IIA} is eightfold more active in the Mn^{2+}- than in the Mg^{2+}-supported adenylylation assay.

The studies of Brown et al. [23] illustrated in Figure 6 first suggested a time-dependent conversion of the regulatory protein (P_{II}) to a form that stimulated the ATase-catalyzed deadenylylation of adenylylated glutamine synthetase. With this time-dependent activation of deadenylylation activity, there was a reciprocal decrease in the ability of P_{II} to stimulate the adenylylation of glutamine synthetase. The transformation of the regulatory protein ($P_{IIA} \rightarrow P_{IID}$) illustrated in Figure 6 depends on the presence of UTP, divalent cation, ATP, α-ketoglutarate, and a crude fraction of ATase, which is now known to have contained also the uridylyltransferase [23, 24]. The stoichiometry of ^{14}C-UMP or (α^{32}P)-UMP incorporation into P_{II} during this transformation was found to be slightly more than one equivalent per 50,000 g of P_{II} (which was of $\sim 50\%$ purity) in the studies of Mangum et al. [24]. Later studies [71] have shown that potentially four molar equivalents of 5'UMP can be attached covalently to specific tyrosyl residues of P_{II}. The regulatory P_{II} protein is composed of four apparently identical subunits containing two tyrosyl residues per subunit of 11,400 mol wt [71].

As indicated in Table 3 and in the scheme of Figure 5, there is a reciprocal relationship between metabolites. When the intracellular level of glutamine increases, the adenylylation of glutamine synthetase is activated [73] while the modification of the regulatory protein is blocked [23]. When ammonia is limiting in the growth medium of E. coli, an increase in the intracellular level of α-ketoglutarate will activate both the uridylylation of the regulatory protein and the deadenylylation reaction catalyzed by ATase + P_{IID}, while inhibiting the adenylylation reaction. UTP is required for the formation of P_{IID} from P_{IIA} [23, 24] and UTP also has the capacity to inhibit adenylylation [37]. Since the form of P_{II} specifies the reaction that ATase catalyzes, the ratio of P_{IIA} to P_{IID}, together with the concentrations of other effectors (Table 3), will determine the average state of adenylylation of glutamine synthetase [13]. Thus, the activity of glutamine synthetase is ultimately controlled by the

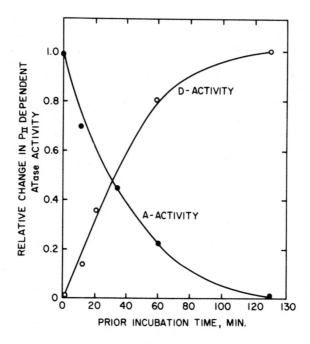

FIG. 6. Data of Brown et al. [23] showing a reciprocal effect of prior incubation of P_{II} with a crude fraction of ATase (containing UTase) on adenylylation (A) and deadenylylation (D) activities of ATase. The enzyme fraction (0.21 mg) and P_{II} (0.10 mg) were incubated at $37^\circ C$ in 0.2 ml containing 0.1 mM ATP, 1.0 mM UTP, 1.0 mM α-ketoglutarate, 1.0 mM $MnCl_2$, 1.0 mM dithiothreitol, and 50 mM 2-methylimidazole (pH 7.2). At the times indicated, 0.02 ml aliquots were assayed for Mg^{2+}-supported A- and D-activities at pH 7.2 (Table 3). Only the relative change in the P_{II}-dependent ATase activity is shown; the A- and D-activities of ATase, pre-incubated without P_{II}, were lower (see text) and are not illustrated here.

various metabolites involved in regulating the activities of the cascade system shown in Figure 5 [66]. Metabolites function in generating either the unmodified or the uridylylated regulatory protein, while simultaneously acting as effectors or substrates of adenylylation and deadenylylation reactions.

IV. SUBUNIT INTERACTIONS

The predominant control of glutamine synthetase activity by adenylylation occurs at the subunit level. This novel regulatory mechanism, however, can provide hybrid molecular species containing both adenylylated and unadenylylated subunits [40, 48]. In fact, it has been calculated [48] that 382 molecular forms are possible, taking into account the symmetrical structure of the enzyme (Fig. 1) and the possible combinations of subunits in molecules containing from 0 to 12 adenylyl groups. It is perhaps of physiological significance that the catalytic potential of each subunit can be affected by subunit interactions. Both homologous and heterologous sub-unit interactions between like subunits and between adenylylated and un-adenylylated subunits, respectively, appear to be involved in the examples considered here.

A. In Biosynthetic Catalysis

The activation of glutamine synthetase in the biosynthetic reaction (1) of Sec. I by Mg^{2+}, Mn^{2+}, or Co^{2+} is intimately linked to the adenylylation state of the enzyme. Adenylylation also affects substrate saturation functions for glutamate, ATP, and NH_4^+ in biosynthetic catalysis, and these functions depend in a complex way upon the divalent cation present [21, 22, 40, 54].

One of the most dramatic changes induced in glutamine synthetase by adenylylation, and one first noted by Wulff et al. [1] and Kingdon et al. [2], is the inactivation of the enzyme in a Mg^{2+}-activated biosynthetic assay. This is illustrated in Figure 7, which also shows that adenylylation produces a reciprocal, although less pronounced, activation by Mn^{2+}. The changed metal ion specificity in the biosynthetic reaction catalyzed by glutamine synthetase upon adenylylation was first observed by Kingdon et al. [2]. Moreover, the pH optimum with Mg^{2+} or Mn^{2+} is different [40, 54]. Since the fully adenylylated form of glutamine synthetase has no activity with Mg^{2+}, Mn^{2+} is concluded to be a specific activator of adenylylated subunits. Conversely, Mg^{2+} or Co^{2+} (Fig. 8) appears to specifically activate unadenylylated subunits of the dodecamer.

Unlike mixtures of unadenylylated and the fully adenylylated enzymes, different preparations of glutamine synthetase exhibit a nonlinear decrease in Mg^{2+}- or in Co^{2+}-dependent activity with increasing adenylylation (Figs. 7 and 8). The increase in Mn^{2+}-activated biosynthetic activity as a function of increasing adenylylation is nonlinear also. Obviously, subunit interactions between adenylylated and unadenylylated subunits in hybrid enzyme molecules must affect the catalytic potential of activated subunits. However, the sigmoidal relationships between activity and state of adenyl-ylation in the cases of Mg^{2+} and Mn^{2+} activations (Fig. 7) are complex and therefore not easily interpretable. In contrast, the concave curve obtained for Co^{2+}-supported activity as a function of adenylylation (Fig. 8) is consistent with the interpretation that within a given hexagonal ring all adenylylated subunits and all unadenylylated subunits in direct contact with

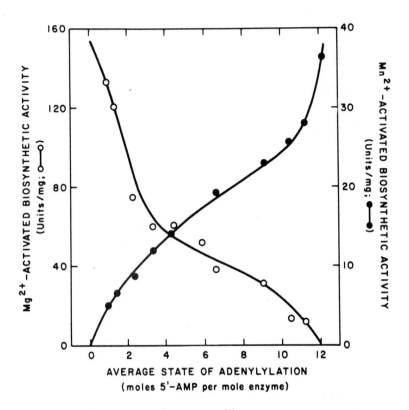

FIG. 7. The variation in Mg^{2+}- and Mn^{2+}-dependent biosynthetic activities (μmoles ADP formed per minute at $37°C$ per milligram protein) as a function of the average extent of adenylylation of the glutamine synthetase preparation. The Mg^{2+}- supported assays at pH 7.6 contained 50 mM $MgCl_2$, 90 mM KCl, 50 mM NH_4Cl, 30 mM L-glutamate, and 5 mM ATP; Mn^{2+}- supported assays at pH 6.5 contained 6 mM $MnCl_2$, 90 mM KCl, 100 mM NH_4Cl, 100 mM L-glutamate, and 5 mM ATP. (The data are from Ginsburg, et al. [40], reprinted by permission of the American Chemical Society.)

adenylylated subunits are inactive [22]. The experimental data of Figure 8 fit quite well the theoretical curve constructed on the assumption that Co^{2+}- activated biosynthetic activity is expressed only by the average number of unadenylylated subunits adjacent to other unadenylylated subunits within a ring of the double hexagon structure [22]. The average number of unadenylylated subunits not in contact with adenylylated subunits within each ring was computed for all unique configurations at each state

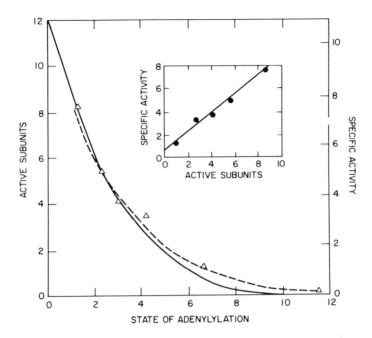

FIG. 8. Heterologous and homologous subunit interactions in hybrid glutamine synthetase molecules that affect Co^{2+}-activation in the biosynthetic reaction. The average number of unadenylylated subunits adjacent only to other unadenylylated subunits within each ring of the double hexagon structure was calculated at different average states of adenylylation (labeled "active subunits") and plotted against the latter quantity in the continuous solid curve. The Co^{2+}-supported biosynthetic activity ($\triangle$---$\triangle$) was determined by measuring inorganic phosphate release per min per mg of enzyme at 37° C in an assay mixture containing 25 mM imidazole-25 mM triethanolamine buffer at pH 7.0, 50 mM L-glutamate, 50 mM NH_4Cl, 7.5 mM ATP, and 7.5 mM $CoCl_2$. In the inset, the Co^{2+}-supported activity is plotted against the calculated active subunits. (From Ref. 22, courtesy of Academic Press.)

of adenylylation, assuming a random distribution of adenylylated subunits within the enzyme molecule. (A nonrandom distribution of adenylylated subunits is not consistent with experimental observations.) A poorer fit of the data of Figure 8 was obtained when the average number of unadenylylated subunits next to like subunits within the same and adjacent rings was calculated for different extents of adenylylation [22]. The interring subunit contact therefore might be less important than the intraring subunit contact in heterologous interactions.

The fact that zero is not intercepted in the inset of Figure 8 could mean that unadenylylated subunits adjacent to adenylylated subunits have some finite activity, but if they are adjacent to other unadenylylated subunits the activity is markedly increased due to homologous interactions. Whatever the mechanism, the marked decrease in the Mg^{2+}- or Co^{2+}-supported activity at the initial stages of adenylylation could be physiologically significant. The covalent attachment of relatively few 5'-adenylyl groups to the glutamine synthetase molecule is accompanied by a greater decrease in catalytic potential in these cases than would occur from a linear response. Since the affinity of the adenylyltransferase for glutamine synthetase appears to decrease with increasing adenylylation [37], the most active enzyme forms would tend to be modified first. It should be pointed out that the data of Figures 7 and 8 were obtained with enzyme forms isolated from different batches of E. coli [15, 29], and with enzyme preparations reisolated after adenylylation in vitro [40]. In either case, hybrid enzyme forms at intermediate stages of adenylylation appear to be present.

Studies of Segal and Stadtman [21] with pairs of divalent cations have demonstrated that two different divalent cations can occupy sites that modulate the expression of biosynthetic activity at the subunit level and also affect subunit interactions in the dodecamer. For example, Cd^{2+} cannot by itself support biosynthetic activity but when present with Mg^{2+} or Co^{2+}, the pH optima and the saturation functions for the substrate glutamate are shifted from those observed with either Mg^{2+} or Co^{2+} alone. The Mg^{2+}-supported activity is very sensitive to inhibition by the addition of Cd^{2+}, Mn^{2+}, or Ca^{2+} [21].

When only a single species of metal ion is present, it seems probable that two metal ion binding sites per activated subunit need to be saturated for the expression of biosynthetic activity. As indicated in Sec. II, C, recent studies [43] show an obligatory saturation of two metal ion binding sites per subunit in the expression of γ-glutamyl transferase activity. However, ATP, an obligatory component of the biosynthetic reaction, has a high affinity for metal ions [74], which complicates the interpretation of kinetically determined saturation functions for divalent cations in the biosynthetic assay. Nevertheless, recent studies [75, 76] indicate that at least two equivalents of Mg^{2+} are bound per subunit — one presumably at the activating site (Sec. II, B) and the other at the catalytic site which accommodates ATP.

Saturation functions for L-glutamate, ammonium ion, and ATP vary according to which divalent cation is used in the biosynthetic reaction. For example, saturation functions for L-glutamate are hyperbolic with Mg^{2+} [54], anticooperative with Mn^{2+} [40, 54], and cooperative with Co^{2+} [22]. Moreover, enzyme forms at different adenylylation states have varying apparent affinities for substrates [40, 54]. If an absolute specificity of unadenylylated subunits for Mg^{2+} or Co^{2+}, or of adenylylated subunits for

Mn^{2+}, is maintained throughout all stages of adenylylation, then heterologous subunit interactions are indicated by such $K'_m ([S]_{0.5})$ shifts observed kinetically with different enzyme forms. In some cases, substrate-induced homotropic interactions [77-82], as well as heterologous interactions between adenylylated and unadenylylated subunits, appear to influence substrate interactions with glutamine synthetase.

As is illustrated by the Hill plots in Figure 9, there is an apparent negative interaction between ATP and the Mn^{2+}-activated enzyme in both equilibrium binding and kinetic saturation functions [54]. The unit slopes at low and high concentrations of ATP are connected by slopes of less than one in the Hill plots of Figure 9. Since the enzyme preparation contained an average of 2.3 equivalents of covalently bound 5'AMP and all subunits bind ATP (Sec. II, D), the binding results are interpretable in terms of adenylylated and unadenylylated subunits of the dodecamer having differing affinities for ATP [82]. Activity, however, is expressed in the kinetic experiments throughout the entire ATP concentration range, with maximal velocity approached only at the highest ATP levels. The activation of adenylylated subunits by Mn^{2+} therefore apparently results in two affinity types, with the apparent affinity for ATP at the low concentrations of this substrate the same as in direct binding measurements. The slope of the corresponding Hill plot of kinetic data for the much more active, fully adenylylated enzyme is unity; a decrease in the average number of unadenylylated subunits reduces the displacement between low and high ATP concentrations [54]. The marked deviation of kinetic and binding data at high ATP concentrations in Figure 9 is not obliterated by lowering the pH of the assay and is due presumably to the influence of L-glutamate and NH_4^+ in the kinetic measurements under the two conditions. The activated enzyme subunits apparently can undergo conformational changes that affect the binding of ATP (and possibly also the velocity), with heterologous subunit interactions influencing the equilibria between different active subunit conformations in a negative or destabilizing manner as described by Koshland et al. [80].

Saturation of glutamine synthetase with L-glutamate in a Mn^{2+}-supported assay is even a more complex function of the adenylylation state of the enzyme [40], with apparent affinity shifts occurring at both low and high substrate concentrations. Again, a negative type of interaction between the Mn^{2+}-activated enzyme and L-glutamate was indicated by bimodal L-glutamate saturation curves in which the affinity of the enzyme for this substrate is apparently greater at low than at high glutamate concentrations. With increasing adenylylation, higher concentrations of L-glutamate are required to saturate the enzyme. A negative type of interaction between the enzyme and NH_4^+ has also been observed in the Mn^{2+}-supported [54] and Co^{2+}-supported [22] biosynthetic assays. There is a synergistic effect between NH_4^+ and L-glutamate in the Co^{2+}-supported assay,

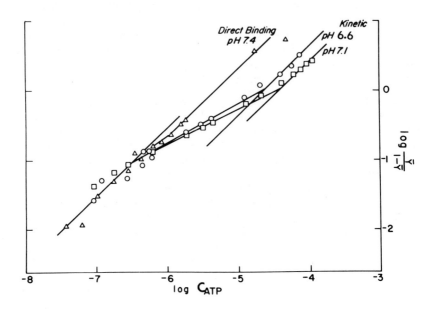

FIG. 9. Hill plots of equilibrium binding and kinetic data obtained for ATP saturation functions in the presence of 5 mM $MnCl_2$ with an unadenylylated glutamine synthetase preparation containing an average of 2.3 moles covalently bound AMP per mole enzyme ($GS_{\overline{2.3}}$). In binding data ($\triangle$) with ^{14}C-ATP in the presence of 5 mM $MnCl_2$, 0.02 M imidazole (pH 7.4) and 0.1 M KCl at $4°$, $\overline{y} = \overline{v}/12$, where $\overline{v}$ is the moles ATP bound per mole enzyme at the indicated free concentration of ATP; kinetic data are expressed as $\overline{y} = v/V_{max}$, where velocities at a given constant ATP concentration were obtained in spectrophotometric biosynthetic assays at $\sim 22°$ containing 5 mM $MnCl_2$ at pH 7.1 or at pH 6.6, and V_{max} is the extrapolated apparent maximum velocity at $v/c_{ATP} = 0$. (From Ref. (54), courtesy of the American Chemical Society.)

although the glutamate saturation function remains sigmoidal at both high and low concentrations of NH_4^+ [22].

In another study of the Co^{2+}-supported activity of glutamine synthetase [22] at subsaturating concentrations of L-glutamate (25 mM) and NH_4^+ (50 mM), low concentrations of L-alanine (8 mM) activated the unadenylylated enzyme 33% but slightly inhibited the adenylylated enzyme. The decrease in the percentage activation by L-alanine under these assay conditions with increasing adenylylation is shown in Figure 10. The percentage activation by this amount of L-alanine was found by Segal and Stadtman [22]

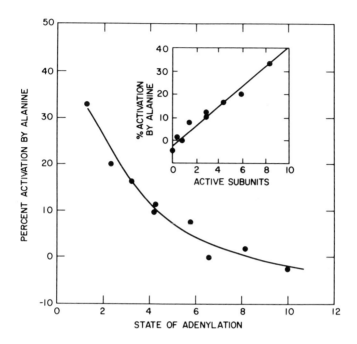

FIG. 10. Percentage activation by L-alanine of glutamine synthetase vs. the average extent of adenylylation of the enzyme preparation. Biosynthetic assays at pH 7.0 contained 7.5 mM $CoCl_2$, 50 mM NH_4Cl, 25 mM L-glutamate, 7.5 mM ATP, with or without 8 mM L-alanine added. In the inset, the same data are replotted as the percentage activation by alanine vs. the average number of unadenylylated subunits adjacent only to unadenylylated subunits within the same ring of the double hexagon (active subunits); see Fig. 8. (From Ref. 22, courtesy of Academic Press.)

to be a linear function of the number of unadenylylated subunits adjacent to like subunits (inset, Fig. 10) in an analysis similar to that used in constructing the inset of Figure 8. In this complex system, L-alanine appears to augment the Co^{2+}-activation of unadenylylated adjacent subunits by promotion of homologous interactions among these subunits. Heterologous interactions between dissimilar subunits appear to diminish the activating effect of L-alanine. At subsaturation of L-glutamate and NH_4^+ in the Co^{2+}-supported assay, glycine or AMP activation of unadenylylated subunits is similar to that produced by L-alanine [22]. At saturating concentrations of L-glutamate and NH_4^+, however, unadenylylated glutamine synthetase is inhibited by L-alanine, glycine, or AMP. High concentrations of these metabolites also inhibit the expression of Co^{2+}-supported activity, regardless of the substrate concentrations [22]. The L-alanine effect shown in Figure 10 is therefore a special case, being quite dependent upon the biosynthetic assay conditions employed.

B. In Binding Effectors

The affinity of the activating metal ion sites for Mn^{2+} that are involved in the tightening process (Sec. II, B) decreases with increasing adenylylation of the enzyme. With 75% adenylylation, the stability constant for Mn^{2+} decreases by about one order of magnitude from that observed with the unadenylylated enzyme [42]. Nevertheless, the stoichiometry of binding remains at one Mn^{2+} per activated subunit, with each metal ion binding site apparently equivalent or noninteracting [42]. These results suggest that adenylylation provokes a conformational change that equivalently affects activating Mn^{2+} binding sites in both adenylylated and unadenylylated subunits of the dodecamer. Heterologous subunit interactions must be involved in such a conformational transition. The stability constant for Mg^{2+} at the activating metal ion binding sites, however, is not significantly affected by adenylylation [28].

The binding of the feedback inhibitor, L-tryptophan, is cooperative, as illustrated in the Hill plot of Figure 11 [55]. Saturation occurred at 12 moles of L-tryptophan bound per mole enzyme (Sec. II, D), and the data of Figure 11 are normalized to one-twelfth or to maximally one L-tryptophan bound per enzyme subunit. The transition is quite sharp, with a Hill coefficient of about 2.5 and an apparent activation energy calculated according to Wyman [78] of 700 cal/mol. With each subunit of the dodecamer apparently having a binding site for L-tryptophan, homologous interactions between these sites on different subunits lead to a cooperative type of transition. This transition is independent of either the adenylylation state of the enzyme or whether Mn^{2+} or Mg^{2+} is present [55]. This is not representative of the response of glutamine synthetase to inhibition by L-tryptophan under assay conditions, since the sensitivity of the enzyme to L-tryptophan inhibition is a function of both the adenylylation state of the enzyme [25] and the metal ion present [20]. Nevertheless, the L-tryptophan binding characteristics are mentioned here as a type of subunit interaction that can occur in glutamine synthetase. The measurements of L-alanine binding to glutamine synthetase in the presence of glutamine and Mn^{2+}, although inaccurate because of the low affinity constant for this effector, appeared to involve a cooperative transition also [55].

The equilibrium binding of another feedback inhibitor, AMP, was to the extent of 12, with the AMP binding sites apparently equivalent or noninteracting [55]. Thus, each enzyme subunit appears to have a binding site for AMP. Although the presence of either L-tryptophan or L-glutamate

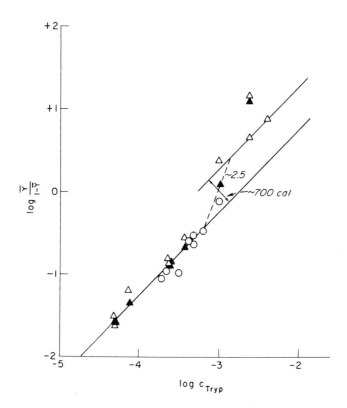

FIG. 11. Hill plot of the equilibrium binding of L-[^{14}C-methylene]trypto-
phan to glutamine synthetase (GS$\overline{9}$ and GS$\overline{2.3}$) at 4° and pH 7.4. The data
were obtained with GS$\overline{9}$ (0) in 5 mM MnCl$_2$, 0.02 M imidazole, and 0.1 M
KCl, and GS$\overline{2.3}$ (10-12 mg/ml) in 1 mM MnCl$_2$ (▲) or 1 mM MgCl$_2$ (△),
0.02 M imidazole chloride, and 0.1 M KCl. The fractional saturation,
$\overline{y}/1-\overline{y}$ was calculated from $\overline{y} = \overline{v}/12$, where $\overline{v}$ equals the moles of L-
tryptophan bound per mole enzyme at the free concentration of L-tryptophan
measured. The Hill coefficient of ~ 2.5 and the interaction energy of
~ 700 cal were estimated by the graphical method of Wyman [78]. (From
Ref. 55, courtesy of the American Chemical Society.)

decreased the apparent affinity of the enzyme for AMP [55], it has been
established that L-tryptophan uniformly and equivalently decreases the
affinity of the enzyme subunits for AMP [83]. Again, the direct measurements of AMP binding to the enzyme were not representative of kinetically
determined inhibition curves [8, 20, 55].

A fluorescent derivative of ATP $(1, N^6$-ethenoadenosine 5'-triphosphate or ε-ATP) has been used recently to enzymatically attach the corresponding monophosphate to adenylylation sites of glutamine synthetase [49, 50]. The ε-adenylylated enzyme derivatives exhibit similar catalytic properties to the corresponding natural enzyme forms isolated from E. coli, and therefore the covalently bound ε-AMP serves as an internal fluorescent probe of ligand interactions with the enzyme affecting the adenylylation site. Excitation of ε-adenylylated glutamine synthetase at 3150 Å yields a fluorescence emission spectrum with a maximum at 4000 Å. A comparison of Mn^{2+} to Mg^{2+} binding to the enzyme derivative indicated that the Mn^{2+}-enzyme complex had a decreased ε-adenosine fluorescence intensity and a blue shift in this emission spectrum. This suggests either that Mn^{2+} binds closer to the adenylylated tyrosyl residue than does Mg^{2+} or that Mn^{2+} causes a more extensive change in the protein conformation than does Mg^{2+}; in either case, Mn^{2+} imposes some further constraint on the covalently bound ε-AMP than does Mg^{2+}. Fluorescence measurements therefore appear to be more sensitive than ultraviolet spectral measurements (Sec. II, B, Fig. 2) in detecting conformational differences between Mg^{2+}- and Mn^{2+}-activated enzyme forms. A possible role of subunit interactions in such conformational differences has yet to be established, however. The binding of ATP or ADP to the catalytic sites of the Mn^{2+}-activated enzyme also cause a shift in the ε-adenosine fluorescence spectrum similar to that provoked by Mn^{2+} [49, 50].

When covalently attached to glutamine synthetase, the ε-adenosine group has a severely quenched fluorescence. Interactions between ε-AMP groups on different subunits and conformational changes induced by adenylylation have not been assessed. Ultraviolet absorption measurements at 2600 Å suggest that there is little or no interaction between 5'-adenylyl groups on different subunits [Eq. (2), Sec. II, A]. However, the ultraviolet absorbency at 2600 Å is the same with Mn^{2+} or Mg^{2+} bound to the enzyme, whereas the fluorescence of covalently bound ε-adenosine is affected differently by the binding of these divalent cations.

C. In Stability

Under various disaggregation conditions tried, inactivation of glutamine synthetase has been found to parallel dissociation of the dodecamer; (see review Ref. 26). Heterologous interactions between adenylylated and unadenylylated subunits in hybrid molecular species, which are indicated by the various catalytic parameters of biosynthetic catalysis (Sec. IV, A), also have a role in the stabilization properties outlined here.

Recently, Ciardi and Shifrin [84] have determined that metal ion-free preparations of unadenylylated enzyme are less stable at pH 8.4 and low ionic strength than is the relaxed adenylylated enzyme. At times when

most of the activity of the unadenylylated enzyme had been lost at this pH
(i.e., when the enzyme was disaggregated to one-twelfth of its size), the
adenylylated enzyme had as much as 50% activity remaining. At more
alkaline pH than pH 8.8, the kinetics of the dissociation of unadenylylated
and adenylylated enzyme forms were similar. In reassociation experiments
[84], however, much lower enzyme activity was recovered with base-
disaggregated adenylylated enzyme than with the dissociated unadenylylated
enzyme. For this reason, the hybridization experiments outlined below
(Sec. IV, D.) were not attempted with base-disaggregated enzyme species.
The observation that unadenylylated enzyme is more susceptible to dissociation at slightly alkaline pH and low ionic strength is exactly opposite to the
stability characteristics of these enzyme forms in urea with ADP present
(see Fig. 12) or at pH 8 in buffers at high ionic strength [42].

In other studies, the inactivation of glutamine synthetase produced by
urea as a function of the state of adenylylation of the enzyme preparation
was investigated [27, 48]. At high urea concentrations (4 to 7 M), the
kinetics of inactivation (which is accompanied by dissociation) of the metal
ion-free enzyme [15, 19, 27, 31, 32] is independent of the extent of adenylylation of the enzyme [27]. With 7 M urea at pH 8.7 (Table 4), the half-
time of inactivation of the enzyme in the presence of EDTA is about 5 min
at 0°C [27]. However, under special conditions (Fig. 12), in which the
unadenylylated ($GS_{\overline{0.8}}$) or adenylylated (GS_{12}) enzymes are incubated at
37°C in a mixture containing 4.0 M urea, ADP, Mn^{2+}, arsenate, and
glutamine, only the adenylylated enzyme is inactivated. For this effect,
ADP is an essential component of the 4 M urea inactivation mixture; in the
absence of ADP both enzyme forms were rapidly inactivated [27].

The differential inactivation of different glutamine synthetase forms by
incubation with 4 M urea under the conditions described in the legend of
Fig. 12 was used by Ciardi et al. [27] and Stadtman et al. [48] to detect
hybrid enzyme forms containing both adenylylated and unadenylylated subunits. As noted in Sec. II, C, heterologous interactions within hybrid
glutamine synthetase molecules are uncoupled in reaction (4). Thus, the
state of adenylylation [Eq. (5), Sec. II, C] indicated by the circled values
in Figure 12 could be determined by the γ-glutamyl transferase assay
method [48] at different times during the incubation of glutamine synthetase
in the special 4 M urea inactivation mixture [27, 48]. As illustrated in
Figure 12, a mixture of the fully adenylylated and unadenylylated enzymes
('Mix $GS_{\overline{7.2}}$') is inactivated to the extent and at the rate expected for the
amount of adenylylated enzyme in the 'Mix $GS_{7.2}$'. Checks on the state of
adenylylation after 30 and 60 min in the inactivation mixture showed that
only the unadenylylated enzyme was present at these times. In contrast,
the native enzyme ($GS_{\overline{7.6}}$), isolated directly from E. coli cells by the
method of Woolfolk et al. [15], was inactivated under the same conditions
at a slower rate than was the artificial 'Mix $GS_{\overline{7.2}}$'. Even more striking

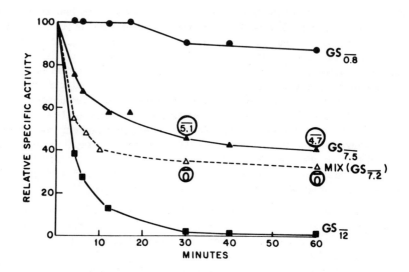

FIG. 12. A differentiation of glutamine synthetase hybrids from mixtures of adenylylated and unadenylylated enzymes. The enzymes designated $GS_{\overline{0.8}}$ and $GS_{\overline{7.5}}$ were isolated directly from different batches of E. coli cells [29] and contained an average of 0.8 and 7.5 molar equivalent covalently bound AMP, respectively. The fully adenylylated enzyme GS_{12} was prepared by enzymatic adenylylation of $GS_{\overline{7.5}}$ [40]. The Mix($GS_{\overline{7.2}}$) was prepared by mixing $GS_{\overline{0.8}}$ and GS_{12} in the proper proportions. The various enzyme preparations (0.24 mg/ml) were incubated at 37° C in the following mixture: 0.187 M buffer at pH 7.0 (composed of equal concentrations of imidazole, 2,4-dimethylimidazole, and 2-methylimidazole), 0.5 mM ADP, 25 mM L-glutamine, 25 mM potassium arsenate, and 4.0 M urea. The enzyme activity and the state of adenylylation were determined by γ-glutamyl transfer assay methods (Sec. II, C.) on aliquots removed at various times from the inactivation mixture. The circled numbers indicate the average state of adenylylation of $GS_{\overline{7.5}}$ (large circles) and of Mix $GS_{\overline{7.2}}$ (small circles) after 30 and 60 min exposure to 4 M urea, as indicated. (From the paper of Stadtman et al. [48], courtesy of Pergamon Press, Inc., using the data of Ciardi et al. [27]).

is the fact that after 30 and 60 min (Fig. 12) the $GS_{\overline{7.5}}$ preparation contained appreciable amounts of adenylylated subunit activity. Since all adenylylated subunits in either GS_{12} or in the 'Mix $GS_{\overline{7.2}}$' are completely inactivated at these times, it is concluded that heterologous subunit interactions in hybrid molecules of the native $GS_{\overline{7.5}}$ preparation must stabilize adenylylated subunits against inactivation by 4 M urea under these conditions. Thus, the resistance of adenylylated subunits to urea-induced inactivation under the conditions described in Figure 12 can be used to detect hybrid species of glutamine synthetase.

D. Formation of Hybrid Molecules from Subunits

Using the criteria of Figure 12 and the conditions of Table 4, Ciardi et al. [27] showed that hybrid enzyme molecules containing both types of subunits could be synthesized in vitro by reassociating a dissociated mixture of adenylylated (GS_{12}) and unadenylylated ($GS_{\overline{0.8}}$) glutamine synthetase. Hybrid enzyme formation was demonstrated more directly in the experiments outlined in Figures 13 and 14. For these studies [27, 85], succinylated derivatives of adenylylated subunits were hybridized with unadenylylated subunits to obtain electrophoretically separable variants [86, 87]. The physical separability afforded by succinylation is not furnished by adenylylation, since the adenylylated (GS_{12}) and unadenylylated ($GS_{\overline{0.8}}$) enzymes are not separated in electrophoretic or isoelectric focusing experiments [27]. By the latter technique, both enzyme forms have acidic isionic points at pH 4.9 [27].

The introduction of about 36 succinyl groups into the adenylylated enzyme molecule increases the net negative charge by about 72 [88] and produces a 75% loss in specific activity of the enzyme in catalysis of reaction (4). Except for the loss in catalytic potential, other catalytic parameters were unaltered by succinylylation. The succinylylated enzyme is therefore well suited for hybridization studies. Figures 13a-e illustrate the mobilities (from left to right) of the reassociated unadenylylated enzyme (Fig. 13a), of the reassociated succinylated, adenylylated (GS_{12}) enzyme (Fig. 13b), and of the forms produced by reassociation of a dissociated equimolar mixture of the enzyme forms of Figures 13a and 13b by the procedure of Table 4. The electrophoresis of any combination of protein (Figs. 13a-c) produced the expected patterns (Figs. 13d and 13e). The broad diffuse protein band of intermediate mobilities formed after dissociation and reassociation of the mixed unadenylylated and succinylated, adenylylated enzymes was further analyzed in the experiment illustrated in Figure 14, equal amounts of succinylated, $5'-[^{14}C]$-adenylyl glutamine synthetase (GS_{12}) and of unadenylylated enzyme ($GS_{\overline{0.8}}$) were dissociated and reassociated (Table 4) prior to polyacrylamide electrophoresis. After electrophoresis (see Fig. 13c), the gel was sliced and the different fractions analyzed for $[^{14}C]$-adenylyl subunit content by radioactivity and for unadenylylated subunits by measurements of γ-glutamyl transferase activity [48] in the presence of Mn^{2+} plus 60 mM Mg^{2+} at pH 7.15 [term (b) in Eq. (5) of Sec. II, C]. As expected for a heterogeneous mixture of hybrids, there was a higher proportion of $[^{14}C]$-adenylylated subunits in the faster migrating region of the protein band; conversely, the trailing end of the protein band was enriched proportionally in unadenylylated subunits. From the

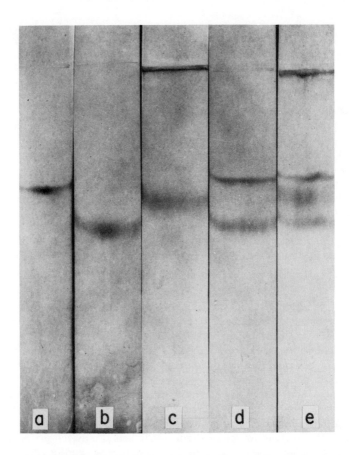

FIG. 13. Separation of hybrid glutamine synthetase forms from the succinylated-adenylylated $GS_{\overline{12}}$ and unadenylylated ($GS_{\overline{0.8}}$) enzymes from
which they were prepared. Succinylated $GS_{\overline{12}}$ (containing an average of
3 succinlyated amino groups per enzyme subunit), unmodified $GS_{\overline{0.8}}$, and
a mixture containing equal amounts of both were each subjected to dissociation (7 M urea) and reassociation conditions (Table 4). Two to 6 μg of
protein were applied to cellulose acetate strips and subjected to electrophoresis in 50 mM potassium phosphate buffer (pH 6.0), using a constant
voltage of 200 volts for 1 hr. Protein then was fixed and stained with
Ponceau S and Nigrosin. The bands of the strips, from left to right, are:
a) $GS_{\overline{0.8}}$, b) succinylated-$GS_{\overline{12}}$, c) hybrids formed from dissociating and
reassociating a mixture of (a + b), d) a mixture of $GS_{\overline{0.8}}$ and succinylated-
$GS_{\overline{12}}$, and e) a mixture of (c +d) composed of all three glutamine synthetase
preparations. (From Ref. 27, courtesy of the American Chemical Society.)

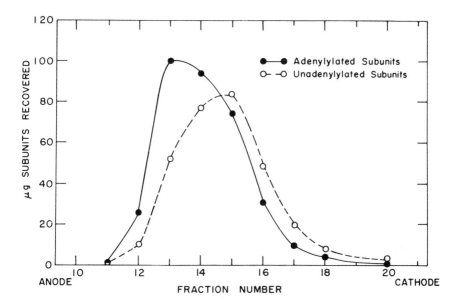

FIG. 14. Distribution of succinylated-adenylylated and unadenylylated subunits in hybrid dodecameric enzymes formed after reversible dissociation. Succinylated-[^{14}C-adenylyl)$_{11}$ -glutamine synthetase (with succinylated GS$_{12}$ having 25% residual γ-glutamyl transferase activity) was mixed with an equal amount of unadenylylated enzyme (GS$_{\overline{0.8}}$). The mixed enzymes were dissociated into subunits and the subunits subsequently reassociated (Table 4). The resultant, reconstituted enzyme mixture (see Fig. 13c) was subjected to electrophoresis on a polyacrylamide disc gel at pH 7.2. The gel was then sliced into 1.5 mm sections and the enzyme eluted with 10 mM imidazole HCl-1 mM MnCl$_2$ buffer (pH 7.0). Aliquots of each fraction were assayed for Mg^{2+}- supported γ-glutamyl transfer activity as a measure of unadenylylated subunits (Sec. II, C), and radioactivity was measured by liquid scintillation counting to estimate adenylylated subunits. From these measurements, the amount of each type of subunit was calculated for each fraction. (From Ref. 27, courtesy of the American Chemical Society.)

distribution of the two types of subunits in Figure 14, it was calculated that the average state of adenylylation varied from 9.0 to 4.0 moles 5'-adenylyl groups per mole enzyme in fractions 12-17 [27]. Using the γ-glutamyl transfer assay method with Mn^{2+} + Mg^{2+} (Sec. II, C), the average state of adenylylation of the molecular hybrids in the same electrophoretic gradient varied from 8.0 to 4.0 [27]. Judging from the composition of molecular

TABLE 4

Dissociating and Reassociating Conditions [a]

Dissociation	Reassociation[b]
1. Add Enzyme (0.5-1.0 mg/ml) to solution containing: 50 mM tris-HCl (pH 8.7), 1 mM EDTA, 143 mM 2-mercaptoethanol, and 7 M urea. 2. Incubate 1-3 hr at 0° C.	1. Dilute dissociation mixture tenfold at 0°C into solution containing: 50 mM tris-HCl (pH 7.5), 400 mM KCl, 10 mM $MnCl_2$, 10 mM $MgCl_2$, 0.4 mM ADP, and 72 mM 2-mercaptoethanol. 2. Incubate 30 min at 4°C and then 2-3 hr at 25°C. 3. Dialyze solution at 4° C against 10 mM imidazole (pH 7)- 1 mM $MnCl_2$ buffer. [c]

[a] Conditions of Ciardi et al. [27], with either unadenylylated or adenylylated glutamine synthetase forms.

[b] Recovered activity varies from 64% to 80% [27].

[c] Dialysis to remove residual urea is necessary to stabilize activity [19, 26, 27, 31, 32]. After reconstitution, the sedimentation properties and appearance in the electron microscope after negative staining (see Fig. 1a) are identical to those of the native enzyme [27].

hybrids derived from mixtures of succinylated, adenylylated, and unadenyl-ylated subunits, the hybridization process appears to be a random process. Generally, it was found [27] that the average states of adenylylation of hybrids formed by reassociation were the same as those of the initial dis-sociated mixtures. The studies of Ciardi et al. [27] demonstrate unequiv-ocally the formation in vitro of molecular hybrids of glutamine synthetase.

V. CONCLUDING REMARKS

Various structural features and some examples of subunit interactions within the dodecameric glutamine synthetase molecule of E. coli have been summarized here. The enzymatically catalyzed adenylylation and deadenylylation of glutamine synthetase in E. coli results in adenylylated, unadenylylated, and hybrid species containing both unadenylylated and adenylylated subunits. The adenylylation-deadenylylation of glutamine synthetase is the basis of a noval, regulatory mechanism in which a single site modification affects most dramatically the catalytic properties of the modified subunit and, more subtly, intersubunit interactions of the dodecamer.

Figure 15 schematically shows one subunit of the dodecameric glutamine synthetase structure. The arrangement of binding sites is arbitrary in Figure 15 since the topography of binding sites located at the surface of each subunit is unknown. Two essential binding sites for divalent cation are shown. One is needed to stabilize an active configuration of the enzyme; the other is presumably involved in activity expression. Shown also is a possibly specific site for a monovalent cation. In addition, substrate and inhibitor sites are shown. Besides these sites, the subunit must have three intersubunit contact sites, which are involved in the association of adjacent subunits in the same and in adjoining hexagonal rings. An inferred site of interaction for the adenylyltransferase that catalyzes the adenylylation and deadenylylation of each subunit is indicated by the open circle in Figure 15 [13]. The specific tyrosyl residue to which the 5'-adenylic acid group may be covalently bound through a phosphodiester linkage upon adenylylation is shown at the subunit center with 5'AMP attached. Since adenylylation can affect the interactions of all the ligands with the enzyme, the interaction of the subunit with the adenylylating-deadenylylating proteins, and subunit interactions within the dodecamer, dashed lines are drawn between these sites (peripheral closed triangles of Fig. 15) and the site of adenylylation (center closed circle of Fig. 15).

Somehow each subunit of 50,000 mol wt must physically accommodate all the sites depicted in Figure 15. It is difficult to imagine a polypeptide chain of this size specifically folded to provide this number of separate sites. Nevertheless, there is substantial evidence that each subunit of glutamine synthetase of E. coli contains a separate site for each different ligand as shown in Figure 15. With the exception of the divalent cation, nucleotide substrates, and subunit contact sites, most of the other sites have relatively poor affinities for ligands. Perhaps this reflects a deficiency in the number of amino acid residues available in the subunit contour for forming separate binding clusters for so many different ligands. In this respect, it should be mentioned that there is considerable antagonism between different effectors of glutamine synthetase that apparently arise from conformational changes induced by ligands, rather than from competition at common binding sites. Some overlapping of binding sites cannot be excluded, however.

Glutamine synthetase in E. coli is involved in nearly every aspect of intermediary metabolism. This enzyme has an important role in the

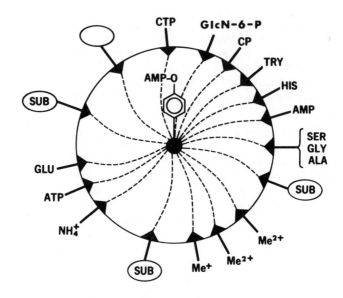

FIG. 15. Diagrammatic representation of a single subunit of E. coli glutamine synthetase showing multiple binding sites. Each closed triangle on
the circumference represents separate sites for the inhibitors (CTP,
glucosamine 6-P, carbamyl phosphate, L-tryptophan, L-histidine, and
AMP, with a possible single site for L-serine, glycine, and L-alanine), for
activating and stabilizing divalent and monovalent cations (Me^{2+}, Me^{+}),
and for the substrates (L-glutamate, ATP, and NH_4^{+}). The encircled
"SUB" indicates interaction sites between adjacent subunits. The open
circle shows a binding site for the enzyme system that catalyzes adenylylation and deadenylylation. The solid center circle represents the site
composed of the tyrosyl residue that undergoes adenylylation and deadenylylation. Interactions between ligand binding sites and the adenylylation
site are indicated by the dashed lines. (From Ref. 13, courtesy of
Academic Press.)

assimilation of nitrogen on the one hand, and on the other, it has a primary
function in the synthesis of amino acids, purine and pyrimidine nucleoside
phosphates, complex polysaccharides via glucosamine 6-phosphate, and
vitamins. Effective regulation of this enzyme therefore requires a capacity to sense fluctuations in metabolite demands for a multiplicity of cellular
functions. To this end, the adenylylating-deadenylylating cascade type of
enzyme regulation system [66, 89] is well adopted since it increases the
capacity for allosteric control of glutamine synthetase. This increased
capacity is due to the fact that each modifying enzyme in a cascade system
is an independent target for allosteric interactions that will affect ultimately

the activity of the last enzyme in the series, which in this case is glutamine synthetase [66, 89]. Thus, the allosteric input is multiplied by regulatory sites present on the proteins of the adenylylating–deadenylylating
enzyme system. The limitations imposed by a subunit of 50,000 mol wt in
providing sufficient allosteric sites for regulating such an important cellular activity may therefore be surmounted by the inclusion of a cascade system in the overall control. Then, in the final analysis, the regulation of
glutamine synthetase activity in E. coli is not restricted by the number of
allosteric sites on each subunit of a dodecameric protein but involves also
metabolite control of the cascade system of regulation which is poised to
shift dramatically the activity of glutamine synthetase according to cellular
demands.

REFERENCES

1. K. Wulff, D. Mecke, and H. Holzer, Biochem. Biophys. Res.
   Commun., 28, 740 (1967).
2. H. S. Kingdon, B. M. Shapiro, and E. R. Stadtman, Proc. Natl.
   Acad. Sci., U. S., 58, 1703 (1967).
3. L. Levintow and A. Meister, J. Biol. Chem., 209, 265 (1954).
4. R. A. Alberty, J. Biol. Chem., 243, 1337 (1968).
5. A. Meister, in The Enzymes, Vol. 6 (P. D. Boyer, H. Lardy, and
   K. Myrback, eds.), Academic Press, New York, 1962, p. 193.
6. D. W. Tempest, J. L. Meers, and C. M. Brown, Biochem. J., 117,
   405 (1970).
7. R. E. Miller and E. R. Stadtman, J. Biol. Chem., 247, 7407 (1972).
8. B. M. Shapiro and E. R. Stadtman, Annu. Review Microbiol., 24,
   501 (1970).
9. H. Holzer, Advan. Enzymol., 32, 297 (1969).
10. H. Holzer, D. Mecke, K. Wulff, K. Liess, and L. Heilmeyer, Jr.,
    Advan. Enzyme Regulation, 5, 211 (1967).
11. E. R. Stadtman, B. M. Shapiro, H. S. Kingdon, C. A. Woolfolk,
    and J. S. Hubbard, Advan. Enzyme Regulation, 6, 257 (1968).
12. E. R. Stadtman, B. M. Shapiro, A. Ginsburg, H. S. Kingdon, and
    M. D. Denton, Brookhaven Symp. Biol., 21, 378 (1968).
13. E. R. Stadtman, A. Ginsburg, W. B. Anderson, A. Segal, M. S.
    Brown, and J. E. Ciardi, in Molecular Basis of Biological Activity,
    Vol. 1 (K. Gaede, B. L. Horecker, and W. J. Whelan, eds.),
    PAABS Symp., Academic Press, New York, 1972, p. 127.
14. A. Ginsburg and E. R. Stadtman, in The Enzymes of Glutamine
    Metabolism, (S. Prusiner and E. R. Stadtman, eds.), Amer. Chem.
    Soc. Symp., Academic Press, New York, 1973, p. 9.
15. C. A. Woolfolk, B. M. Shapiro, and E. R. Stadtman, Arch. Biochem.
    Biophys., 116, 177 (1966).

16. C. A. Woolfolk and E. R. Stadtman, Biochem. Biophys. Res. Commun., 17, 313 (1964).

17. C. A. Woolfolk and E. R. Stadtman, Arch. Biochem. Biophys., 118, 736 (1967).

18. H. S. Kingdon, J. S. Hubbard, and E. R. Stadtman, Biochemistry, 7, 2136 (1968).

19. B. M. Shapiro and A. Ginsburg, Biochemistry, 7, 2153 (1968).

20. H. S. Kingdon and E. R. Stadtman, J. Bacteriol., 94, 949 (1967).

21. A. Segal and E. R. Stadtman, Arch. Biochim. Biophys., 152, 367 (1972).

22. A. Segal and E. R. Stadtman, Arch. Biochem. Biophys., 152, 356 (1972).

23. M. S. Brown, A. Segal, and E. R. Stadtman, Proc. Natl. Acad. Sci., U. S., 68, 2949 (1971).

24. J. H. Mangum, G. Magni, and E. R. Stadtman, Arch. Biochem. Biophys., 158, 514 (1973).

25. B. M. Shapiro, H. S. Kingdon, and E. R. Stadtman, Proc. Natl. Acad. Sci., U. S., 58, 642 (1967).

26. A. Ginsburg, Advan. Protein Chem., 26, 1 (1972).

27. J. E. Ciardi, F. Cimino, and E. R. Stadtman, Biochemistry, 12, 4321 (1973).

28. J. B. Hunt and A. Ginsburg, Biochemistry, 11, 3723 (1972).

29. B. M. Shapiro and E. R. Stadtman, in Methods in Enzymology, Vol. 17A (H. Tabor and C. W. Tabor, eds.), Academic Press, New York, 1970, p. 910.

30. B. M. Shapiro and E. R. Stadtman, J. Biol. Chem., 242, 5069 (1967).

31. C. A. Woolfolk and E. R. Stadtman, Arch. Biochem. Biophys., 122, 174 (1967).

32. R. C. Valentine, B. M. Shapiro, and E. R. Stadtman, Biochemistry, 7, 2143 (1968).

33. D. Eisenberg, E. C. Heidner, P. Goodkin, M. N. Dastoor, B. H. Weber, F. Wedler, and J. D. Bell, Cold Spring Harbor Symp., Quant. Biol., 36, 291 (1971).

34. A. K. Lahiri, C. Noyes, H. S. Kingdon, and R. L. Heinrikson, Fed. Proc. Fed. Amer. Soc. Exp. Bid., 31, 473 Abstr. (1972).

35. B. M. Shapiro and E. R. Stadtman, J. Biol. Chem., 243, 3769 (1968).

36. R. L. Heinrikson and H. S. Kingdon, J. Biol. Chem., 246, 1099 (1971).

37. S. B. Hennig and A. Ginsburg, Arch. Biochem. Biophys., 144, 611 (1971).

38. D. H. Wolf, E. Ebner, and H. Hinze, Eur. J. Biochem., 25, 239 (1972).

39. F. Cimino, W. B. Anderson, and E. R. Stadtman, Proc. Natl. Acad. Sci., U. S., 66, 564 (1970).

40. A. Ginsburg, J. Yeh, S. B. Hennig, and M. D. Denton, Biochemistry, 9, 633 (1970).

41. R. E. Miller, E. Shelton, and E. R. Stadtman, Arch. Biochem. Biophys., 163, 155 (1974).

42. M. D. Denton and A. Ginsburg, Biochemistry, 8, 1714 (1969).

43. J. B. Hunt, P. Z. Smyrniotis, A. Ginsburg, and E. R. Stadtman, Arch. Biochem. Biophys., 166, 102 (1975).

44. J. B. Hunt, P. D. Ross, and A. Ginsburg, Biochemistry, 11, 3716 (1972).

45. R. B. Timmons and P. B. Chock, unpublished data.

46. R. W. Henkens, G. D. Watt, and J. M. Sturtevant, Biochemistry, 8, 1874 (1969).

47. D. Mecke, K. Wulff, and H. Holzer, Biochim. Biophys. Acta, 128, 559 (1966).

48. E. R. Stadtman, A. Ginsburg, J. E. Ciardi, J. Yeh, S. B. Hennig, and B. M. Shapiro, Advan. Enzyme Regulation, 8, 99 (1970).

49. P. B. Chock, C. Y. Huang, R. B. Timmons, and E. R. Stadtman, Proc. Natl. Acad. Sci., U. S., 70, 3134 (1973).

50. R. B. Timmons, C. Y. Huang, E. R. Stadtman, and P. B. Chock, in Third International Symposium on Metabolic Interconversion of Enzymes (E. H. Fischer, E. G. Krebs, H. Neurath, and E. R. Stadtman, eds.), Springer, New York, 1974, p. 209.

51. S. G. Rhee and P. B. Chock, personal communication.

52. S. Chaberek and A. E. Martell, in Organic Sequestering Agents, John Wiley and Sons, Inc., 1959, p. 78.

53. G. G. Hammes, R. W. Porter, and C.-W. Wu, Biochemistry, 9, 2992 (1970).

54. M. D. Denton and A. Ginsburg, Biochemistry, 9, 617 (1970).

55. A. Ginsburg, Biochemistry, 8, 1726 (1969).

56. P. D. Ross and A. Ginsburg, Biochemistry, 8, 4690 (1969).

57. S. R. Tronick, J. E. Ciardi, and E. R. Stadtman, J. Bacteriol., 115, 858 (1973).

58. T. F. Deuel, A. Ginsburg, J. Yeh, E. Shelton, and E. R. Stadtman, J. Biol. Chem., 245, 5195 (1970).

59. W. B. Anderson, S. B. Hennig, A. Ginsburg, and E. R. Stadtman, Proc. Natl. Acad. Sci., U. S., 67, 1417 (1970).

60. W. B. Anderson and E. R. Stadtman, Arch. Biochem. Biophys., 143, 428 (1971).

61. C. P. Heinrich, F. A., Battig, M. Mantel, and H. Holzer, Arch. Mikrobiol., 73, 104 (1970).

62. M. Mantel and H. Holzer, Proc. Natl. Acad. Sci., U. S., 65, 660 (1970).

63. E. R. Stadtman, in The Enzymes, 3rd ed., Vol. 8 (P. D. Boyer, ed.), Academic Press, New York, 1973, p. 1.

64. R. M. Wohlhueter, Eur. J. Biochem., 21, 575 (1971).
65. W. B. Anderson and E. R. Stadtman, Biochem. Biophys. Res. Commun., 41, 704 (1970).
66. S. P. Adler, J. H. Mangum, G. Magni, and E. R. Stadtman, in Third International Symposium on Metabolic Interconversion of Enzymes (E. H. Fischer, E. G. Krebs, H. Neurathy, and E. R. Stadtman, eds.), Springer-Verlag, New York, 1974, p. 221.
67. E. Ebner, D. H. Wolf, C. Gancedo, S. Elsasser, and H. Holzer, Eur. J. Biochem., 14, 535 (1970).
68. S. B. Hennig, W. B. Anderson, and A. Ginsburg, Proc. Natl. Acad. Sci., U. S., 67, 1761 (1970).
69. C. E. Caban and A. Ginsburg, Fed. Proc. Fed. Amer. Soc. Exp. Biol., 33, 1426 Abstr. (1974).
70. B. M. Shapiro, Biochemistry, 8, 659 (1969).
71. S. P. Adler, D. Purich, and E. R. Stadtman, Fed. Proc. Fed. Amer. Soc. Exp. Biol., 33, 1427, Abstr. (1974).
72. E. J. Oliver and A. Ginsburg, unpublished data.
73. R. M. Wohlhueter, H. Schutt, and H. Holzer, in The Enzymes of Glutamine Metabolism (S. Prusiner and E. R. Stadtman, eds.), Amer. Chem. Soc. Symp., Academic Press, New York, 1973, p. 45.
74. D. D. Perrin and V. S. Sharma, Biochim. Biophys. Acta, 127, 35 (1966).
75. R. E. Miller, P. B. Chock, and E. R. Stadtman, unpublished data.
76. R. B. Timmons, S. G. Rhee, D. L. Luterman, and P. B. Chock, Biochemistry, 13, 4479 (1974).
77. D. E. Koshland, Jr., Proc. Natl. Acad. Sci., U. S., 44, 98 (1958).
78. J. Wyman, Advan. Protein Chem., 19, 223 (1964).
79. J. Monod, J. Wyman, and J.-P. Changeux, J. Mol. Biol., 12, 88 (1965).
80. D. E. Koshland, Jr., G. Némethy, and D. Filmer, Biochemistry, 5, 365 (1966).
81. M. M. Rubin and J. P. Changeux, J. Mol. Biol., 21, 265 (1966).
82. J. Wyman, J. Amer. Chem. Soc., 89, 2202 (1967).
83. M. A. Berberich and A. Ginsburg, unpublished data.
84. J. E. Ciardi and S. Shifrin, personal communication.
85. J. E. Ciardi and F. Cimino, Fed. Proc. Fed. Amer. Soc. Exp. Biol., 30, 1175 Abstr. (1971).
86. E. A. Meighen, and H. K. Schachman, Biochemistry, 9, 1163 (1970).
87. E. A. Meighen, V. Pigiet, and H. K. Schachman, Proc. Natl. Acad. Sci., U. S., 65, 234 (1970).
88. I. M. Klotz, in Methods in Enzymology, Vol. 11 (C. H. W. Hirs, ed.), Academic Press, New York, 1967, p. 576.
89. E. R. Stadtman and A. Ginsburg, in The Enzymes, 3rd ed., Vol. 10 (P. D. Boyer, ed.), Academic Press, New York, 1974, p. 755.

Chapter 3

UDP-GALACTOSE-4-EPIMERASE

Othmar Gabriel*   Robert A. Darrow
Herman M. Kalckar

Biochemical Research Laboratory   Charles F. Kettering
Harvard Medical School and the   Research Laboratory
Massachusetts General Hospital   Yellow Springs, Ohio
Boston, Massachusetts

I. INTRODUCTION   86

II. GENERAL MOLECULAR PROPERTIES OF UDP-GALACTOSE-
  4-EPIMERASE AND ITS REACTION MECHANISM   87

III. UDP-GALACTOSE-4-EPIMERASE FROM YEAST, SUBUNITS,
  AND PROSTHETIC GROUPS   90
  A. Introduction   90
  B. General Properties of Purified Candida 4-Epimerase   90
  C. Heterogeneities of Enzyme Protein Observed by Poly-
   acrylamide Gel Electrophoresis   91
  D. Subunits, Monomers, Dimers, Tetramers, Apoenzymes —
   Some Definitions   92
  E. Properties of the Prosthetic Group in Candida 4-Epimerase   99
  F. NAD$^+$ Exchange   100
  G. Studies of Peptide Hydrogen Exchanges and Their Relation
   to Conformational Changes of the Enzyme Protein   102
  H. Isoelectric Point   102
  I. Asymmetry of the Native Enzyme Protein   105

---

*Presently affiliated with the Department of Biochemistry, Georgetown University Schools of Medicine and Dentistry, Washington, D. C.

IV. UDP-GALACTOSE-4-EPIMERASE FROM E. COLI   105
     A. Main Structural Features of the Enzyme   106
     B. Properties of the Catalytically Active Enzyme Protein   108
     C. Reaction Mechanism   108
     D. Reductive Inhibition   112
     E. Side Reactions of the Catalytically Active Enzyme   116
     F. Subunits   119

 V. UDP-GALACTOSE-4-EPIMERASE FROM OTHER
     SOURCES   120
     A. Mammalian Sources   120
     B. Plant Sources   121
     C. Torula Yeast   121

VI. OVERALL VIEW OF REACTIONS CATALYZED BY UDP-
     GALACTOSE-4-EPIMERASE AND RELATIONS TO OTHER
     SUGAR TRANSFORMATIONS   122
     A. Reactions Catalyzed by Active Enzyme   122
     B. Possible Regulatory Function of Enzyme-NADH   122
     C. Subunit Structure of TDPG-Oxidoreductase: Mechanism
         of Reaction, Molecular Properties   124
     D. Common Mechanistic and Structural Properties of Enzymes
         Involved in Enzyme-NAD$^+$ Mediated Hydrogen Transfer
         Reactions   126

I. INTRODUCTION

It is the objective of this chapter to provide to the reader the presently
available body of knowledge concerning the mode of action of UDP-galac-
tose-4-epimerase and to relate the function of this important metabolic
enzyme to its unique protein structure.

To accomplish this, we feel it necessary to present first a generalized
picture of some of the properties of UDP-galactose-4-epimerase in terms
of its mechanism and structure. The molecular properties discussed
first are applicable to all the different epimerases, isolated from various
sources. In this way, we hope to avoid redundancy and at the same time
emphasize at the outset the considerable variations encountered when
carrying out detailed studies on epimerases isolated from different species.
As a matter of fact, ambiguity was introduced into the literature when
results of studies on epimerases isolated from different sources were
used interchangeably.

This general discussion (Sec. II) will be followed by a detailed description of individual epimerases according to their source of isolation. Consequently, the induced Candida 4-epimerase yeast enzyme, UDP-galactose-

4-epimerase, epimerase from E. coli, and mammalian epimerases will be described separately. This will contrast the properties common to all 4-epimerases from their species variations.

We wish to indicate that some of the principles concerning relations between structure and function of UDP-galactose-4-epimerase are applicable to several other enzymes involved in quite different sugar transformations. The feature common to this whole group of enzymes is the fact that the initiation of enzyme catalysis appears to involve identical intermediates and transition states. Moreover, these similarities appear to extend to similarities in protein structure such as subunits and coenzymes as well as conformational similarities at the active site.

Finally, we wish to emphasize the complementary role studies on the mode of action of enzymes in mechanistic terms can provide for a better understanding of protein structure and function.

II. GENERAL MOLECULAR PROPERTIES OF UDP-GALACTOSE-4-EPIMERASE AND ITS REACTION MECHANISM

UDP-galactose-4-epimerase was first described by Leloir [1] and was found to catalyze the conversion of glucose to galactose as their UDP-derivatives. In the early days of this work, the enzymes of this type were also referred to as galactowaldenase although, as we will discuss below, the reaction mechanism is not that of a Walden inversion.

The interest of several investigators was attracted by the fact that, at least on paper, the conversion of UDP-glucose to UDP-galactose appears to be extremely simple, involving the inversion of a single center of asymmetry at carbon 4 of the hexose moiety of the sugar nucleotide according to scheme (1).

However, as we will discuss, it was not until very recently that experimental verification for an oxidoreductase mechanism was finally accomplished, although such a mechanism had been suspected since 1956 when the presence of NAD^+ was found to be essential for liver-4-epimerase activity [2]. So far, in all cases studied, UDP-galactose-4-epimerase appears to be composed of two protein subunits. A pyridine nucleotide, nicotinamide adenine dinucleotide (NAD), in its oxidized or reduced form was found to be an essential part of the catalytically active enzyme protein, although the degree of binding of the coenzyme to the protein varies considerably in different species. The nature of this binding of coenzyme to the protein, in some instances very strong, by noncovalent bonds is still the subject of speculation. The stoichiometry of protein subunits to coenzyme was found to be two subunits per one mole of pyridine nucleotide. The binding of

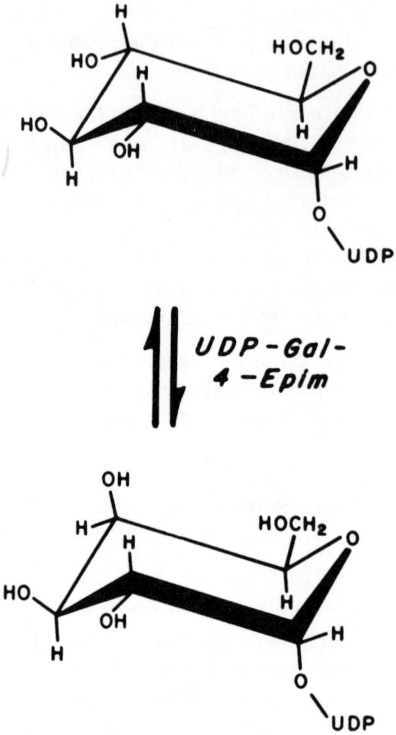

SCHEME 1

coenzyme to the subunits will be the subject of the discussion below and
poses some interesting questions concerning the asymmetrical properties
of the functional enzyme protein.

From the extensive studies in our laboratories as well as in those of
several other investigators, the following general reaction mechanism appears to be applicable to all the UDP-galactose-4-epimerases so far
examined [scheme (2)].

As can be seen, the reaction involves an intramolecular hydrogen
transfer mediated by the enzyme-bound pyridine nucleotide. The reaction
is initiated by a hydride abstraction from the 4-position of the substrate and
transfer to the enzyme-bound pyridine nucleotide, resulting in the formation of a 4-ulose intermediate. The existence of this intermediate was,
until recently, the subject of speculation. One reason for it is the fact
that, under normal catalysis conditions, this intermediate is not released
to the medium but remains enzyme-bound as indicated in the reaction
scheme by brackets. Concomitant with the formation of UDP-D-xylo-4-
hexosulose, enzyme NAD^+ is converted to enzyme-NADH. It is this

REACTION MECHANISM OF UDP-GAL-4-EPIMERASE

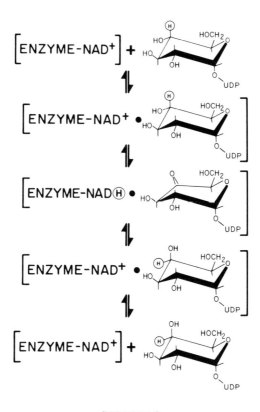

SCHEME 2

conversion of the coenzyme from its oxidized to its reduced form that appears to be responsible for the tight binding of the 4-ulose intermediate to
the enzyme protein. As we will discuss below, there is increasing experimental evidence to suggest that the conversion enzyme-NAD⁺ to enzyme-
NADH is accompanied by a conformational change of the enzyme protein.
Generally, in all cases studied, enzyme-NADH appears to be a much
"tighter" conformation of the polypeptide structure than enzyme-NAD⁺.

To complete the reaction sequence, restoration of enzyme-NAD⁺ is accomplished by the stereospecific donation of hydride to the 4-ulose intermediate. This results in conversion of enzyme-NADII to enzyme-NAD⁺
accompanied by release of the reaction product of the 4-epimeric UDP-
hexose to the medium. Enzyme-NAD⁺ can now undergo a new cycle of this
sequence of events.

The present state of information concerning the reaction mechanism is not addressing itself to the question of how this hydrogen transfer is actually accomplished. Considering the fact that this intramolecular hydrogen transfer involves the "B" side of the nicotinamide moiety of the coenzyme only, many important questions concerning conformational changes of enzyme protein, coenzyme, and substrate as well as their relative positions at the active site during enzyme catalysis remain unanswered. The transfer of an axial hydrogen (glucose) to an equational position (galactose) may be facilitated by the formation of the 4-hexosulose intermediate, since introduction of the keto group will result in a flattening of the hexose ring at the 4-position.

III. UDP-GALACTOSE-4-EPIMERASE FROM YEAST, SUBUNITS, AND PROSTHETIC GROUPS

A. Introduction

UDP-galactose-4-epimerase from Candida 4-epimerase (Candida pseudotropicalis or Saccharomyces fragilis) was found to contain bound NAD^+ and NADH, the latter in a highly fluorescent form [3, 4].

It should be noted that native purified epimerase contains between 15-20% of the coenzyme in its reduced form, enzyme-NADH, while the rest of the enzyme occurs as enzyme-NAD^+. Titration of the available sulfhydryl groups with p-CMB* releases enzyme-bound NAD^+ as well as NADH [5]. A further study of the reaction with p-CMB showed that the release of 1 mol of NAD^+ gave rise to formation of two subunits [6, 7]. This type of asymmetry, one NAD^+ per dimer, has not only been found for the Candida 4-epimerase but also for the E. coli 4-epimerase [8] and for the E. coli TDPG oxidoreductase [9]. Moreover, Yoshida and Hoagland [10] have more recently found that human glucose-6-phosphate dehydrogenase has one tightly bound NADP molecule per active dimer.

B. General Properties of Purified Candida 4-Epimerase

Candida 4-epimerase was isolated from cells induced during their growth by galactose and purified by various ammonium sulfate fractionation steps and DEAE cellulose chromatography according to Darrow and Rodstrom [7]. The purified native epimerase showed a characteristic blue fluorescence emission with excitation maximum at 3500 Å and emission

---

*p-CMB is the abbreviated form for p-chloromercuribenzoate.

maximum at 4350 Å. This fluorescence of the enzyme is due to the proportion of enzyme-NADH present [4, 5, 11].

The purity and homogeneity of purified epimerase was investigated by various biochemical and physicochemical criteria [7]. The catalytic activity in the presence of an optimal concentration of cations [12] reached a specific activity of the order of 80 μmol of UDP-galactose being converted into UDP-glucose per minute per milligram protein at 23°C. The values of bound NAD$^+$ reached a final value of 8 nmol per milligram active protein [average figure of five preparations ranging from 7.3 to 9.0 (cf. Ref. 7)] corresponding to one mole of coenzyme per mole of enzyme protein.

The Schlieren pattern obtained in the analytical ultracentrifuge [13] was not strictly symmetrical, possibly reflecting an equilibrium of dimeric and monomeric forms of enzyme protein at low ionic strength [13, 14]. However, calculation of the apparent diffusion coefficient according to Schachman [15] gave values which were constant with time, indicating only a low degree of heterogeneity. Moreover, the dependence of the sedimentation coefficient on the protein concentration extrapolated to zero protein concentration gave a value of S = 6.95 $\pm$ 0.10, in reasonable agreement with the average value of 6.49 S obtained in sucrose gradient centrifugation [6, 13]. It will become evident later in this chapter that heterogeneities seen in these preparations are a consequence of different molecular aggregation of epimerase itself, whereas foreign impurities could be present only in amounts barely discernible by biochemical and physicochemical criteria.

C. Heterogeneities of Enzyme Protein Observed by Polyacrylamide Gel Electrophoresis

The rate of migration of the enzyme protein during disc gel electrophoresis was inversely proportional to the polyacrylamide gel concentration. In addition to the main band, there were several minor components which became more apparent at higher concentrations of acrylamide gel. The inclusion of 4 M urea in 7.5% acrylamide gel made the main component more diffuse but did not resolve it into more than one band [7].

In an attempt to understand the multiple banding, reelectrophoresis of individual bands in 7.5% polyacrylamide gel was carried out [7]. In each instance, a new set of multiple bands was obtained, indicating formation of components during the process of electrophoresis itself. Whether this was due to establishment of an equilibrium between different forms of enzyme [7] and/or to artifact formation by residual harmful components which remain present after gel polymerization [16] has not been resolved.

The use of SDS-polyacrylamide* gel electrophoresis also revealed multiband patterns, especially upon prior incubation of epimerase at 37° C for several hours with dilute SDS solutions [17]. In contrast, when epimerase is first mixed with SDS at 95° C for 30 min and then subjected to polyacrylamide gel electrophoresis, the multibands are faint or nondetectable [17].

Epimerase purified by electrofocusing, followed by examination by SDS acrylamide gel electrophoresis, resulted in the appearance of a single intense band.

These findings indicate to us the presence of traces of a yeast protease responsible for the appearance of proteolytic fragments of the enzyme protein in preparations not purified by electrofocusing. The proteolytic action is most apparent after denaturation of the enzyme protein by preincubation with dilute SDS solutions.

D. Subunits, Monomers, Dimers, Tetramers, Apoenzymes — Some Definitions

We wish to discuss experimental conditions that will result in the conversion of native Candida 4-epimerase into specific and defined molecular entities. To facilitate the following presentation concerning various quaternary structures of 4-epimerase, a simplified summary of different molecular species of epimerase is shown in Table 1.

Candida 4-epimerase subunits can be obtained by denaturation of the enzyme with 6 M guanidine hydrochloride yielding units with a molecular weight of 60,000. Subunits obtained by this treatment lose their bound NAD^+. The denaturation process caused by guanidine hydrochloride is irreversible; hence, there is no possibility of retrieving active "holoenzyme."

As we have already mentioned, subunits can also be obtained by treatment with p-CMB (14 SH groups per mole of enzyme titratable [5]). Subunits prepared in this way have lost their prosthetic group; we like to refer to them as apoepimerase [6]. This type of apoepimerase has a molecular weight of 60,000 [13]. Under certain conditions, removal of the mercuric reagent by excess mercaptoethanol gave rise to a new type of apoepimerase which seemed to have a molecular weight of 2 x 60,000 since its sedimentation corresponded to 6 S rather than 4 S [6]. Usually, however, dimer formation from the monomer-apoepimerase required not only removal of the mercuric reagent by excess mercaptoethanol but also addition of NAD^+

_____

*SDS is the abbreviation used for sodium dodecylsulfate.

TABLE 1

Summary of Molecular Forms of _Candida_ 4-Epimerase and Their Relation to Molecular Weight, Bound Coenzyme, and Catalytic Activity[a]

Molecular form(s)	Mode of treatment	Molecular weight and sedimentation coefficient values [S]	Bound NAD$^+$	Catalytic activity
(a) Monomeric subunits	Denaturation with guanidine-HCl and urea	60,000 (3.5–4S)	0	0
(b) Monomer-apo-epimerase	Titration of –SH by p–CMB	60,000 (3.5–4S)	0	0
(c) Monomer-apo-epimerase[b]	Removal of p–CMB	60,000 (3.5–4S)	0	0
(d) NAD$^+$-monomer+apomonomer	Low ionic strength (0.001 M)	60,000 (3–4S)	1	0
(e) NAD$^+$-dimer-epimerase	Addition of NAD$^+$ to (c)	120,000 (6S)	1	100%
(f) NAD$^+$-dimer-epimerase	Native epimerase in 0.01 M tris–HCl buffer	120,000 (6S)	1	100%
(g) NAD$^+$-tetramer	Addition of cations to (f)	240,000 (11S)	2	100%

[a] For references and other details, see text.

[b] An inactive dimer-apo-epimerase can be formed in glycine buffer in presence of spermine in the absence of NAD$^+$.

[14]. This treatment resulted in full restoration of enzymatic activity, and the enzyme was found to be largely-NAD^+ consisting of two subunits [13]. A third way to prepare enzyme subunits consists of exposure of the native enzyme to low ionic strength (0.001 M). By this treatment, one can obtain a mixed population of subunits, monomers still linked to NAD^+, and apoepimerase, a monomer free of NAD^+ [14]. When this population of monomers is again transferred into a milieu of high ionic strength [6, 13, 14], a mixture of dimeric and tetrameric holoepimerase is formed. Both molecular species contain the same proportion of coenzyme, namely 1/2 mol of NAD^+ per mole of subunit. As we shall see, the tetrameric form is characterized by having a much higher affinity for the substrate UDP-galactose [13].

1. Subunits Formed by Titration of Candida 4-Epimerase with p-CMB or by Treatment with Guanidine Hydrochloride

Studies of the sedimentation properties of epimerase in a sucrose gradient containing glycine buffer pH 8.7 revealed that the prevailing component had a sedimentation coefficient of approximately 6 S [6]. Epimerase fully titrated with p-CMB (14 eq. per mole native 6 S epimerase [5]) showed a sedimentation pattern corresponding to 4 S and had lost catalytic activity as well as its bound NAD^+ [5]. The catalytic activity can be fully restored by addition of mercaptoethanol and NAD^+ [5, 14]. This restoration is accompanied by a reconversion of 4 S particles to 6 S [14].

Measurements of the intrinsic viscosity of Candida 4-epimerase have been carried out by Ottesen and Johannsen [18], who obtained values of 3.3 ml per gram for native epimerase and 3.8 ml per gram protein treated with p-CMB. These figures are within the usual range for globular proteins [19] and are a further indication that the decrease in sedimentation velocity after p-CMB treatment is not merely due to unfolding of the peptide chains.

The molecular weight of the 4S particles was difficult to ascertain because of the tendency of the p-CMB-treated protein to aggregate. However, the high speed equilibrium technique by Yphantis [20], which separates aggregates by centrifugal fractionation, permitted an estimate of the molecular weight of the 4 S particles at about 60,000 [13]. As has already been mentioned, this range of molecular weight was also found after denaturation of epimerase with 6 M guanidine hydrochloride or 8 M urea [14]. High speed equilibrium sedimentation established the molecular weight of these denatured preparations as ranging between 59,300 and 73,700, taking into account corrections for the partial specific volume under these circumstances [14]. When epimerase was chromatographed on BioGel in 6 M guanidine hydrochloride, the Bertlands found an apparent average molecular weight of 59,130 [14]. A summary of data obtained in these experiments is provided in Table 2. It should be added that in these experiments heterogeneities were detectable only under special conditions and then only in traces [7, 14]. The problem of heterogeneities arising from traces of yeast protease after preincubation of enzyme with dilute SDS was mentioned above.

2. Formation of Dimers and Tetramers

As we indicated before, Candida 4-epimerase requires certain cations for maximum catalytic activity, and sucrose gradient sedimentation analyses revealed that in buffers of higher cation concentration the sedimentation coefficient increased from a value of 6.5 S to 11 S [6, 13].

If the purified epimerase is dissolved in 0.1 M glycine buffer, adjusted to pH 8.7 by NaOH, the epimerase appears as an almost pure dimer population. The sodium ion concentration in this buffer is only about 0.01 M; addition of UDP-galactose and NAD^+ may bring the sodium concentration up to 0.013 M. In the presence of 0.1 mM UDP-galactose, Darrow and Rodstrom [6, 13] found that addition of various cations (such as NaCl, Na_2SO_4, $MgCl_2$, spermine, or spermidine) increased the catalytic activity of the Candida 4-epimerase markedly. This effect is apparently due to a lowering of the K_m for UDP-galactose. In the case of spermine, 10 mM of this organic cation will lower the K_m tenfold [13].

Similar experiments performed in the presence of various salts established cations as the active entity responsible for the increased affinity between enzyme and substrate. This type of effect was found to be proportional to the concentration of the cation, up to a maximum value, as well

TABLE 2

Molecular Weight Determinations of Yeast UDP-Galactose-4-
Epimerase in 6M Guanidine Hydrochloride by Sedimentation
Equilibrium (35,600 rpm, 20°, $\bar{v} = 0.73$) [4]

Time of exposure in 6 M guanidine hydrochloride	Molecular weight
1 hour	59,300-73,700
48 hours	61,000-72,200
5 days	68,700
22 days	71,800

as to the number of positive charges per ion at any given concentration. At concentrations of 10 mM or below, spermine was the most effective cation. Inhibition observed at high spermine concentrations may be due to an interference with the indicator enzyme UDP-glucose dehydrogenase [21] rather than with epimerase. For these reasons a two-step assay [22] was used for the measurement of epimerase activity. Michaelis-Menten kinetics were observed in each case. At 0.2, 0.5, and 10 mM spermine concentrations, the K_m for UDP-galactose was determined to be 0.6, 0.4, and 0.1 mM respectively.

The effect of cations on the affinity of the Candida 4-epimerase toward its substrate appears to be accompanied by changes in the quaternary structure of the enzyme. As mentioned above, the sucrose density gradient pattern obtained in glycine buffer at low cationic strength (about 0.01 M) showed a sedimentation coefficient of 6 S [6], corresponding to a molecular weight of 125,000 [7]. In the presence of 0.1 M spermine under conditions of optimal affinity, the sedimentation coefficient changes to a value around 11 S.

To look more carefully into the correlation of molecular size and enzymatic activity, a series of density gradient runs were performed at various spermine concentrations, using microgram amounts of epimerase, to approximate the concentrations used in the enzymatic assay. For example, at a spermine concentration of 5×10^{-3} M, the existence of both epimerase forms with sedimentation coefficient values of 6 S and 11 S was observed in the same gradient.

The molecular weight of the 11 S species was determined by Darrow and Rodstrom [13] by equilibrium centrifugation in 0.1 M tris-HCl pH 7.5 corresponding to a cation concentration of 0.08 M. This concentration of univalent cations is close to optimal, both for the enhancement of enzymatic activity (low K_m) and to the formation of 11 S. At low speed equilibrium centrifugation, the apparent average molecular weight was found to be 225,000. Thus, the 11 S species evidently represents a dimer of the 6 S species.

Samples of purified epimerase were found to contain variable amounts of enzymatically active protein with higher affinity for the substrate but incapable of forming 11 S particles in the presence of excess cations. It is possible that endogenous yeast proteases, which may still occur in traces in the purified Candida 4-epimerase preparations, may convert part of the 6 S population to a modified 6 S protein. This 6 S protein, although still able to have high affinity for the substrate, has lost its ability to form tetramers. Thus, it appears that the ability to form 11 S tetramers may be a more delicate indicator of the structural integrity of epimerase than the catalytic activity and changes in the K_m would indicate.

3. Subunit Studies at Low Ionic Strength

In an additional study concerning the nature of Candida 4-epimerase subunits, the Bertlands [14] attempted to correlate general ionic effects with the aggregation state of the enzyme. These authors decided to examine the effect of low ionic strength on NAD^+-epimerase without resorting to any chemical modification. Dissociation of native NAD^+-epimerase into apoenzyme and NAD^+ was found to take place during gel filtration of the enzyme in tris buffers of low ionic strength (0.001 to 0.003 M) at a neutral pH [14]. When NAD^+-epimerase in 0.001 M tris-HCl (pH 7.35) (conductivity 50μmhos or less) was examined by ultracentrifugation, it was found to sediment as a mixture of two components, a light one with S values ranging from 2.8 to 3.6 and a slightly heavier one with S values of 4.0 to 4.7 [14]. High speed sedimentation equilibrium ultracentrifugation according to Yphantis (4° C, 0.001 to 0.003 M tris-HCl in 1% sucrose, 18 hr, ca. 26,000 rpm) indicated a molecular weight of 53,000 to 61,000 for these subunits [14]. Charge effects may be responsible for distorting the S values to a moderate degree. Similar sedimentation studies were carried out by treatment of enzyme-NAD^+ with p-CMB. When examined at the same low ionic strength, a single protein peak with a sedimentation coefficient of 3.7 was found, as compared to a value of 4 S found by Darrow and Rodstrom [13] at somewhat higher ionic strength, corresponding to a molecular weight of about 60,000. These details are emphasized because it is well known that charged macromolecules subjected to ultracentrifugation in a very diluted ionic atmosphere can behave differently as compared to conditions providing higher concentrations of counter ions [23]. Exposure of charged groups in the protein could well account for moderate differences in S values, e.g., 3 S or 3.7 S versus 4 S. It is, however, doubtful that the sedimentation equilibrium data performed at low ionic strength would show a 60,000 mol wt where, in fact, the true molecular weight still remained 120,000. It should also be noted that the molecular weight determinations at low ionic strength ranging around 60,000 were measured over a wide range of protein concentrations (from 6.7 mg per ml to 0.5 mg per ml).

A summary of the sedimentation properties of Candida 4-epimerase at low ionic strength under various experimental conditions [14] is presented in Table 3.

Examination of native epimerase in 0.001 M tris HCl buffer at pH 8.08 by chromatography on Sephadex G-150 resulted in a more complex elution profile, but the major fraction appeared as a peak with an apparent molecular weight of about 60,000. The final part of the study concerning the state of aggregation of epimerase at various ionic strengths concerned the reassociation of monomers to form dimers.

TABLE 3

Sedimentation Properties of Yeast UDP-Galactose-4-Epimerase in
0.001 M Tris HCl Buffer (pH 7.4) at 4° (Conductivity 50μmhos) [14]

Type of enzyme examined	Molecular weight
NAD^+-epimerase	52,700-60,500
Apo-epimerase with bound p-CMP	61,000
Apo-epimerase after p-CMB removal	54,800
NADH-epimerase	53,300

When enzyme kept at low ionic strength and dissociated to its monomeric
form is transferred into a milieu of high ionic strength, formation of dimers
can occur without the addition of NAD^+, as was demonstrated by Bertland
and Bertland [14]. These results are shown in Table 4.

When p-CMB subunits (obtained by treatment of native enzyme with
p-CMB) are exposed in 0.01 M buffer to mercaptoethanol to remove bound
p-CMB, the addition of NAD^+ is required to form enzymatically active
dimers.

As the above studies demonstrate, monomeric enzyme subunits can be
obtained by various experimental techniques such as treatment with
guanidine-hydrochloride, titration with p-CMB, or by exposure to low buffer concentration (0.001 M). Within the limits of experimental errors,
subunits obtained by any of the above-mentioned methods were found to have
the identical molecular weight of about 60,000.

TABLE 4

Dependence of Sedimentation Properties of Yeast UDP-Galactose-4-
Epimerase Upon Buffer Concentration [14]

Concentration of tris-HCl buffer	Molecular weight
0.001 M	57,300- 61,500
0.01 M	76,400- 86,700
0.1 M	115,000-118,000

E. Properties of the Prosthetic Group in Candida 4-Epimerase

Native purified samples of epimerase (even those only partially purified) when exposed to ultraviolet light at 3400-3500 Å display a striking
blue fluorescence. The fluorescence of these speciments is reminiscent
of NADH, albeit with a moderate blue shift of the fluorescence emission to
4350 Å [5, 11, 24]. Most likely, the Candida 4-epimerase is a mixture of
dimeric molecules, 15-20% of which are carrying one NADH imbedded in a
lipophilic region, and 80-85% of which contain one NAD^+. The latter fraction represents the catalytically active epimerase [5]. At higher ionic
strength, this fraction would consist of tetramers carrying two NAD^+
groups [13]. Incubation of epimerase with 5'UMP and D-glucose or
D-galactose or with 5'UMP and L-arabinose (25) results in almost quantitative conversion to NADH-epimerase showing greatly enhanced fluorescence and low catalytic activity [25, 26]. This conversion of enzyme-
NAD^+ to enzyme-NADH, referred to by us as reductive inhibition, can be
achieved in several ways. For example, prolonged incubation of UDP-
galactose with E. coli epimerase or UDP-glucose with Candida 4-epimerase
will render highly fluorescent epimerase [26, 27, 28]. A detailed discussion of the phenomenon of the process of reductive inhibition will be presented in the section concerned with the E. coli epimerase.

Liberation of the prosthetic group from the fluorescent NADH-epimerase can be accomplished by treatment with p-CMB. The liberated coenzyme was tested with lactic acid dehydrogenase and an excess of pyruvate
and it was clearly shown to behave like typical NADH [25]. In contrast,
p-CMB titration of NAD^+-epimerase released a coenzyme which
did not behave like NAD^+ [29]. Denaturation of NAD^+-epimerase by
ethanol did give rise to release of NAD^+, as ascertained by its ability to
act as coenzyme in the UDP-glucose dehydrogenase test and identical
mobility with authentic NAD^+ upon thin layer chromatography on poly-
ethyleneimine (PEI) [30]. By these specific methods, the stoichiometry
of one NAD^+ (or NADH) per dimer was once more confirmed.

The strong binding of NAD^+ and NADH in the dimeric and tetrameric
forms of epimerase is evident from several facts. Repeated chromatography runs on Sephadex columns do not decrease the nicotinamide nucleotide content to any detectable degree [31]. Moreover, the above-
mentioned enzymatic tests for NAD^+ or NADH are negative as long as the
dimer structure remains intact. Likewise, addition of a highly active
NAD^+-glycohydrolase [32] to native epimerase does not destroy the bound
NAD^+ [31], whereas the NAD^+ liberated by denaturation of epimerase
was destroyed by this enzyme [31].

F. NAD$^+$ Exchange

Another experimental approach to examine the problem of monomer formation at low ionic strength was carried out. This approach concerns the fate of NAD$^+$ in relation to the dimeric enzyme protein under different experimental conditions. If the NAD$^+$ (or NADH) of the dimer is dependent on its strongly bound state by a hydrophobic milieu created by aggregation of two monomers, one would expect that NAD$^+$ would be liberated whenever dissociation into monomers takes place. Curiously enough, this does not always happen. As mentioned above, when a potent NAD$^+$-glycohydrolase purified from Fuasrium solani [32] is incubated with the native dimeric NAD$^+$-epimerase, no bound NAD$^+$ is hydrolyzed. If the same experiment is performed at low ionic strength under conditions where monomers are known to be formed, only 20-25% of the NAD$^+$ is hydrolyzed although the NAD$^+$ glycohydrolase present is able to split more than one thousandfold the amount of free NAD$^+$ under the same conditions [31].

We are, therefore, faced with the problem of finding out whether NAD$^+$ remains at random bound to monomers or alternatively, we have to postulate that the subunits are nonidentical (see also Sec. II). NAD$^+$ would then be expected to have higher affinity to one type of monomer. Another reason for higher affinity of NAD$^+$ to one subunit could be conformational change of the protein caused by the binding of NAD$^+$ to the subunit. Of course, one could also cite the outcome of the above experiment as a criterion against the notion of monomer formation at low ionic strength. Different experiments were performed to study the process of NAD$^+$-dissociation from enzyme subunits at low ionic strength. For instance, when enzyme-NADH is incubated at low buffer concentration (0.001 M), a marked decrease of the characteristic NADH fluorescence is observed. Upon rapid addition of concentrated buffer to a final concentration of 0.1 M, fluorescence is restored again to its original value [30].

More convincing evidence concerning the affinity of NAD$^+$ to subunits at different ionic strengths was obtained by NAD$^+$ exchange experiments, performed by Bugge and Kalckar [30]. Epimerase was stored overnight in 0.001 M tris-hydrochloride buffer pH 8.0 at 4°C. After this period of time, the sample was divided into two equal portions. Sample A was incubated for 72 hours at 4°C in the presence of 40 μmoles [^{14}C] NAD$^+$. Sample B was first adjusted with tris-HCl, pH 8.0 to a final buffer concentration of 0.1 M, followed by the addition of the identical amount of [^{14}C] NAD$^+$. Again the incubation was carried out at 4°C for 72 hours.

After the completed incubation period, the individual samples were subjected to chromatography on a Sephadex G-50 column to separate enzyme-bound NAD$^+$ from excess free [^{14}C] NAD$^+$. The data obtained in this experiment are shown in Figure 1. As can be seen, incubation of epimerase at low ionic strength (left diagram in Fig. 1) results in a

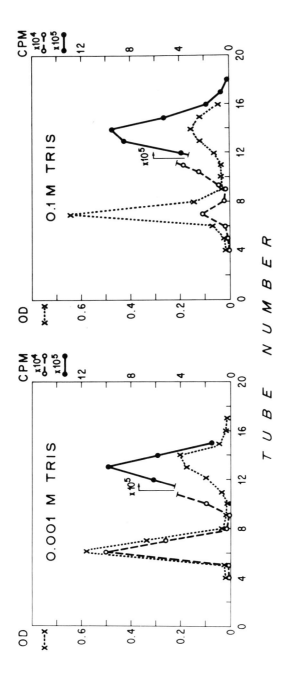

FIG. 1. The experimental details are described in the text [30]. The diagram at the left side of the figure represents the elution profile obtained for sample A (exchange study at 0.002 M tris hydrochloride), while the diagram on the right side shows the elution profile for sample B (exchange study at 0.1 M tris hydrochloride).

marked exchange of protein-bound NAD^+ with $[^{14}C] NAD^+$ present in the medium. In contrast, at high ionic strength (right diagram in Fig. 1) only negligible exchange with labeled NAD^+ takes place. From these experiments, it appears certain that NAD^+ is more accessible to exchange at low ionic strength than at high ionic strength, although it is evident that NAD^+ remains largely protein-bound, even at low buffer concentrations. These observations are consistent with data mentioned earlier, concerning the fact that only a small proportion (about 25%) of NAD^+ is hydrolyzed by NAD^+-glycohydrolase at low ionic strength [31].

G. Studies of Peptide Hydrogen Exchanges and Their Relation to Conformational Changes of the Enzyme Protein

Exchange studies of the tritiated peptide hydrogens of dimeric enzyme-NAD^+ show that more than half the peptide hydrogens exchange very rapidly, too fast to be measured by the back exchange method, whereas about 400 peptide hydrogens show measurable exchange by this method [29]. The exchange feature of this class of peptide hydrogens is illustrated in Figure 2, from which it is also clear that enzyme-NADH shows a different pattern as compared to enzyme-NAD^+. Enzyme-NADH retains more peptide tritium than enzyme-NAD^+, indicating that the reduced form is more taut than the native enzyme.

When the same tritium back exchange method was applied to monomers formed by titration of the SH groups by p-CMB, shown in Figure 3, many more peptide hydrogens exchanged, indicating that the monomer formed by p-CMB is more easily accessible to proton exchange with the medium than the native enzyme protein.

H. Isoelectric Point

In a discussion on the effect of low ionic strength on the properties of epimerase, it is relevant to mention various attempts to determine the isoelectric point of the enzyme.

Starch gel electrophoresis under conditions which should preserve the dimeric structure of the enzyme (at 8-25 mM buffer concentrations) indicated the isolelectric point of the protein to be at the acidic side of the pH scale [33]. Numerous experiments using the method of gel electrofocusing indicated an isoelectric point of 4.5 to 5.0 [31]. The ampholines used for electrofocusing may alter the isoelectric point to some extent. Another problem that has not been eliminated concerns the low ionic strength used in electrofocusing (of the order of 0.001 M). Under these experimental conditions, it is likely that native epimerase migrates in its dissociated form and that the isoelectric point of the subunits (or one of

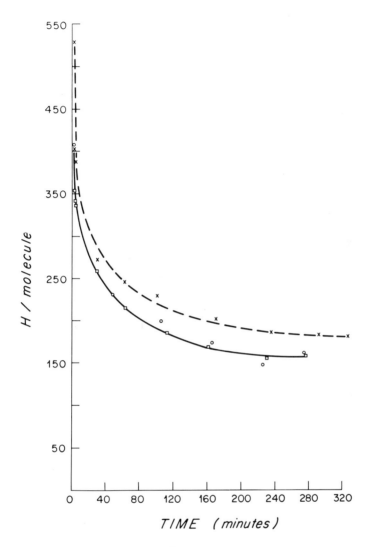

FIG. 2. Time dependence of [^{3}H]-exchange (out). Native epimerase, (enzyme-NAD$^+$): Open squares, full line. Epimerase plus L-fucose and 5'-UMP, (enzyme-NAD$^+$): Open circles, full line. Epimerase plus L-arabinose and 5'-UMP, (enzyme NADH): Crosses, dashed line. (From Ref. 29, p. 262, courtesy of Munksgaard and Academic Press.)

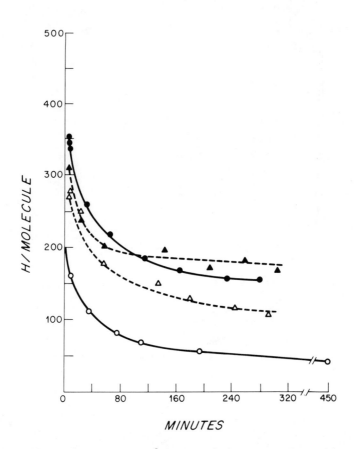

FIG. 3. Time dependence of [³H] exchange (out). Native epimerase:
full circles, full line. p-CMB-epimerase: Open circles, full line. Reactivated p-CMB-epimerase (reactivated in presence of [³H] - H₂O: full
triangles, dashed line. Reactivated p-CMB-epimerase ([³H] - labeled
after reactivation): Open triangles, dashed line. (From Ref. 29, p. 264,
courtesy of Munksgaard and Academic Press.)

them) is around pH 4.5. The protein band located in the electrofocusing
gel showed catalytic activity, indicating that the enzyme was not denatured
by the procedure. A further positive proof for the maintenance of the
integrity of the functional enzyme protein was the appearance of a blue
fluorescent band that was observed after incubation of the gel with 5'UMP
and L-arabinose in 0.1 M tris-HCl buffer [30].

I. Asymmetry of the Native Enzyme Protein

Neither electrofocusing nor electrophoresis performed at low ionic strength was a method capable of resolving dissociated monomers into two individual distinct proteins. And yet, the monomers obtained at low ionic strength may differ in several ways.

According to several independent lines of experimental evidence, it has been established that there is only one pyridine nucleotide per two enzyme subunits [5-8]. After dissociation of the enzyme into subunits, based on the NAD^+ exchange data reported in Sec. III, F, the coenzyme remains largely bound to the subunits, either at random or to only one specific type of subunit. In either case, only about half the subunits should be linked to NAD^+, the other half has to be free of NAD^+.

Another approach to establish the occurrence of two different subunits was provided by preliminary experiments by Bertland [34]. This author determined C-terminal amino acids in epimerase by a modified method of Matsuo et al. [35, 36, 37]. The principle of this method is based upon the selective tritiation of C-terminal amino acids by formation of an oxazolone derivative followed by incubation in $[^3H] H_2O$.

According to Bertland [34], when C-terminal amino acid analysis was performed on Candida 4-epimerase, evidence for two different amino acids was found: Leucine and isoleucine were recognized as one C-terminal amino acid, while either phenylalanine or tyrosine was found to be the second C-terminal amino acid. From these data it appears that Candida 4-epimerase does not contain two identical subunits in spite of the fact that electrophoretic methods failed to provide experimental evidence for two different subunits.

It should be emphasized, however, that with the data presently available, the question of identical or nonidentical subunits for the Candida 4-epimerase enzyme has not been satisfactorily resolved. One attractive possibility that has not been ruled out is the existence of two identical protein subunits which, upon binding of coenzyme to either one of these subunits, causes a conformational change of the subunit engaged in the linkage. Thus, after binding of NAD^+ to one subunit, the subunits behave like two different entities.

IV. UDP-GALACTOSE-4-EPIMERASE FROM E. COLI

The first reactions of galactose metabolism in bacterial systems involve catalysis by galactokinase, galactose-1-phosphate-uridyltransferase, and

UDP-galactose-4-epimerase [38, 22]. In E. coli, these enzymes are coordinately induced [39] since their structural genes are located adjacent
to the galactose operon [39, 40].

A. Main Structural Features of the Enzyme

Cultures of E. coli induced for these enzymes either by the addition of
D-fucose (6-deoxy-D-galactose) or, in the case of kinase-less mutants, by
endogenous induction, were used for purification and isolation of UDP-
galactose-4-epimerase.

Wilson and Hogness [41] have described a method by which they obtained an almost homogeneous preparation which, according to Schlieren
patterns, indicated better than 95% purity. The catalytic activity of this
purified E. coli 4-epimerase is as high as 14×10^3 units per mg (1 unit or
1 μmole UDP-galactose converted per hour). The sedimentation coefficient
of this preparation was found to be $S = 5.02$, and sedimentation equilibrium
centrifugation indicated a molecular weight of 79,000. The E. coli enzyme
did not have any requirements for NAD^+ since it contains firmly bound
NAD^+. Enzymatic determination of NAD^+ released from the native enzyme
using the UDP-glucose dehydrogenase assay method indicated 1 mol of
NAD^+ per 7.9×10^4 g epimerase. The K_m for UDP-galactose was found to
be 1.6×10^{-4} M [41].

An important structural feature of the enzyme concerns the occurrence
of three different groups of half cysteine residues in a total of eight [8]:

1. Two residues with -SH groups are capable of reacting with Ellman's
    reagent [5,5'-dithiobis (2-nitrobenzoic acid)] in the native state of
    the enzyme.
2. Two residues with -SH groups are accessible to Ellman's reagent
    only after denaturation with 3.5 M guanidine hydrochloride.
3. The four remaining -SH residues are incapable of reacting either
    with Ellman's reagent or with N-ethylmaleimide under denaturing
    conditions. The sulfhydryl groups of the enzyme react readily
    with mercurial compounds causing inactivation of the enzyme and
    dissociation into subunits. This process will be discussed in more
    detail later.

The frequency of amino acid residues per mole of active E. coli 4-epi-
merase was also determined by Wilson and Hogness [8]. One mole of
epimerase contains 720 amino acids or 360 amino acids per subunit, assuming identical subunits. The sum of lysine and arginine residues per
mole of enzyme totaled 56. Assuming that E. coli epimerase is a symmetrical dimer composed of two identical subunits, there should be 56/2 =
28 loci susceptible to attack by trypsin, and therefore, a total of 29 tryptic

peptides should be discernible. Both native epimerase and reduced alkylated enzyme were subjected to trypsin digestion. Separation of the peptides, according to Helinski and Yanofsky [42], resulted in resolution of
27-29 ninhydrin positive spots.

The molecular weight of E. coli 4-epimerase in solutions of guanidine
hydrochloride (4.65 M or 6.14 M) was determined using sedimentation
equilibrium centrifugation in double sector cells [8].

Assuming a partial specific volume of 0.73 ml per gram, the molecular
weights for the subunits was found to be 40,000 [8]. From the amino acid
composition and number of total amino acids, the molecular weight of the
active E. coli epimerase was assessed to be 80,000 [8] in excellent
agreement with the value found by sedimentation equilibrium studies of
native E. coli epimerase [8]. The subunit found in guanidine hydrochloride has half the molecular weight of the native enzyme, a fact consistent
with the existence of two identical subunits.

In the same study, these authors [8] carried out a quantitative determination of the amino terminal residues using the carbamylation methods of
Stark and Smyth [43]. Two moles of aspartic acid per mole of active enzyme was the only N-terminal amino acid found. The authors, however,
emphasized that the method does not distinguish between aspartic acid and
asparagine. Consequently, one or both N-terminal amino acids could be
asparagine.

From the above experimental evidence, Wilson and Hogness [8] concluded that native E. coli epimerase is most likely composed of two
identical subunits containing 360 amino acids each, and one mole of NAD^+
held together by noncovalent bonds. In analogy with the Candida 4-epimer-
ase, the E. coli enzyme has identical stoichiometry between coenzyme and
subunits, namely 1 mol of NAD^+ (or NADH) per dimer. In contrast to the
Candida 4-epimerase, however, it appears that the NAD^+ molecule is attached to a symmetrical dimer, a very intriguing type of steric
configuration.

Some of the problems concerning the questions of symmetry between
one NAD^+ and two subunits have already been discussed for the Candida
4-epimerase and are also applicable to the E. coli enzyme. For example,
does the occurrence of dimers with only one mole of coenzyme preclude
additional uptake of one more mole of NAD^+ (in presence of excess NAD^+)
due to negative cooperativity of the subunits linked to one mole of NAD^+?

A more detailed discussion of subunit formation from E. coli epimerase
will follow later in this review.

B. Properties of the Catalytically Active Enzyme Protein

In contrast to 4-epimerases isolated from other sources, the E. coli enzyme has a rather broad specificity. The sugar nucleotide can have variations in the sugar moiety as well as in the pyrimidine base of the molecule as indicated in Table 5. Although the broad substrate specificity for UDP-galactose-4-epimerase in terms of both the pyrimidine base and the sugar moiety has been reported [47], we have chosen to provide experimental evidence that TDP-glucose is indeed a substrate for the enzyme since the above data were not obtained on homogeneous enzyme preparations. For this purpose, four independent lines of evidence were provided:

1. The ratio of enzymatic activities between UDP-glucose and TDP-glucose remained unchanged throughout enzyme purification.
2. Induction of cultures by the addition of D-fucose showed the same extent of induction for UDP-glucose and TDP-glucose [45, 46].
3. E. coli mutants with absent UDP-galactose-4-epimerase activity are also inactive when tested with TDP-glucose.
4. TDP-galactose is a competitive inhibitor for UDP-galactose epimerization. The K_m for TDP-galactose at 37° C was found to be 1.2×10^{-3} M, while the K_i for both UDP-glucose and UDP-galactose was 1.8×10^{-4} M in agreement with K_m values previously determined [41].

Considering the above range of substrate specificity, it is apparent that uracil can be replaced by thymine or cytidine, but even a purine base like adenine is acceptable as a substrate.

Variation in the sugar moiety is possible in the 6-position where 6-deoxy-, 6-aldo-derivatives, or complete absence of carbon 6 (pentoses) can serve as substrates so long as the sugar is homologous to the parent hexose of the D-family. One notable exception is the 2-epimer of D-glucose, D-mannose, which cannot serve as a substrate [47].

C. Reaction Mechanism

The apparently simple conversion, involving only one center of asymmetry of a hexose, was explored by the use of radioactive isotopes. Earlier reports in the literature established that hexoses tritiated at carbon 4 did not exchange hydrogen into the medium during epimerization, nor was ^{3}H from ^{3}[H] H_2O incorporated during the enzymatic reaction [48-50].

Addition of UDP-galactose to the E. coli enzyme caused appearance of an NADH-like absorption with a maximum at 345° Å [41], suggestive of

TABLE 5

Substrate Specificity of E. coli-4-Epimerase

UDP-D-glucose	[41]
UDP-D-galactose	[41]
UDP-D-xylose	[44]
UDP-L-arabinose	[44]
UDP-D-fucose	[45]
UDP-6-deoxy-D-glucose	[45]
TDP-D-glucose	[46]
TDP-D-galactose	[46]
TDP-6-deoxy-D-galactose	[46]
deoxy-UDP-D-glucose	[45]
UDP-D-galactose-hexodialdose (1.6)	[47]
CDP-D-glucose	[47]
ADP-D-glucose	[47]

the formation of enzyme-NADH. At full saturation with UDP-galactose (the K_m being 1.6 x 10^{-4} M), the increase of density at 345° Å in the difference spectrum indicated that close to 20% of the E. coli epimerase was in the NADH form. This degree of reduction was reached within one minute at saturating concentrations of UDP-galactose [41].

A detailed representation concerning the enzyme-bound coenzyme and its role in the catalytic process involved in the oxidoreductase mechanism is outlined in scheme [2]. One possible approach to gain insight into the transfer of hydrogen from the substrate to the coenzyme was to study rate-limiting steps involved in this process. When a preparation of UDP-[4-^{3}H]glucose was first used as a substrate of the enzyme, an inverse isotope effect was reported [51]. That is to say that the tritiated substrate was used preferentially over the nontritiated compound. More recently, the same experiment was repeated, and a small positive isotope effect for UDP-[4-^{3}H]glucose was published [52]. We decided to reinvestigate these experiments in order to locate the position of initial substrate oxidation. For this purpose, it is necessary to consider first two possible reaction sequences, both of them involving the participation of enzyme-bound NAD$^+$. In the first reaction sequence, catalysis would occur according to scheme (2): removal of hydrogen at carbon 4 of the substrate, resulting in the formation of the corresponding 4-hexosulose intermediate. This 4-hexosulose-derivative would then, in turn, serve as hydrogen acceptor to result in the formation of the 4-epimeric product.

A second possible mechanism outlined in scheme (3) proceeds by the initial substrate oxidation attacking at carbon 3, followed by a ketoenediol transformation between carbons 3 and 4, causing loss of the center of asymmetry at carbon 4. A reversal to the keto-function at carbon 3 will then lead to formation of the 4-epimer at carbon 4. In the last step of this

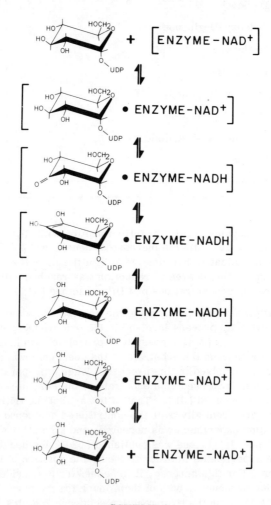

SCHEME 3

sequence, stereo-specific reduction of the keto-group at carbon 3 would terminate the reaction sequence and release the product from the enzyme.

To distinguish between these two mechanisms, kinetic data were obtained using substrates tritiated specifically at carbon 3 or carbon 4 respectively. It would be expected that a substrate with carbon-bound tritium involved in the initiation of the enzymatic reaction and resulting in breaking the linkage between carbon and tritium will show a kinetic isotope effect. The outcome of these experiments should permit to distinguish between the positions in which initial oxidation of the substrate took place, and thus permit differentiation between the two outlined mechanisms. When TDP-[3-^{3}H] glucose and TDP-[4-^{3}H] glucose were used as substrates in these experiments [45, 46] only the 4-tritiated substrate showed a small but significant positive isotope effect, while the 3-tritiated sugar nucleotide behaved like unlabeled material. These data were interpreted as indicating that the initial removal of hydrogen by enzyme catalysis occurs at carbon 4 of the substrate and initial attack at carbon 3 is, therefore, unlikely.

However, it should be emphasized that the absence of any exchange of tritium with the medium should not be considered as proof for a hydride mechanism according to scheme (2) since it is well established that enzymatic reactions involving proton transfer as indicated in scheme (3) do not necessarily lead to hydrogen exchange with the medium [53].

These data are consistent with earlier observations establishing that isotope exchange did not occur with the medium during epimerization of 4-tritiated hexoses nor was tritium from ^{3}HOH incorporated [48-50]. All the experimental facts reported so far appear to be in agreement with the reaction mechanism shown in scheme (2) but do not provide direct evidence for a 4-hexosulose intermediate. In addition, no direct proof for transfer of hydrogen from the 4-position of the substrate to the coenzyme had been demonstrated. Tentative evidence for the involvement of a 4-ulose intermediate was provided by Nelsestuen and Kirkwood [47]. These investigators subjected E. coli epimerase to reduction with [^{3}H] NADH. The NADH bound to the enzyme was shown to stereospecifically reduced on the B side of the nicotinamide ring. When TDP-6-deoxy-D-xylo-4-hexosulose was added to this preparation of enzyme-[^{3}H] NADH, a mixture of TDP-[4-^{3}H]-6-deoxy-D-glucose and TDP-[4-^{3}H]-6-deoxy-D-galactose was formed. Unfortunately, this is not unambiguous proof for the participation of a 4-hexasulose intermediate in the enzymatic reaction since 2-ketoglucose [29] and inosose-2 [54] will also oxidize enzyme-NADH.

Direct hydride transfer from the 4-position of the substrate to the enzyme-bound coenzyme was demonstrated in the following way: TDP[4-^{3}H]-glucose was incubated with epimerase until enzyme-NADH was detectable by fluorescence [26]. The enzyme was then isolated by

Sephadex chromatography, and the protein fraction was found to contain a significant amount of radioactivity. The coenzyme was released from the isolated enzyme and was characterized as $[^3H_1]$ NADH. When an identical experiment was carried out with TDP-$[3-^3H]$ glucose as the substrate, no radioactivity was located in the enzyme-bound reduced coenzyme (Table 6) [55].

Complementary evidence consistent with this reported hydride transfer was the demonstration of UDP-D-xylohexosulose by Maitra and Ankel [56]. These investigators incubated the enzyme with UDP-galactose and upon addition of $[^3H]$ NaBH$_4$, UDP-$[4-^3H]$ glucose and UDP-$[4-^3H]$ galactose were identified. These results were interpreted to indicate the occurrence of UDP-D-xylo-4-hexosulose as an intermediate in the reaction.

D. Reductive Inhibition

Provided that the reaction mechanism for UDP-galactose-4-epimerase presented in scheme (2) is valid, one important prediction can be made: Only enzyme-NAD$^+$ can be catalytically active to initiate enzyme catalysis. In contrast, enzyme-NADH is enzymatically inactive since it cannot act as hydrogen acceptor.

Several groups of compounds have been described that are capable of causing enzyme inhibition according to:

$$[^3H] \text{ inhibitor} + \text{enzyme NAD}^+ \rightarrow [\text{(inhibotor)}_{\text{oxidized}}^- $$
$$\text{enzyme} - [^3H_1] \text{ NADH}] \tag{1}$$

We refer to this process as <u>reductive inhibition.</u> Thus, a reductive inhibitor has the ability to donate hydrogen to the enzyme-bound coenzyme but cannot accept it, resulting in accumulation of enzyme-NADH and concomitant loss of catalytic activity. One criterion used to recognize reductive inhibition is the reversal of this process by oxidation of enzyme-NADH. One suitable intermediate for this purpose is thymidine diphospho-6-deoxy-D-xylo-4-hexosulose according to:

$$[\text{(inhibitor)}_{\text{oxidized}} - \text{enzyme } [^3H_1] \text{ NADH}] + \text{TDP-6-deoxy-D-}$$
$$\text{xylo-4-hexosulose} \rightarrow [\text{enzyme-NAD}^+] + \text{(inhibitor)}_{\text{oxidized}}^+ $$
$$\text{TDP-6-deoxy-}[4-^3H] \text{ glucose} + \text{TDP-6-deoxy-}[4-^3H] \tag{2}$$
$$\text{galactose}$$

We wish to distinguish between several different types of compounds causing reductive inhibition:

TABLE 6

Reaction of TDP-[3-^{3}H] glucose and TDP-[4-^{3}H] glucose
with NAD$^+$-epimerase [55]

Substrate	% newly generated NADH-epimerase	cpm [^{3}H] per nmol NADH-epimerase
TDP-[3-^{3}H] glucose	24	<12
TDP-[4-^{3}H] glucose	20	450

1. The first observation concerning reductive inhibition was reported by Kalckar and coworkers [11, 26]. Upon incubation of free sugars such as D-galactose or L-arabinose in the presence of 5'UMP, a marked increase of fluorescence was observed and related to the formation of enzyme-NADH. As expected, the enzyme preparations subjected to this process showed decreased enzymatic activity. The mechanism of this reaction was elucidated by incubation of UDP-galactose-4-epimerase with [1-^{3}H] galactose in the presence of 5'UMP, resulting in tritium transfer to enzyme-NAD$^+$ yielding [^{3}H$_1$] NADH [57]. The product of the oxidation was found to be the corresponding sugar acid (D-galactonic acid). Davis and Glaser [58] reported an isotope effect of this reductive inhibition upon reaction with [3-^{2}H]-glucose but not with [4-^{2}H] glucose, thus implying the 3-position of the sugar to be the point of initial oxidation and suggesting that the 3-carbon position is also the point of initial attack in the natural substrate. However, the same authors reported lack of ^{3}H-transfer to enzyme-NAD$^+$ when this experiment was repeated with [3-^{3}H]glucose. Ketley and Schellenberg reinvestigated this question again [59], by examination of the ability to produce (enzyme-[^{3}H$_1$] NADH) by incubation of [1-^{3}H] glucose, [4-^{3}H]galactose], [1-^{3}H] galactose and [4-^{3}H]glucose in the presence of 5'UMP. According to these authors, only the first three compounds were capable of transferring tritium, while [4-^{3}H] glucose did not transfer any label. Ketley and Schellenberg concluded from these experiments that only α-anomers of free sugars can act as hydrogen donors in reductive inhibition. However, the fact that the natural substrate of the enzyme, UDP-glucose transfers hydrogen from the 4-position to yield enzyme-NADH remains unexplained.

We would like to provide a different interpretation of Davis and Glaser's [58] and Ketley and Schellenberg's [59] experiments.

The main difference between the natural substrate and the action of
5'UMP and free sugar is the missing covalent linkage to keep the
sugar properly positioned in relation to the active site of the enzyme.
This, in turn, will cause the free sugar to find the best fit for the
active site, and probably more than one possible alignment with
free sugar and the active site is possible. Considering the circum-
scribed stereospecificity of the enzyme, it is not surprising that
only homologous sugars like D-galactose and L-arabinose work
best. One reason for this could be that in the C-1 conformation
both of these sugars have one axial OH-group at carbon 4. This
OH-group could be instrumental in facilitating the proper alignment
with the active site. As a space filling model of these sugars re-
veals, when D-galactose is turned around such that carbon 1 of the
molecule is exposed to the position at the active site usually occupied by carbon 4 of UDP-galactose, the axial OH-group of the free
sugar at carbon 4 sits now in the identical position taken up by the
oxygen at the anomeric carbon of the sugarnucleotide. Unfortunately,
no experimental data for a specific probability of free sugar to expose either 4-position or 1-position to participate in the dehydrogen-
ation of the reductive inhibition has been reported, although this
may very well be one of the reasons for the low specific activity
found in the hydrogen transfer from [1-^{3}H] galactose [57].
Similarly, the data reported by Davis and Glaser [58] could be
understood to be related to preferential binding of the free sugar at
a specific position. Consequently, the results of these "model
experiments" with 5'UMP and free sugars and their bearing on the
actual reaction mechanism of the natural substrate have to be
interpreted with great care.

2.   A second type of reductive inhibition was also initially described by
      Kalckar and coworkers [5, 60] and involved the use of NaBH$_4$.
      As was demonstrated in these early experiments, the presence of
      5'UMP is essential for the stability of enzyme-NADH produced.
      Later, the same observations were confirmed by Nelsestuen and
      Kirkwood [47]. It is essential to emphasize that enzyme-NADH
      prepared by this procedure in the absence of 5'UMP provides
      unstable enzyme-NADH. From these experiments, it appears that
      5'UMP changes the conformation of enzyme-NADH such that enzyme-
      NADH is much less accessible to autoxidation. However, it should
      be pointed out here that this is a quantitative difference, and we
      found that enzyme-NADH prepared by any method undergoes slow
      spontaneous autoxidation, a fact that as we will see, has not been
      considered sufficiently, to account for a number of experimental
      observations reported in the literature.

3.   The third class of compounds capable of causing reductive inhibition
      is a group of sugar nucleotides characterized by a phosphate linkage
      to the terminal carbon of the sugar moiety as well as a free reduc-
      ing end at carbon 1. One example of such a reductive inhibitor is

5-(adenosine-5'-pyrophosphoryl)-D-ribose, [ADP-ribose-(5)], a
degradation product of NAD^+. The labile nature of NAD^+ to mild
alkaline hydrolysis resulting in the specific cleavage into free
nicotinamide and ADPR-(5) has been recognized for a long time
[61]. Examination of commercial samples of NAD^+ revealed in
some of them the presence of up to 10% ADP-ribose-(5). The
ambiguity introduced concerning interpretation of kinetic data obtained with NAD^+- requiring enzymes using impure NAD^+ preparation have been discussed [62]. Purified NAD^+ free of ADP-ribose-
(5) contamination does not inhibit epimerase. Several derivatives,
analogous of 5-(adenosine 5'-pyrophosphoryl)-D-ribose were prepared by organic synthesis with the objective to test these compounds for their ability to act as reductive inhibitors. For example,
reduction of ADP-ribose-(5) with $NaBH_4$ yielded 5-(adenosine-5'-
pyrophosphoryl)-ribitol. Substitution of the adenosine moiety with
thymidine led to synthesis of 5-(thymidine-5'-pyrophosphoryl)-D-
ribose. In addition, 6-(thymidine-5'-pyrophosphoryl)-D-glucose
was also prepared. It should be mentioned that the latter two compounds have a free reducing end on their sugar moiety. Incubation
of UDP-galactose-4-epimerase with these sugar nucleotides gave
the results indicated in Table 7 [63]. For the identification of the
reaction intermediates in the process of reductive inhibition caused
by sugars with free reducing end, 6-(thymidine-5'-pyrophosphoryl)-
D-glucose was chosen as the inhibitor. This nucleotide was prepared [$U^{14}C$]-labeled in the glucose moiety as well as with
[$1-^3H$] glucose. Using these labeled compounds, the following
reaction mechanism was established for the process of reductive
inhibition [63]:

$$[enzyme-NAD^+] + 6-(thymidine-5'-pyrophosphoryl) -$$

$$[1-^3H]-D-glucose \rightarrow [enzyme-[^3H_1] NADH-6- \qquad (3)$$

$$(thymidine-5'-pyrophosphoryl)-D-gluconic acid]$$

Reactivation of the enzyme-NADH conjugate obtained according
to the above reaction (3) was accomplished by addition of TDP-6-
deoxy-D-xylo-4-hexosulose and was found to proceed as follows:

$$[enzyme-[^3H_1] NADH-6-(thymidine-5'-pyrophosphoryl)-D-$$

$$gluconic acid] + TDP-6-deoxy-D-xylo-4-$$

$$hexosulose \rightarrow [enzyme-NAD^+] + 6-(thymidine- \qquad (4)$$

$$5'-pyrophosphoryl)-D-gluconic acid + TDP-[4-^3H] -$$

$$6-deoxy-D-galactose + TDP-[4-^3H]-6-deoxy-D-$$

$$glucose.$$

From these experiments (63), the reaction mechanism of reductive inhibition by sugar nucleotides with a free reducing end appears to be analogous to that of free sugar and 5'UMP, as described before.

E. Side Reactions of the Catalytically Active Enzyme

Several other features of UDP-galactose-4-epimerase, apparently not directly related to its catalytic activity under normal reaction conditions, have to be discussed. As we have already mentioned above, there is substantial evidence to support the view that enzyme-NADH is subject to autoxidation. Attachment of nucleotides to enzyme-NADH has a stabilizing effect, as was demonstrated in the instance of formation of enzyme-NADH by sodium borohydride:

$$[\text{enzyme-NADH}] \xrightarrow[\text{fast}]{\text{autoxidation}} [\text{enzyme-NAD}^+] \qquad (5)$$

$$[\text{enzyme-NADH-5'UMP}] \xrightarrow[\text{slow}]{\text{autoxidation}} \begin{array}{l}[\text{enzyme-NAD}^+] \\ +5'\text{UMP}\end{array} \qquad (6)$$

It should be noted, however, that we have evidence to support the view that slow autoxidation of enzyme-NADH occurs irrespective of the nature of the nucleotide attached.

$$[\text{enzyme-NADH-sugar nucleotide}] \xrightarrow{\text{autoxidation}} \begin{array}{l}[\text{enzyme-NAD}^+] \\ + \text{sugar nucleotide}\end{array}$$
$$(7)$$

TABLE 7

Reductive Inhibition of UDP-Galactose-4-Epimerase [63]

Compound	M	Preincubation time in minutes	% inhibition of UDP-galactose-4-epimerase
ADP-ribose-(5)	5×6^{-5}	45	50
ADP-ribitol	5×10^{-4}	45	0
TDP-ribose-(5)	1×10^{-8}	45	35
TDP-glucose-(6)	1×10^{-5}	5	50

[a] Before assaying, the enzyme was preincubated at $37°$ C for the time interval indicated in the table.

This property of enzyme-NADH to be subject to autoxidation results in each instance in the formation of enzyme-NAD$^+$ and release of bound nucleotide from the enzyme protein.

Another feature of UDP-galactose-4-epimerase was elucidated by the elegant experiments of Wee and Frey [64], as well as by our own observations [65] resolving a number of misconceptions concerning the reaction mechanism. Accepting the validity of the reaction mechanism for epimerase shown in scheme (2), the hydrogen transfer of the reaction has to be intramolecular. This was indeed experimentally verified earlier [66]. Therefore, any accumulation of enzyme-NADH during the course of the reaction violates the basic principle of the reaction mechanism since it will exclude intramolecular hydrogen transfer from taking place. Consequently, reaction conditions leading to accumulation of enzyme-NADH involve intermolecular hydrogen transfer and, therefore, must be reactions not operative during normal catalysis.

Wee and Frey [64] demonstrated that the catalytically active enzyme-NADH complex with UDP-D-xylo-4-hexosulose will slowly exchange with excess substrate (or end product) according to:

$$
\begin{array}{lll}
[\,\text{enzyme-NADH-UDP-D-} & +\,\text{UDP-hexose} & [\,\text{enzyme-NADH-} \\
\quad \text{xylo-4-hexosulose}\,] & \underset{+\,\overline{\text{UDP-D-xylo-}}}{\overset{\longrightarrow}{\rightleftharpoons}} & \quad \text{UDP-hexose}\,] + \qquad (8) \\
& \qquad \text{4-hexosulose} & \quad \text{D-xylo-4-hexosulose}
\end{array}
$$

As indicated, this process is reversible and leads to a slow exchange of the enzymatically active nucleotide-4-hexosulose intermediate to the enzymatically inactive enzyme-NADH-UDP-hexose conjugate. This process will take place when the enzyme is exposed for prolonged periods of time to an excess of substrate and product. Enzyme-NADH-UDP-hexose is a catalytically inactive form of epimerase and has two different ways for slow restoration of catalytic activity:

1. Autoxidation according to reaction (7) will lead to formation of enzyme-NAD$^+$ and release of UDP-hexose. Similarly, autoxidation of the catalytically active enzyme-NADH-UDP-D-xylo-4-hexosulose will yield active enzyme-NAD$^+$ and will cause release of 4-keto-intermediate.

2. Exchange of enzyme-NADH-UDP-hexose with UDP-D-hexosulose formed in (1) will restore the catalytically active species, enzyme-NADH-UDP-D-xylo-4-hexosulose, which is in rapid equilibrium with product formation. It should be noted that the affinity of the 4-hexosulose intermediate for the enzyme is at least one hundred-fold stronger than for the substrate (or product). Therefore, the UDP-4-hexosulose intermediate can successfully compete with

UDP-hexose at concentrations around 10^{-7} M. One reason for this high affinity is the similarity of UDP-D-xylo-4-hexosulose or its close relation to a transition state of the enzymatic reaction.

As a matter of fact, this is another argument that can be used to implicate UDP-D-xylo-4-hexosulose as an intermediate of the enzymatic reaction. According to the transition state theory as it pertains to enzymatic catalysis, it was predicted and experimentally verified that transition state intermediates (or analogues thereof) bind to the enzyme much more tightly than the substrate itself [67]. Consequently, prolonged incubation of UDP-galactose-4-epimerase with substrate approaches a level about two-thirds enzyme-NADH and one-third enzyme-NAD$^+$, since both of the above mentioned processes 1 and 2 prevent complete inactivation of the enzyme.

It is, therefore, our interpretation of an earlier report by Wee and coworkers [68] that autoxidation of enzyme-NADH-UDP-D-4-hexosulose according to reaction (7), as well as exchange of enzyme-bound UDP-4-hexosulose intermediate with UDP-hexose [reaction (8)] is responsible for accumulation of 4-keto-intermediate in the medium which can be trapped by [^{3}H] NaBH$_4$ reduction as the 4-epimeric UDP-[4-^{3}H] hexoses. Consequently, we do not think that these data invalidate the earlier report by Maitra and Ankel [56] suggesting UDP-D-xylo-4-hexosulose as the intermediate of enzymatic catalysis. Further corroboration of this view was provided later by Maitra and Ankel [69] demonstrating an inactivation time dependent change in the ratio of trapped UDP-hexoses following [^{3}H] NaBH$_4$ reduction (ratio of UDP- [4-^{3}H] glucose : UDP-[4-^{3}H] galactose changing from 5.6 to 2.6). From these data one can conclude that the 4-hexosulose intermediate is initially enzyme-bound leading to preferential formation of UDP-[4-^{3}H] glucose while at later time intervals the ratio of 2.6 approaches the value found for sodium borohydride reduction of 4-hexosulose to the 4-epimeric sugars obtained in free solution [70, 71].

Another aspect relating to the above-mentioned problem of prolonged incubations and the validity of data obtained in these experiments concerns the demonstration of the direct hydrogen transfer from substrate to enzyme-bound coenzyme. As we mentioned earlier, the intramolecular hydrogen transfer according to scheme (2) occurs under normal short-time enzyme catalysis. During longer incubation periods the steady state intermediate enzyme-NADH-UDP-D-xylo-4-hexosulose exchanges slowly with excess UDP-hexoses according to reaction (8). Under these conditions, as was shown by Wee and Frey [64], intermolecular hydrogen transfer can be experimentally verified. Only during this prolonged incubation in the presence of substrate do two events occur which are not part of the catalytic process shown in scheme (2): (1) UDP-D-xylo-4-hexosulose is released from the enzyme protein by the processes discussed above and (2) concomitant with process (1) accumulation of enzyme-NADH-UDP-hexose occurs. It is this form of the reduced enzyme protein which is not the steady state intermediate of the catalytic process that can be isolated due to its stability and slow reactivation process. Consequently, all attempts reported so far in the literature [46, 47, 55] that claim the demonstration of direct hydrogen transfer from the substrate to the coenzyme do not represent isolation of the catalytic active steady state intermediate enzyme-NADH-UDP-D-xylo-4-hexosulose conjugate but instead enzyme-NADH-UDP-hexose accumulates. Nevertheless, the arguments concerning the mechanism of hydrogen transfer originating from carbon 4 of the substrate to the coenzyme are still valid.

F. Subunits

UDP-galactose-4-epimerase from E. coli has the same general structural features as were described in detail for the yeast enzyme. The catalytically active enzyme contains two apparently identical subunits linked together by 1 mol of NAD^+ [41]. As has already been mentioned earlier, subunits can be obtained by treatment with guanidine hydrochloride. The molecular weight of subunits obtained in this way was determined by ultracentrifugation to be 40,000 daltons [8].

Another way to obtain subunits, accompanied by enzyme inactivation is treatment with compounds like p-CMB. One important feature of this dissociation process is the possibility for its complete reversibility. Addition of NAD^+ and 2-mercaptoethanol will quantitatively restore enzymatic activity. Unfortunately, data concerning the dissociation of the enzyme into subunits, as well as on their reassociation to the holoenzyme, are limited and controversial.

Preparation of enzyme-$[^{14}C]$ NAD^+ using the above mentioned approach of dissociation-reassociation was described by Davis and Glaser [58]. First, the enzyme was treated with p-CMB (10^{-3} M) for one hour at $30°C$. This was followed by reassociation of the enzyme in the presence of $[^{14}C]$ NAD^+ and addition of 2-mercaptoethanol. The radioactive enzyme was isolated by Sephadex chromatography. When we tried to repeat the same experiment, our findings were at variance. After inactivation of the enzyme in the identical way as described by Glaser, addition of 2-mercaptoethanol alone in the absence of exogenous NAD^+ was sufficient to restore fully enzymatic activity. From these data it appears likely that in our experiment NAD^+ had not become dissociated from the enzyme protein. To verify this, inactivated enzyme was incubated with $[^{14}C]$ NAD^+ in the presence of 2-mercaptoethanol and no incorporation of radioactivity into the active enzyme could be demonstrated [72]. These results raise a number of questions concerning the apparent variant nature of the linkage between protein and coenzyme in different enzyme preparations as well as the identity of subunits, somewhat similar to the findings reported for the yeast enzyme.

An even more startling discrepancy between our results and those reported by Glaser [73] concerns kinetic studies on the reactivation of apoenzyme under conditions where the reactions of the apoenzyme and NAD^+ were studied at different NAD^+ concentrations. At given time intervals, further association to active holoenzyme was prevented by addition of ADP-ribose-(5) at 5×10^{-4} M final concentration. The extent of enzyme-NAD^+ formation is then determined by measurement of enzymatic activity. This experiment is difficult to reconcile with our own findings in which ADP-ribose-(5) was found to act at this concentration as a powerful reductive inhibitor [63]. We found this to be true for all our preparations tested including epimerases isolated from other species [63]. It has to be assumed that Glaser's preparation was not sensitive to ADP-ribose-(5) inhibition. More studies are required to resolve these differences and to learn more about the nature of the subunits and the linkage between the coenzyme and the enzyme-protein subunits.

V. UDP-GALACTOSE-4-EPIMERASE FROM OTHER SOURCES

Although UDP-galactose-4-epimerases occurring in microorganisms are the best-studied examples, there are a number of important 4-epimerases occurring in other sources which we wish to discuss in this presentation.

A. Mammalian Sources

In contrast to the enzymes isolated from microbiological systems, the calf liver enzyme has an absolute requirement for NAD^+ [74]. During the purification procedure of this enzyme, the coenzyme is removed; to restore the functional enzyme again, addition of coenzyme is required. A further consequence of the rather low interaction between enzyme and coenzyme is the finding that exogenous NADH is a strong inhibitor for the enzyme. As expected from the reaction mechanism [scheme (2)], enzyme-NADH is catalytically inactive, but due to the low affinity of NAD^+ to the enzyme, NADH acts as a strong inhibitor [74, 75].

A somewhat intermediary affinity between enzyme and coenzyme was reported by Tsai and coworkers [76] in bovine mammary UDP-galactose-4-epimerase. Purification of 4-epimerase isolated from the mammary gland resulted in partial retention of the coenzyme. However, for maximal activity the mammary enzyme had to be supplemented with additional NAD^+ [76].

Tsai and associates [76] examined a number of nucleotides and their stabilizing effect upon diluted solutions of the mammary enzyme. NAD^+, $NADP^+$, and NADH, as well as a number of uridine nucleotides with pyrophosphate linkage, were reported to act as stabilizers. Glycerol has also been found to have a stabilizing effect on the enzyme.

It is apparent that UDP-galactose-4-epimerase preparations isolated from different sources display wide variations in their affinity for the coenzyme, ranging from very tight binding as in the E. coli enzyme to a very loose association resulting in complete dependence on exogenous NAD^+ for catalytic activity, as described for the liver enzyme. The structural elements involved in the noncovalent binding of the coenzyme to the protein still await a careful examination and are at present not fully understood. The susceptibility of UDP-galactose-4-epimerase to inhibition by NADH can be used as an indication to assess the extent of binding between coenzyme and 4-epimerase: The liver enzyme with very loose binding to NAD^+ is strongly inhibited by exogenous NADH [74] while the activity of the E. coli enzyme is insensitive to NADH addition. Whether the proportion between NAD^+ and NADH has physiological significance for the activity of mammalian 4-epimerase remains to be seen.

B. Plant Sources

UDP-galactose-4-epimerase was also reported in higher plants [77, 78]. The enzyme from germinating Phaseolus aureus was characterized by its ability to accept UDP and TDP sugar nucleotides as substrates [79]. Some enzymes, sparingly studied but potentially using the same catalytic principle, may involve the catalysis of the 4-epimerization of UDP-N-acetyl-D-glucosamine [80] and UDP-galactouronic acid [81].

C. Torula Yeast

Kawai and Tochikura [82, 83] described an interesting type of yeast, UDP-galactose-4-epimerase, isolated from Torulopsis Candida (strain 1F00768) grown on a 5% lactose growth medium. This epimerase depends on addition of NAD^+ for catalytic activity [83] just like the 4-epimerase from calf liver [74] or bovine mammary gland [76]. The K_m for NAD^+ in the Torula epimerase was found to be 1.4×10^{-4} M [82]; NAD^+ could not be replaced by $NADP^+$. The K_m for UDP-galactose was found to be 1.2×10^{-3} M. The Torula epimerase was also found to be inhibited by preincubation with 5'UMP and D-galactose; in the presence of 1×10^{-3} M 5'UMP and 0.1 M galactose about 70% inhibition was observed. Glucose was also found to be an inhibitor if incubated with 5'UMP. The observations, as well as the high degree of specificity for 5'UMP, correspond to those reported for the Candida 4-epimerase.

VI. OVERALL VIEW OF REACTIONS CATALYZED BY UDP-GALACTOSE-4-EPIMERASE AND RELATIONS TO OTHER SUGAR TRANSFORMATIONS

A. Reactions Catalyzed by Active Enzyme

Our present view concerning the reactions catalyzed by UDP-galactose-4-epimerase in the presence of substrate is summarized in scheme (4). The actual catalytic process involving intramolecular hydrogen transfer from carbon 4 of the substrate leading to formation of enzyme-NADH-UDP-D-xylo-4-hexosulose is shown in the center of scheme (4). This steady state intermediate of the reaction is converted to the 4-epimeric product and releases enzyme-NAD$^+$ to undergo a new cycle of catalysis.

In addition to this catalytic process, the enzyme-NADH-UDP-D-xylo-4-hexosulose conjugate can undergo several alternate fates shown in scheme (4). Especially under conditions of prolonged incubation periods, several side reaction products will accumulate. As we discussed in Sec. IV, E, enzyme-NADH conjugates are characterized to undergo either autoxidation [upper part of scheme (4)] or they can be subject to exchange between the nucleotide bound to the enzyme and excess substrate [lower part of scheme (4)]. To our knowledge, this scheme is consistent with all experimental facts reported so far. The scheme also illustrates the necessity to distinguish between the intermediate actually participating in the catalytic transformation of glucose into galactose nucleotides and other products not directly related to this enzymatic process.

B. Possible Regulatory Function of Enzyme-NADH

In the metabolic conversion of glucose to galactose, UDP-galactose-4-epimerase plays an important part, and we therefore entertained the possibility for a regulatory role of this enzyme. The ratio of enzyme-NAD$^+$ to enzyme-NADH could be of regulatory significance since only enzyme-NAD$^+$ can serve as hydrogen acceptor to initiate enzyme catalysis, while enzyme-NADH is catalytically inactive. We also considered exploring the possibility that ADP-ribose-(5), a degradation product of NAD$^+$, could act in vivo as a reductive inhibitor and thereby lead to increased levels of enzyme-NADH. When crude extracts of E. coli B are examined for the proportion of enzyme-NAD$^+$ to enzyme-NADH, about 50% of the reduced enzyme can be found [84]. The process of autoxidation may be responsible for the fact that in purified enzyme preparations the amount of enzyme-NADH is diminished to about 10-20% [84]. We examined E. coli B and a mutant strain E. coli MU-2a [85], containing about a tenfold level of NAD$^+$, and found an essentially unchanged ratio of reduced and oxidized forms of the epimerase. Determination of the in vivo steady state levels of ADP-ribose-(5) in both strains revealed that only insignificant amounts of

ADP-ribose-(5) were present [84] . The method employed was capable of
detecting nanomoles of ADP-ribose-(5) and will be published elsewhere
[86] . These data appear to eliminate the possibility that the process of
reductive inhibition has a regulatory function in vivo.

SIDE REACTIONS OF UDP-GALACTOSE-4-EPIMERASE

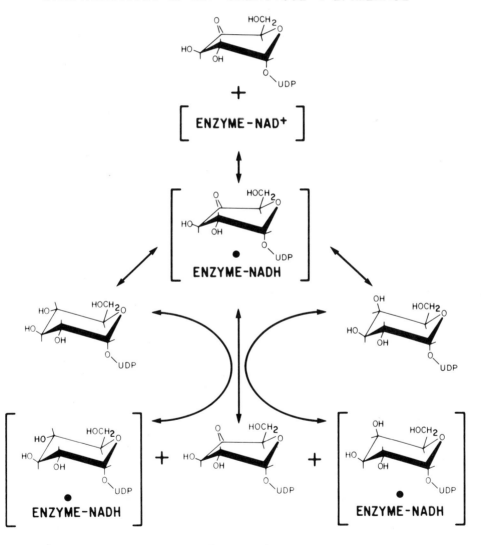

SCHEME 4

C. Subunit Structure of TDPG-Oxidoreductase:
Mechanism of Reaction, Molecular Properties

The detailed study of UDP-galactose-4-epimerase was rewarding in
more than one way. Analogous principles elucidated for the epimerase
were apparently applicable to several other enzymes involved in sugar
transformations. One example studied extensively by us [9, 71, 87, 89] is
the reaction of TDPG-oxidoreductase, the enzyme initiating 6-deoxy-
hexose biosynthesis. By using TDP-[4-^{3}H] glucose as a substrate we
were able to demonstrate quantitative conversion to TDP-[6-^{3}H]-6-deoxy-
D-xylo-4-hexosulose by this enzyme. The reaction occurred without exchange of tritium with the medium. The tritium originally present at
carbon 4 of the substrate was transferred to carbon 6 of the 4-hexosulose
derivative (71). These findings are consistent with a reaction mechanism
shown in scheme (5), analogous to the reaction sequence described for
UDP-galactose-4-epimerase. Initiation of enzyme catalysis occurs by the
attack of enzyme-NAD$^+$ on TDP-glucose to remove the axial hydrogen at
carbon 4, resulting in formation of enzyme-NADH-TDP-D-xylo-4-
hexosulose. This process is identical to the initiation of enzyme catalysis
described for epimerase. The enzyme-bound 4-hexosulose intermediate
rearranges by β-elimination of water between carbons 5 and 6 to yield a
5,6-glucoseen derivative. This latter compound served in turn as hydrogen
acceptor for enzyme-NADH to restore enzyme-NAD$^+$ and resulted in release of end product.

TDPG-oxidoreductase from E. coli was purified to homogeneity [9],
and several molecular properties of this enzyme were found to be similar
to UDP-galactose-4-epimerase.

A twelve hundredfold increase in purity was obtained by conventional
methods of protein purification, and the enzyme was finally isolated in
crystallized form [9]. The preparation behaved like a single component
in the ultracentrifuge and on disc gel electrophoresis. Determination of
molecular weight by sedimentation equilibrium gave a value of 88,000.
Fluorometric measurements established that 1 mol of NAD$^+$ was bound
per mole of enzyme. Treatment of the native enzyme which compounds
like p-chloro-mercuri-phenylsulfonate results in the release of enzyme-
bound NAD$^+$ and the formation of two protein subunits with an apparent
molecular weight of 40,000 each (88). Incubation of these inactive subunits in the presence of mercaptoethanol and excess NAD$^+$ leads to reassociation and complete restoration of enzymatic activity [9].

One other similarity to UDP-galactose-4-epimerase is the fact that
TDPG-oxidoreductase is susceptible to reductive inhibition. When TDP-
6-deoxy-D-glucose, a close structural analog of the natural substrate
TDPG, is incubated with the enzyme, the events shown in reaction (9) take
place.

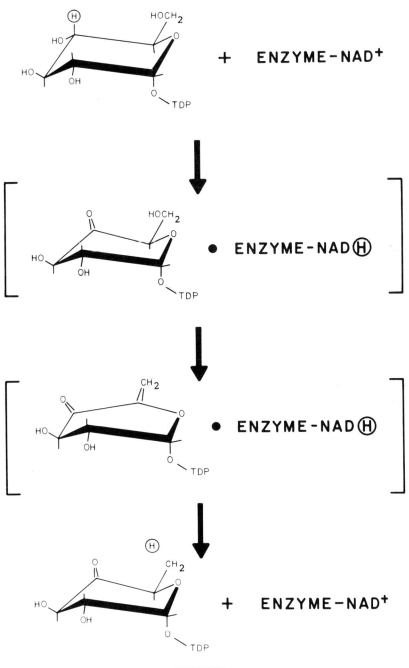

SCHEME 5

TDP-6-deoxy-D-glucose [enzyme-NAD$^+$] $\rightleftharpoons$ [enzyme-NADH-TDP-

6-deoxy-D-xylo-4-hexosulose]   (9)

The stability of the enzyme-NADH conjugate was evidenced by its isolation
by Sephadex chromatography and direct spectrophotometric demonstration
of enzyme-NADH [89]. In contrast to the reaction of the enzyme with its
natural substrate TDP-glucose, the substrate analogue TDP-6-deoxy-D-
glucose cannot undergo β-elimination of water followed by molecular rearrangement. Consequently, there is no acceptor for enzyme-NADH with
a resultant accumulation of enzyme-NADH, accompanied by enzyme
inactivation.

Stereospecificity of the hydrogen transfer to the coenzyme with respect
to the nicotinamide moiety was established to have "B" specificity for
both accepting and donating hydrogen [89].

A summary of all the findings for both enzymes, TDPG-oxidoreductase
and UDP-galactose-4-epimerase, is shown in Table 8.

D.   Common Mechanistic and Structural Properties of Enzymes Involved in Enzyme-NAD$^+$ Mediated Hydrogen Transfer Reactions

From a mechanistic viewpoint, UDP-galactose-4-epimerase and TDP-
glucose-oxidoreductase are best described as oxidoreductases in which
intramolecular hydride transfer is mediated by enzyme-NAD$^+$. Different
enzymatic end products are a consequence of different ways of stabilization of the 4-hexosulose intermediates. During the molecular rearrangement initiated by the formation of the 4-hexosulose, the reaction inter-
mediate(s) are held enzyme-bound until at the final step enzyme-NADH
donates hydrogen to the last intermediate, and enzyme-NAD$^+$ releases
the end-product from the enzyme. There is increasing evidence that
transformation of enzyme-NAD$^+$ to enzyme-NADH must cause a conformational change of enzyme protein since during normal catalysis no intermediate is released once enzyme-NADH is formed.

The initiation of enzyme catalysis is identical for both enzymes:
Hydrogen at position 4 of the hexose moiety is removed and 4-tritiated
substrates are rate limiting and display a kinetic isotope effect of about
2.5. We interpret these observations to indicate identical initiation for
enzyme catalysis with a close resemblance of the transition state(s) leading to the 4-hexosulose intermediate. Another resemblance occurring at
the active site of these enzymes is the stereospecificity of the hydrogen
acceptance and donation at the nicotinamide moiety of the coenzyme.

TABLE 8

Comparison of TDPG-oxidoreductase [71, 9, 88] and UDP-galactose-4-epimerase [29, 41, 46, 47, 63] from E. coli B

I. Similarities

a. Molecular weight (about 80,000)

b. 1 mole of active enzyme contains 2 subunits

c. 1 mole of active enzyme contains 1 mole of NAD^+ firmly bound

d. Initial reaction step involves substrate oxidation at carbon 4 caused by removal of axial hydrogen at carbon 4 and conversion to 4-hexosulose intermediate

e. Reduction of enzyme-NAD^+ to enzyme-NADH

f. Oxidoreductase mechanism-intramolecular hydride shift

g. Enzyme-NADH does not release intermediate(s)

h. Reoxidation of enzyme-NADH to enzyme-NAD^+ by hydrogen transfer to enzyme bound intermediate is last step resulting in product release

i. Identical isotope effects for 4-tritiated substrates

j. Similar or identical substrates

k. "B"-stereospecificity for hydrogen transfer with respect to nicotinamide moiety of coenzyme

II. Differences

TDGP-oxidoreductase	UDP-galactose-4-epimerase
a. 4 → 6 hydride shift	4 → 4 hydride shift
b. Rearrangement by β-elimination	No rearrangement

The similarities found for the reaction mechanism for UDP-galactose-4-epimerase and TDPG-oxidoreductase are complemented by the apparent identical gross structural features described in Table 8 for the enzyme proteins. These enzymes consist of two subunits held together by 1 mol of NAD^+. In both enzymes, subunits can be obtained by treatment with mercurial derivatives, while complete reversal to the active enzyme protein by the addition of compounds containing SH-group and NAD^+ can be accomplished.

Examination of the literature made us aware of the fact that there is a growing list of enzymes involved in sugar transformations that appear to have the same general reaction mechanism outlined for UDP-galactose-4-epimerase and TDPG-oxidoreductase and are characterized from a mechanistic viewpoint as oxidoreductases with identical initiation of enzyme catalysis. In all instances, enzyme-NAD^+ is the initial hydrogen acceptor. A detailed presentation of this group of enzymes as well as a discussion of the common features of their reaction mechanisms was recently presented [87].

At first glance, it is surprising to find close similarities concerning minute details of reaction mechanism as well as structural properties of the enzyme proteins, in various metabolic catalysts of which we previously mentioned human glucose-6-phosphate dehydrogenase [10]. We do not believe that it is simply coincidental that these enzymes happen to have the same catalytic principle. We therefore propose as a working hypothesis that the similarities between different enzymes are a consequence of the maintenance of catalytic principles during the evolutionary process of enzyme catalysis. We realize the necessity for further vigorous research to prove or disprove our hypothesis. We feel that irrespective of the final decision leading to acceptance or rejection of this hypothesis, our comparative approach to study this group of enzymes is an important asset to gain new insight into a better understanding of enzyme catalysis.

ACKNOWLEDGEMENTS

We are indebted to Dr. A. U. Bertland II for helpful discussions.

This work was supported by a grant awarded to Dr. O. Gabriel from the National Institute of Allergy and Infectious Diseases (AI-07241). Dr. H. M. Kalckar's support was awarded from the National Institutes of Health (AM-05507 and AM-05639), the National Science Foundation (GB-30785X), and the Wellcome Trust.

ADDENDUM

Material that appeared in press or information obtained after completion of our manuscript is presented in this addendum. The titles refer to corresponding sections in the text of our article.

UDP-Galactose-4-Epimerase from E. Coli

A recent study undertaken jointly by Schlesinger, Niall, and Wilson [1'], has provided new and important information concerning the primary structure of UDP-galactose-4-epimerase. Highly purified E. coli epimerase in its native form, as well as after oxidation with performic acid, was subjected to Edman degradation. The sequential, stepwise degradation starting from the amino-terminal end of the protein was followed by isolation and identification of the resulting phenylthiohydantoin amino acids by thin-layer chromatography and gas chromatography. Only one amino-terminal amino acid derivative was released, and its parent amino acid was identified as methionine, thus amending the previously reported amino-terminal aspartate [2'].

Altogether, 28 amino acids, starting from the amino-terminal end were identified: H_2N-methionine-arginine-valine-leucine-valine... The determination of this amino acid sequence provides further experimental evidence that the two subunits of epimerase are identical. It also indicates that the translation initiation signal methionine is preserved as the N-terminal unit in both subunits of E. coli epimerase.

Another interesting fact emerging from this study concerns a comparison of the above reported amino acid sequence with the E. coli galactose operon messenger RNA. The structural gene from E. coli epimerase maps at the operator end of the galactose operon. A partial sequence of the E. coli galactose operon messenger RNA has been elucidated [3']. A comparison of the mRNA sequence with the amino acid sequence found for epimerase strongly suggests that the first 26 bases of the mRNA are not involved in the translation process of epimerase, but the triplets starting with bases 27, 28, 29, and the next 16 (or more) correspond to the amino acid sequence of epimerase [1'].

The reductive inhibition of UDP-galactose-4-epimerase by $NaBH_4$, which is short-lived unless 5'UMP is present, [4'] was subjected to a renewed, careful study by Davis et al. [5']. They confirm earlier observations [6'] that the instability of reduced epimerase is due to autoxidation, since the reduced enzyme complex remains stable when kept under nitrogen gas. Replacement of $NaBH_4$ with the more stable $NaCNBH_4$ as the reducing agent in rate studies revealed a dependence of [enzyme-NADH] formation upon availability of 5'UMP. Hence, 5'UMP also plays a role in [epimerase-NADH] formation rather than merely stabilizing reduced epimerase toward autoxidation.

The rate at which $NaCNBH_3$ reacts with [epimerase-NAD^+] in the presence of 5'UMP is substantially faster than the rate at which it reacts with free NAD^+. It should be noted that the experimental evidence presented by Davis et al. [5'] does not eliminate a possible stabilizing role by 5'UMP on the [epimerase-NADH] complex toward autoxidation. This interpretation appears to be consistent with earlier observations of the same group [7'] studying the reoxidation of [UDP-hexose-enzyme-NADH] produced by $NaBH_4$ in presence of substrate. For these experiments the authors found that addition of cyclohexanone as well as of cyclohexanol to [UDP-hexose-enzyme-NADH] results in appreciable restoration (about 40% within 90 minutes) of enzymatic activity. Since cyclohexanone was not reduced to cyclohexanol in these experiments, its role as hydrogen acceptor was eliminated and it appears to act solely by causing the release of UDP-hexose from [UDP-hexose-enzyme-NADH]. The resulting [enzyme-NADH] is then subject to autoxidation leading to [enzyme-NAD$^+$] concomitant with restoration of enzymatic activity.

ADP-ribose-(5) has been shown to be an effective "reductive inhibitor" of E. coli UDP galactose-4-epimerase at concentrations as low as 10^{-6} M [8']. When comparative studies with the yeast epimerase were carried out, incubation of the enzyme with ADP-ribose-(5) at 3×10^{-3} M did not result in loss of enzymatic activity. Similarly, reactivation of reductively inhibited yeast epimerase by TDP-4-keto-6-deoxyglucose required at least tenfold higher concentrations than the E. coli enzyme to restore enzymatic activity [9'].

In a related study, we compared UDP-galactose-4-epimerase isolated from yeast and from E. coli for their ability to become reductively inhibited by incubation with L-arabinose and 5'dTMP or 5-iodo-2'deoxy-uridine-5'monophosphate. The E. coli enzyme was inhibited at nucleotide concentrations of 10^{-4} M, while the yeast enzyme required at least tenfold higher nucleotide concentrations for effective reductive inhibition to take place.

The above mentioned observations indicate to us considerable differences between the active sites of the yeast and E. coli enzymes. This difference is reflected in the rather broad substrate specificity for the E. coli enzyme [6'] and the more stringent substrate specificity of the yeast enzyme [10']. Although the two enzymes have many features in common, for instance two subunits and one coenzyme per active enzyme unit, it should be pointed out that they differ considerably in many other respects.

Liver UDP-Galactose-4-Epimerase

Examination of UDP-galactose-4-epimerase in goat and beef livers revealed at least two distinct forms of the enzyme, A and B, to be present

[11']. Fractionations by ammonium sulfate, followed by chromatography on calcium phosphate gel or DEAE cellulose gave rise to the separation of the two enzyme forms. Epimerase A was found to be inhibited by galactose-6-phosphate and higher concentrations of substrate, whereas the B form was found to be stimulated by galactose-1-phosphate and galactose-6-phosphate. Both forms showed an absolute requirement for NAD$^+$.

A study on calf liver UDP-galactose-4-epimerase was carried out by Langer and Glaser [12']. The enzyme was purified by ammonium sulfate precipitation, DEAE chromatography and affinity chromatography using the 3-picolylamine analogue of NAD$^+$ linked to Sepharose. The molecular weight of the enzyme was determined to be 70,000 composed of two subunits; maximal enzymatic activity was dependent upon availability of NAD$^+$. The enzyme appears to be capable of binding NAD$^+$ analogues, which contain the ADP-ribose portion of the NAD$^+$ molecule.

REFERENCES

1. L. F. Leloir, Arch. Biochem. Biophys., 33, 186 (1951).
2. E. S. Maxwell, J. Amer. Chem. Soc., 78, 1074 (1956).
3. E. S. Maxwell, H. de Robichon Szulmajster, and H. M. Kalckar, Arch. Biochem. Biophys., 78, 407 (1958).
4. E. S. Maxwell and H. de Robichon Szulmajster, J. Biol. Chem., 235, 308 (1960).
5. C. R. Creveling, A. Bhaduri, A. Christensen, and H. M. Kalckar, Biochem. Biophys. Res. Commun., 21, 624 (1965).
6. R. A. Darrow and R. Rodstrom, Proc. Natl. Acad. Sci., U. S., 55, 205 (1966).
7. R. A. Darrow and R. Rodstrom, Biochemistry, 7, 1645 (1968).
8. D. B. Wilson and D. S. Hogness, J. Biol. Chem., 244, 2132 (1969).
9. S. F. Wang and O. Gabriel, J. Biol. Chem., 244, 3430 (1969).
10. A. Yoshida and V. D. Hoagland, Jr., Biochem. Biophys. Res. Commun., 40, 1167 (1970).
11. A. Bhaduri, A. Christensen, and H. M. Kalckar, Biochem. Biophys. Res. Commun., 21, 631 (1965).
12. R. A. Darrow and C. R. Creveling, J. Biol. Chem., 239, PC362 (1964).
13. R. A. Darrow and R. Rodstrom, J. Biol. Chem., 245, 2036 (1970).
14. L. H. Bertland and A. U. Bertland II, Biochemistry, 10, 3145 (1971).
15. H. K. Schachman, Methods in Enzymol., 4, 32 (1957).

16.   O. Gabriel, Methods in Enzymol., 22, 565 (1971).

17.   D. Stathakos, unpublished data.

18.   M. Ottesen and J. T. Johansen, unpublished data.

19.   H. K. Schachman, Cold Spring Harbor Symp. Quant. Biol., 28, 409
       (1963).

20.   D. A. Yphantis, Biochemistry, 3, 297 (1964).

21.   R. A. Darrow, unpublished data.

22.   H. M. Kalckar, K. Kurahashi, and E. Jordan, Proc. Natl. Acad.
       Sci., U. S., 45, 1776 (1959).

23.   K. O. Pedersen, J. Phys. Chem., 62, 1282 (1958).

24.   A. U. Bertland II, Biochemistry, 9, 4649 (1970).

25.   A. U. Bertland II, Y. Seyama, and H. M. Kalckar, Biochemistry,
       10, 1545 (1971).

26.   H. M. Kalckar, A. U. Bertland II, and B. Bugge, Proc. Natl. Acad.
       Sci., U. S., 65, 1113 (1970).

27.   Y. Seyama and H. M. Kalckar, Biochemistry, 11, 40 (1972).

28.   A. Bhaduri, R. A. Darrow, H. M. Kalckar, and E. Randerath,
       Fed. Proc., 24, 478 Abstr. (1965).

29.   H. M. Kalckar, A. U. Bertland, J. T. Johansen, and M. Ottesen,
       Alfred Benzon Symp., Munksgaard, Copenhagen, No. 1 (1969),
       p. 247.

30.   B. Bugge and H. M. Kalckar, unpublished data.

31.   D. Stathakos and B. Bugge, unpublished data.

32.   D. Stathakos, I. Isaakidou, and H. Thomou, Biochim. Biophys.
       Acta, 302, 80 (1973).

33.   A. U. Bertland II, unpublished data.

34.   A. U. Bertland II, unpublished data.

35.   H. Matsuo, Y. Fuyimoto, and T. Tatsuno, Tetrahydron Letters, 39,
       3465 (1965).

36.   H. Matsuo, Y. Fuyimoto, and T. Tatsuno, Biochem. Biophys. Res.
       Commun., 22, 69 (1966).

37.   G. N. Holcomb, S. A. James, and D. N. Ward, Biochemistry, 7,
       1291 (1968).

38.   H. M. Kalckar, Advances in Enzymol., 20, 111 (1958).

39.   G. Buttin, J. Mol. Biol., 7, 164, 183 (1963). (In French).

40.   M. B. Yarmolinsky and H. Wiesmeyer, Proc. Natl. Acad. Sci.,
       U. S., 46, 1626 (1960).

41.   D. B. Wilson and D. S. Hogness, J. Biol. Chem., 239, 2469 (1964).

42.   D. R. Helinski and C. Yanofsky, Biochim. Biophys. Acta, 63,
       10 (1962).

43.   G. R. Stark and D. G. Smyth, J. Biol. Chem., 238, 214 (1963).

44.   H. Ankel, and U. S. Maitra, Biochem. Biophys. Res. Commun.,
       32, 526 (1968).

45.   W. L. Adair, R. W. Gaugler, and O. Gabriel, Fed. Proc., 30,
       376 Abstr. (1971).

46. W. L. Adair, Jr., O. Gabriel, D. Ullrey, and H. M. Kalckar, J. Biol. Chem., 248, 4635 (1973).

47. G. L. Nelsestuen and S. Kirkwood, J. Biol. Chem., 246, 7533 (1971).

48. L. Anderson, A. M. Landel, and D. F. Diedrich, Biochim. Biophys. Acta, 22, 573 (1956).

49. A. Kowalsky and D. E. Koshland, Jr., Biochim. Biophys. Acta, 22, 575 (1956).

50. H. M. Kalckar and E. S. Maxwell, Biochim. Biophys. Acta, 22, 588 (1956).

51. R. D. Bevill III, E. A. Hill, F. Smith, and S. Kirkwood, Can. J. Chem., 43, 1577 (1965).

52. G. L. Nelsestuen and S. Kirkwood, Biochim. Biophys. Acta, 220, 633 (1970).

53. I. A. Rose, Ann. Rev. Biochem., 35, 23 (1966).

54. J. N. Ketley and K. A. Schellenberg, Biochim. Biophys. Acta, 284, 549 (1972).

55. H. M. Kalckar, O. Gabriel, D. Stathakos, and D. Ullrey, Anales Assoc. Quim., Argentina, 60, 129 (1972).

56. U. S. Maitra and H. Ankel, Proc. Natl. Acad. Sci., U. S., 68, 2660 (1971).

57. Y. Seyama and H. M. Kalckar, Biochemistry, 11, 36 (1972).

58. L. Davis and L. Glaser, Biochem. Biophys. Res. Commun., 43, 1429 (1971).

59. J. N. Ketley and K. A. Schellenberg, Biochemistry, 12, 315 (1973).

60. R. A. Darrow, C. R. Creveling, and H. M. Kalckar, Fed. Proc., 22, 360 Abstr. (1963).

61. N. O. Kaplan, S. P. Colowick, and C. C. Barnes, J. Biol. Chem., 191, 461 (1951).

62. K. Dalziel, J. Biol. Chem., 238, 1538 (1963).

63. W. L. Adair, Jr., O. Gabriel, D. Stathakos, and H. M. Kalckar, J. Biol. Chem., 248, 4640 (1973).

64. T. G. Wee and P. A. Frey, J. Biol. Chem., 248, 33 (1973).

65. O. Gabriel, unpublished data.

66. L. Glaser and L. Ward, Biochim. Biophys. Acta, 198, 613 (1970).

67. G. E. Lienhard, Science, 180, 149 (1973).

68. T. G. Wee, J. Davis, and P. A. Frey, J. Biol. Chem., 247, 1339 (1972).

69. U. S. Maitra and H. Ankel, J. Biol. Chem., 248, 1477 (1973).

70. O. Gabriel, Carbohydrate Res., 6, 319 (1968).

71. O. Gabriel and L. C. Lindquist, J. Biol. Chem., 243, 1479 (1968).

72. J. Y. Lin and O. Gabriel, unpublished data.

73. L. Glaser, The Enzymes, Vol. 6 (P. D. Boyer, ed.), Academic Press, New York, 1972, p. 355.

74. E. S. Maxwell, J. Biol. Chem., 229, 139 (1957).

75. E. A. Robinson, H. M. Kalckar, H. Troedsson, and K. Sanford, J. Biol. Chem., 241, 2737 (1966).
76. C. M. Tsai, N. Holmberg, and K. E. Ebner, Arch. Biochem. Biophys., 136, 233 (1970).
77. D. F. Fan and D. S. Feingold, Plant Physiol., 44, 599 (1969).
78. T. N. Druzhinina, M. A. Novikova, and V. N. Shibaev, Biokhimya, 34, 110 (1969).
79. E. F. Neufeld, Biochem. Biophys. Res. Commun., 7, 461 (1962).
80. L. Glaser, J. Biol. Chem., 234, 2801 (1959).
81. M. A. Gaunt, H. Ankel, and J. S. Schutzbach, Methods in Enzymol., 28, 426 (1972).
82. H. Kawai and T. Tochikura, Agr. Biol. Chem., 35, 1578 (1971).
83. H. Kawai and T. Tochikura, Agr. Biol. Chem., 35, 1587 (1971).
84. J. Y. Lin, W. L. Adair, and O. Gabriel, Fed. Proc., 32, 667 Abstr. (1973).
85. B. Witholt, J. Bacteriol., 109, 350 (1972).
86. J. Y. Lin, B. Bugge, and O. Gabriel, Carbohydrate Res., 37, 47 (1974).
87. O. Gabriel, Advances in Chemistry Series, 117, 387 (1973).
88. H. Zarkowsky, E. Lipkin, and L. Glaser, J. Biol. Chem., 245, 6599 (1970).
89. S. F. Wang and O. Gabriel, J. Biol. Chem., 245, 8 (1970).

ADDENDUM REFERENCES

1'. D. H. Schlesinger, H. D. Niall, and D. Wilson, Biochem. Biophys. Res. Comm., 61, 282 (1974).
2'. D. B. Wilson and D. S. Hogness, J. Biol. Chem., 244, 2132 (1969).
3'. R. E. Musso, B. de Grombrugghe, I. Pastan, J. Skylar, P. Yot, and S. Weissman, Proc. Nat. Acad. Sci., U. S., 71, 4940 (1974).
4'. A. U. Bertland II, Y. Seyama, and H. M. Kalckar, Biochem., 10, 1545 (1971).
5'. J. E. Davis, L. D. Nolan, and P. A. Frey, Biochim. Biophys. Acta, 334, 442 (1974).
6'. G. L. Nelsestuen and S. Kirkwood, J. Biol. Chem., 246, 7533 (1971).
7'. T. G. Wee and P. A. Frey, J. Biol. Chem., 249, 856 (1974).

8'. W. L. Adair, Jr., O. Gabriel, D. Stathakos, and H. M. Kalckar, J. Biol. Chem., 248, 4640 (1973).

9'. J. S. Myers and O. Gabriel, unpublished observations (1974).

10'. A. Bertland, B. Bugge, and H. M. Kalckar, Arch. Biophys. Biochem., 116, 280 (1966).

11'. M. Ray and A. Bhaduri, Biochim. Biophys. Acta, 302, 129 (1973).

12'. R. Langer and L. Glaser, J. Biol. Chem., 249, 1126 (1974).

Chapter 4

LACTOSE SYNTHETASE: α-LACTALBUMIN AND
β-(1 → 4) GALACTOSYLTRANSFERASE

Kurt E. Ebner*
Steve C. Magee [†]

Department of Biochemistry
Oklahoma State University
Stillwater, Oklahoma

I.	INTRODUCTION	138
II.	LACTOSE SYNTHETASE	139
	A. Requirement for Two Proteins	140
	B. Identification of the B Protein as α-Lactalbumin	141
	C. The A Protein is a Galactosyltransferase	141
III.	STRUCTURAL SIMILARITIES BETWEEN α-LACTALBUMIN AND LYSOZYME	142
IV.	EVOLUTIONARY RELATIONSHIP BETWEEN α-LACTALBUMIN AND LYSOZYME	148
V.	β-(1 → 4) GALACTOSYLTRANSFERASE	149
	A. General Properties	149
	B. Catalytic Properties	155

*Presently affiliated with the Department of Biochemistry, University of Kansas Medical Center, Kansas City, Kansas.

[†] Presently affiliated with The Sigma Chemical Company, Enzyme Research and Development, St. Louis, Missouri

C. Reaction Kinetics   158
D. Relationship to Other Galactosyltransferases   265

VI. INTERACTION BETWEEN α-LACTALBUMIN AND
    GALACTOSYLTRANSFERASE   167
    A. Physical Interactions   167
    B. Chemical Modification of α-Lactalbumin   168

VII. BIOLOGICAL SIGNIFICANCE   170

I. INTRODUCTION

α-Lactalbumin is found in the milk of all mammalian species that are able
to synthesize lactose, the prominent carbohydrate found in most milks.
Bovine skim milk contains about 0.7-1.5 mg of α-lactalbumin per milliliter
and is a major component of the whey or noncasein proteins of milk [1].
α-Lactalbumin is easy to crystallize from milk, and this was done in 1953
by Gordon and Semmett [2], who suggested the name for this protein.
Many of the properties of α-lactalbumin are available in a review by
Gordon [3]. α-Lactalbumin was subjected to many chemical and physical
studies since it was readily available in pure form, but its function was
not discovered until 1966 when Brodbeck and Ebner [4] demonstrated that
the enzyme involved in the biosynthesis of lactose required two proteins
and that α-lactalbumin was one of these proteins [5, 6].

The biosynthesis of lactose involves the transfer of galactose from
UDP-galactose to glucose to form lactose (4-O-β-D-galactopyranosyl-α-
D-glucopyranose).

$$\text{UDP-galactose} + \text{glucose} \longrightarrow \text{lactose} + \text{UDP} \tag{1}$$

This reaction was discovered in 1962 by Watkins and Hassid [7] in particulate fractions prepared from lactating bovine and guinea pig mammary
glands. In 1964 Babad and Hassid [8] found that the enzyme was soluble
in bovine milk and were able to purify the enzyme some seventyfold before
losing activity [9]. They also were able to show that the enzyme would
catalyze the transfer of galactose from UDP-galactose to N-acetylglucosamine at about 25% the rate of transfer to glucose [9].

$$\text{UDP-galactose} + \text{N-acetylglucosamine} \longrightarrow \text{N-acetyl-} \tag{2}$$
$$\text{lactosamine} + \text{UDP}$$

Later work by Brew et al. [10] showed that the other protein (not α-
lactalbumin) required for the biosynthesis of lactose could effect the
formation of N-acetyllactosamine by itself and that α-lactalbumin inhibited

this reaction. Further investigations clearly showed that the biosynthesis of lactose required the presence of both a galactosyltransferase and α-lactalbumin for meaningful enzymatic rates as summarized in several reviews [11-18].

The purpose of this chapter is to review the chemical and physical properties of the galactosyltransferase and α-lactalbumin as well as to review the progress made in the understanding of the interaction of these two proteins as related to lactose biosynthesis and other related reactions.

II. LACTOSE SYNTHETASE

Lactose synthetase (UDP-galactose: D-glucose 1-galactosyltransferase, EC 2.4.1.22) is the name given to the enzymatic system which catalyzes the formation of lactose from UDP-galactose and glucose. In reality, the name is a misnomer, but it is useful for describing the biosynthesis of lactose even though two proteins are involved. One of these proteins is α-lactalbumin. The other is a particulate galactosyltransferase found in most tissues and concerned with glycoprotein biosynthesis rather than with lactose biosynthesis which is restricted to the mammary gland since it has the ability to synthesize α-lactalbumin in response to hormonal stimulation. The elucidation of the mechanism of lactose biosynthesis has led to the discovery of a new type of biological control in which one enzyme, a galactosyltransferase, is modified kinetically by the common milk-whey protein, α-lactalbumin, so that lactose is formed at meaningful rates.

Earlier studies by Gander et al. [19, 20] indicated that a crude enzymatic preparation from bovine mammary tissue could synthesize lactose-1-phosphate from UDP-galactose and glucose-1-phosphate, and the lactose-1-phosphate was subsequently hydrolyzed by a phosphatase to form lactose. These results have not been confirmed and it is now well established that the galactosyl acceptor is glucose [7-9]. In 1964, Babad and Hassid [8] assumed that the reaction of the enzyme prepared by Watkins and Hassid [7] in 1962 was catalyzed by a single protein, but now their loss of activity may be attributed to the separation of the two proteins required for enzymatic activity. The enzyme did not transfer galactose to glucose-1-phosphate but did transfer galactose to N-acetylglucosamine. The reaction required Mn^{2+} though Mg^{2+} was 25% as effective. Later studies with more purified preparations showed that Mg^{2+} was ineffective; the observed result may have been due to the displacement of Mn^{2+} by Mg^{2+} from other proteins present in the preparation. The tissue preparation used by Gander et al. [19, 20] was a soluble enzyme and was active with Mg^{2+} but not Mn^{2+}, and it would appear that it was a different system than that isolated by Hassid and his coworkers [7-9].

A. Requirement for Two Proteins

Initial attempts to solubilize lactose synthetase from mammary gland particles were unsuccessful since there were large losses of activity. Also, initial attempts made to purify the enzyme from skim milk resulted in large losses of activity. Brodbeck and Ebner [4] noted that the activity of the enzyme did not correspond to the protein profile on a molecular sieve column and were able to separate a partially purified soluble lactose synthetase from bovine milk into two protein fractions by chromatography on Bio-Gel P-30 (Fig. 1). Approximately 900 mg of protein from the 60% saturated ammonium sulfate precipitate dissolved in 10 ml of 20 mM tris-HCl-10 mM $MgCl_2$, pH 7.4, were applied to a column (3 x 160 cm) of Bio-Gel P-30. The column was equilibrated and eluted with the above buffer. Fractions, 3.5 ml each, were collected after the first 300 ml passed through the column. The key to the figure is as follows: $\bigcirc$, Protein distribution in eluted fractions (A_{2800} Å) determined upon 1:10 dilution with the same buffer; $\blacktriangle$, lactose synthetase activity of Fraction A in the presence of 0.2 ml (0.5 mg of protein) of Fraction B obtained from tubes 56 and 57; $\blacksquare$, lactose synthetase activity of Fraction B assayed with 0.2 ml (1.1 mg of protein) of Fraction A obtained from tubes 22 and 23. Activity was obtained when both protein fractions were combined. The larger molecular weight protein was called the A protein and the lower molecular weight protein was called the B protein. Neither protein alone exhibited lactose synthetase activity when assayed under the conditions used, and the results showed that two proteins were required for the biosynthesis of lactose from UDP-galactose and glucose. In addition, lactose synthetase was resolved into two protein fractions by chromatography on Bio-Gel P-30 from solubilized bovine mammary tissue [21] and from the milk of the sheep, goat, and human [6]. Subcellular distribution studies

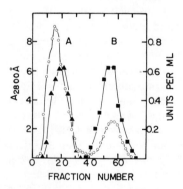

FIG. 1. Gel Filtration of Partially Purified Lactose Synthetase on Bio-Gel P-30

showed that the A protein was associated mainly with a crude microsomal fraction whereas the B protein was found equally distributed between the soluble and particulate fractions [21]. The B protein could be dissociated from particles carefully prepared in 0.25 M sucrose by 20 mM tris-10 mM EDTA, pH 7.4, or by 20% acetone-0.2M ammonium sulfate [21]. Bovine milk microsomes, mainly associated with the cream fraction, had specific activities of the A and B proteins comparable to microsomal particles isolated from mammary tissue. The difficulty in finding activity of lactose synthetase in tissue extracts was attributed to the ease of dissociation of the B protein from particulate fractions during the homogenization and solubilization procedures.

B. Identification of the B Protein as α-Lactalbumin

The lower molecular weight protein, or B protein, was crystallized, and it was demonstrated that five times crystallized α-lactalbumin isolated by a completely different procedure substituted for the B protein in the lactose synthetase assay since it was possible to quantitatively assay for one protein in the presence of saturating amounts of the other protein [5]. Considerations of the similarity in molecular weight, spectral properties, and concentration in milk led to the suspicion that the B protein may be α-lactalbumin. Further work based on identical chemical, physical, and immunological properties showed that the B protein of the lactose synthetase enzyme was identical to α-lactalbumin, the common whey protein found in milk [6]. Impure preparations of the A protein and α-lactalbumin from the milk of the bovine, goat, sheep, and human are qualitatively interchangeable in the enzymatic rate assay [6], and to date α-lactalbumin isolated from the milk of a variety of species all react enzymatically with bovine galactosyltransferase [22].

C. The A Protein is a Galactosyltransferase

The initial studies by Babad and Hassid [9] showed that their preparations from milk would catalyze the transfer of galactose from UDP-galactose to N-acetylglucosamine at about 25% of the rate when glucose was the galactosyl acceptor. Brew et al. [10] observed that the A protein would by itself catalyze the formation of N-acetyllactosamine and that α-lactalbumin would markedly inhibit this reaction. Similar activity was observed with a solubilized preparation obtained from rat liver. This enzyme appeared to be similar to a soluble fraction isolated from goat colostrum and particulate fractions prepared from the lung, brain, spleen, large intestine, mammary gland, and liver of the rat [23]. These preparations catalyzed the formation of N-acetyllactosamine from UDP-galactose and N-acetylglucosamine. In addition, other workers [23] were able

to transfer galactose to a number of acceptors containing β-N-acetylgluco-
saminyl end groups of the type sialic acid → galactose → N-acetylglucosamine → glycoprotein when the sialic acid and galactose were previously
removed by sialidase and β-galactosidase. Further work has shown
clearly that the A protein is a galactosyltransferase, found in most tissues
and involved in glycoprotein biosynthesis in the formation of a β-Gal-(1 → 4)-
β-GlcNAc-linkage.

III. STRUCTURAL SIMILARITIES BETWEEN α-LACTALBUMIN AND LYSOZYME

The similarity in molecular weight, amino acid composition, and number
of disulfide groups was noted for bovine α-lactalbumin and hen's egg-white
lysozyme in 1958 [24]. Brew and Campbell [25] also noted structural
features which were similar between hen's egg-white lysozyme and guinea
pig α-lactalbumin even though guinea pig α-lactalbumin did not have any
detectable lysozyme activity. The similarity in structure between lysozyme
and α-lactalbumin and a biological function for α-lactalbumin led to the
determination of the amino acid sequence and location of the disulfide bonds
of bovine α-lactalbumin [26-30]. Comparison of the linear amino acid
sequence of bovine α-lactalbumin and hen's egg-white lysozyme showed
that 49 amino acids are in the identical or corresponding positions and an
additional 23 are conservative replacements of the total of 123 amino acids
in the sequence [30]. These results indicate a high degree of similarity
between the structures of α-lactalbumin and lysozyme, and indeed Browne
et al. [31] constructed a three-dimensional model of α-lactalbumin based
on the coordinants of hen's egg-white lysozyme. The high degree of
structural relatedness between the two proteins suggests that they are related in an evolutionary sense, and a proposal was made that they both
were derived from a common ancestral protein and evolved by the process
of gene duplication to give rise to the present proteins [27]. Both lysozyme and α-lactalbumin participate in similar types of reactions in that
lysozyme catalyzes the hydrolysis of a β-(1 → 4) glycosidic bond whereas
α-lactalbumin, in conjunction with the galactosyltransferase, is concerned
indirectly with the synthesis of a β-(1 → 4) glycosidic bond. It should be
pointed out that the galactosyltransferase has the catalytic site and α-lactalbumin does not have any catalytic activity but rather acts as a protein
modifier of reactions catalyzed by the galactosyltransferase. Neither
lysozyme nor α-lactalbumin participate in nor interfere with each other's
reaction [31].

   The complete amino acid sequence of human [32], guinea pig [33],
bovine [28], and the partial sequence of kangaroo [34] α-lactalbumin
have been completed and are compared with human leukemic and hen's

egg-white lysozyme in Table 1. Comparison of human α-lactalbumin with bovine α-lactalbumin shows that 72% of the residues are in identical positions in the amino acid sequence and 6% are chemically similar. The corresponding values for a comparison of the human α-lactalbumin/human lysozyme are 39% and 12%, respectively. Both the amino acid sequences of the human and guinea pig α-lactalbumin support the idea of a structural homology between α-lactalbumin and lysozyme. Bovine, human, and guinea pig α-lactalbumins are roughly equally different from hen's egg-white and human lysozyme. Since there are a large number of sequence changes between members of the α-lactalbumin group, it is apparent that this protein has undergone a high rate of evolutionary change which is greater than that observed in the lysozyme family. Calculations of helix probability profiles for denatured bovine α-lactalbumin and hen egg-white lysozyme also support the idea that these two proteins are structurally homologous [35]. Initial attempts on the x-ray crystallography of goat α-lactalbumin have been reported [36, 37], but as yet it has been difficult to prepare adequate isomorphous crystals suitable for continuous study.

The postulate of structural relatedness between α-lactalbumin and lysozyme has led to numerous studies concerned with providing evidence for or against this relationship, and in general the results have shown that gross similarities exist though differences in the fine details occur. Krigbaum and Kugler [38] concluded from small angle x-ray diffraction studies that the molecular conformations are quite different between α-lactalbumin and lysozyme. These results have been criticized [39], and the data actually support the similarity between the two proteins when the percentage dimerization of α-lactalbumin is considered. NMR studies on the denaturation of α-lactalbumin and lysozyme showed that α-lactalbumin has a reduced stability to denaturation when compared to lysozyme and that there are fewer aliphatic-aromatic interactions in α-lactalbumin, leading to a more open structure [40]. The circular dichroism spectra for four lysozymes (hen egg, duck egg, goose egg, and human urinary leukemia) and four α-lactalbumins (bovine, camel, guinea pig, and human) were compared [41]. The egg lysozymes were different from other proteins in the group which all had prominent Cotton effects at about 2700 Å and a positive feature at 2950 Å. These differences were rationalized in terms of aromatic contributions. Sommers et al. [42] examined the fluorescence properties of bovine, goat, human, and guinea pig α-lactalbumin and concluded that the fluorescence properties of these α-lactalbumins may be satisfactorily explained on the lysozyme-α-lactalbumin analogy model.

Rotational diffusion measurements of α-lactalbumin and lysozyme conjugated to 1-dimethylaminonaphthalene-5-sulfonate showed differences in the rotational relaxation times. Such differences were attributed to differences in conformation or hydration of α-lactalbumin and lysozyme [43].

TABLE 1

Amino Acid Sequence of Bovine, Guinea Pig, Human, and Kangaroo (Partial)
α-Lactalbumins Compared to Human Leukemic and Chicken Lysozymes[a]

```
                1                   5                  10                       15

BαLA   Glu-Gln-Leu-Thr-Lys-Cys-Glu-Val-Phe-Arg-Glu-Leu-Lys-          -Asp-Leu-
GPαLA  Lys-Gln-Leu-Thr-Lys-Cys-Ala-Leu-Ser-His-Glu-Leu-Asn-          -Asp-Leu-
HαLA   Lys-Gln-Phe-Thr-Lys-Cys-Glu-Leu-Ser-Gln-Leu-Leu-Lys-          -Asp-Ile-
KαLA   Ile-Asp-Tyr-Arg-Lys-Cys-Gln-Ala-Ser-Gln-Ile-Leu-Lys-Glu-His-Gly-Met-
H Ly   Lys-Val-Phe-Glu-Arg-Cys-Glu-Leu-Ala-Arg-Thr-Leu-Lys-Arg-Ley-Gly-Met-
C Ly   Lys-Val-Phe-Gly-Arg-Cys-Glu-Leu-Ala-Ala-Ala-Met-Lys-Arg-His-Gly-Leu-
                1                   5                  10                       15


                20                  25                 30

BαLA   Lys-Gly-Tyr-Gly-Gly-Val-Ser-Leu-Pro-Glu-Trp-Val-Cys-Thr-Thr-Phe-His-
GPαLA  Ala-Gly-Tyr-Arg-Asp-Ile-Thr-Leu-Pro-Glu-Trp-Leu-Cys-Ile-Ile-Phe-His-
HαLA   Asp-Gly-Tyr-Gly-Gly-Ile-Ala-Leu-Pro-Glu-Leu-Ile-Cys-Thr-Met-Phe-His-
KαLA   Asp-Lys-Val-             -Ile-Pro-Glu-Leu-Val-Cys-Thr-Met-Phe-His-
H Ly   Asp-Gly-Tyr-Arg-Gly-Ile-Ser-Leu-Ala-Asn-Trp-Met-Cys-Leu-Ala-Lys-Trp-
C Ly   Asp-Asn-Tyr-Arg-Gly-Tyr-Ser-Leu-Gly-Asn-Trp-Val-Cys-Ala-Ala-Lys-Phe-
                20                  25                 30


                35                  40                 45                     50

BαLA   Thr-Ser-Gly-Tyr-Asp-Thr-Glu-Ala-Ile-Val-Glu-Asn-             -Asn-Gln-Ser-Thr-
GPαLA  Ile-Ser-Gly-Tyr-Asp-Thr-Gln-Ala-Ile-Val-Lys-Asn-             -Ser-Asn-His-Lys-
HαLA   Thr-Ser-Gly-Tyr-Asp-Thr-Gln-Ala-Ile-Val-Glu-Asn-             -Asn-Gln-Ser-Thr-
KαLA   Ile-Ser-Gly-Leu-Ser-Pro-Gln-Ala-Glu-Val-
H Ly   Glu-Ser-Gly-Tyr-Asn-Thr-Arg-Ala-Thr-Asn-Tyr-Asn-Ala-Gly-Asp-Arg-Ser-Thr-
C Ly   Glu-Ser-Asn-Phe-Asn-Thr-Gln-Ala-Thr-Asn-Arg-Asn-Tyr-        -Asp-Gly-Ser-Thr-
                35                  40                 45                     50
```

```
              50              55              60              65
BαLA   Asp-Tyr-Gly-Leu-Phe-Gln-Ile-Asn-Asn-Lys-Ile-Trp-Cys-Lys-Asn-Asp-Gln-Asp-
GPαLA  Glu-Tyr-Gly-Leu-Phe-Gln-Ile-Asn-Asn-Lys-Asp-Cys-Phe-Cys-Glu-Ser-Thr-Thr-
HαLA   Glu-Tyr-Gly-Leu-Phe-Gln-Ile-Ser-Asn-Lys-Leu-Trp-Cys-Lys-Ser-Ser-Gln-Val-
H Ly   Asp-Tyr-Gly-Ile-Phe-Gln-Ile-Asn-Ser-Arg-Tyr-Trp-Cys-Asn-Asp-Gly-Lys-Thr-
C Ly   Asp-Tyr-Gly-Ile-Leu-Gln-Ile-Asn-Ser-Arg-Trp-Trp-Cys-Asn-Asp-Gly-Arg-Thr-
                  55              60              65

              70              75              80              85
BαLA   Pro-His-Ser-Asn-Ile-Cys-Asn-Ile-Ser-Cys-Asp-Lys-Phe-Leu-Asn-Asn-Asp-
GPαLA  Val-Gln-Ser-Arg-Asp-Ile-Cys-Asp-Ile-Ser-Cys-Asp-Lys-Leu-Asn-Asp-Asn-
HαLA   Pro-Gln-Ser-Arg-Asn-Ile-Cys-Asp-Ile-Ser-Cys-Asp-Lys-Phe-Leu-Asn-Asp-Asn-
H Ly   Pro-Gly-Ala-Val-Asn-Ala-Cys-His-Leu-Ser-Cys-Ser-Ala-Leu-Leu-Gln-Asp-Asn-
C Ly   Pro-Gly-Ser-Arg-Asn-Leu-Cys-Asn-Ile-Pro-Cys-Ser-Ala-Leu-Leu-Ser-Ser-Asp-
                  75              80

              90              95             100
BαLA   Leu-Thr-Asn-Asn-Ile-Met-Cys-Val-Lys-Lys-Ile-Leu-    -Asp-Lys-Val-
GPαLA  Leu-Thr-Asn-Asn-Ile-Met-Cys-Val-Lys-Lys-Ile-Leu-    -Asp-Ile-Lys-
HαLA   Ile-Thr-Asn-Asn-Ile-Met-Cys-Ala-Lys-Lys-Ile-Leu-    -Asp-Ile-Lys-
H Ly   Ile-Ala-Asp-Ala-Val-Ala-Cys-Ala-Lys-Arg-Val-Arg-    -Asp-Pro-Gln-
C Ly   Ile-Thr-Ala-Ser-Val-Asn-Cys-Ala-Lys-Lys-Ile-Val-Ser-Asn-Gly-Asp-
                  95
```

TABLE 1 (cont'd)

	100	105	110	115
BαLA	Gly-Ile-Asn-Tyr-Trp-Leu-Ala-His-Lys-Ala-Leu-Cys-Ser-Glu-Lys-Leu-Asp-
GPαLA	Gly-Ile-Asn-Tyr-Trp-Leu-Ala-His-Lys-Pro-Leu-Cys-Ser-Asp-Lys-Leu-Glu-
HαLA	Gly-Ile-Asn-Tyr-Trp-Leu-Ala-His-Lys-Ala-Leu-Cys-Thr-Glu-Lys-Leu-Glu-
H Ly	Gly-Ile-Arg-Ala-Trp-Val-Ala-Trp-Arg-Asn-Arg-Cys-Gln-Asn-Arg-Asp-Val-
C Ly	Gly-Met-Asn-Ala-Trp-Val-Ala-Trp-Arg-Asn-Arg-Cys-Lys-Gly-Thr-Asp-Val-

	105	110	115	120

	120	
BαLA	Gln-Trp-Leu-	-Cys-Glu-Lys-Leu-
GPαLA	Gln-Trp-Tyr-	-Cys-Glu-Ala-Gln-
HαLA	Gln-Trp-Leu-	-Cys-Glu-Lys-Leu-
H Ly	Arg-Gln-Tyr-Val-Gln-Gly-Cys-	-Gly-Val-
C Ly	Gln-Ala-Trp-Ile-Arg-Gly-Cys-	-Arg-Leu-

	125

aBαLA (bovine α-lactalbumin) [28]; GPαLA (guinea pig α-lactalbumin) [33]; HαLA (human α-lactalbumin) [32]; KαLA (Kangaroo α-lactalbumin) [34]; H Ly (human leukemic lysozyme) [34]; and C Ly (chicken lysozyme) [32]. The top number refers to the sequence of amino acids in bovine α-lactalbumin, and the bottom numbers are the sequence of amino acids in chicken lysozyme. The sequences are aligned to give the highest degree of homology.

The extrinsic fluorescence of 2-p-toluidinyl-naphthalene-6-sulfonate was used as a fluorescence probe of the hydrophobic areas of bovine α-lactalbumin and hen's egg-white lysozyme [44]. Both proteins bound a mole of dye per mole of protein and had similar dissociation constants although the dye binding site on α-lactalbumin had a more pronounced hydrophobic character than that on lysozyme, even though both are considered moderate. On the whole, the work supported the idea of similar conformations with some specific modification of the side chains. Similarities of binding of N-methyl nicotinamide chloride, which forms a charge transfer complex with a planar tryptophanyl group, have been demonstrated for α-lactalbumin and lysozyme; and the results are interpreted to provide support for the proposed model of α-lactalbumin [45, 46].

The four disulfides in lysozyme and α-lactalbumin are in similar positions [27, 30, 31], and their respective stabilities may reflect differences in tertiary structure. The disulfides of α-lactalbumin were more accessible to reduction by dithiothreitol in 4 M urea than those of lysozyme [47], which is in agreement with results obtained by reduction with mercaptoethanol and guanidine hydrochloride [48]. Iyer and Klee [49] have also shown that all the disulfides of α-lactalbumin are readily reduced with dithiothreitol in aqueous buffers at room temperature. These data indicate that the conformations of the disulfides in α-lactalbumin and lysozyme indeed do differ to some extent.

Immunological techniques have shown little [50] if any [48, 51, 52] cross-reactivity between lysozyme and α-lactalbumin. However, immunological cross-reactivity was observed between reduced and carboxymethylated bovine α-lactalbumin and hen egg-white lysozyme in contrast to lack of immunological cross-reactivity of the native proteins [53]. The peptide, 60-83, of lysozyme reacted with antibodies to both reduced and carboxymethylated proteins, and it was concluded that the open-chain peptides in the region of 60-83 of both α-lactalbumin and lysozyme are structurally similar.

The immunological differences between native proteins are a reflection of conformational differences or differences in the amino acid side-chains exposed to the medium. It is of interest to note that antisera to various bird lysozymes will cross-react with other bird lysozymes but that human lysozyme does not cross-react with bird antisera [54]. In a similar manner antisera to ruminant α-lactalbumin will cross-react with ruminant α-lactalbumins but not with human or pig α-lactalbumin [22]. All the α-lactalbumins and lysozymes have their respective catalytic functions, but there are distinct differences in antigenic responses with both proteins indicating probable differences in the location of the catalytic and antigenic sites.

In general, there is a large body of evidence to indicate that α-lactal-bumin and lysozyme have similar structures and in particular the backbone structure. However, certain structural differences are apparent in those residues which are exposed to the solvent, and the final assessment must await the results obtained from the x-ray crystallographic analysis.

IV. EVOLUTIONARY RELATIONSHIP BETWEEN α-LACTALBUMIN AND LYSOZYME

The high degree of structural homology between hen egg-white lysozyme and α-lactalbumin [16] led to the proposal that the structural genes for α-lactalbumin and lysozyme were derived from a common ancestral gene and by the process of gene duplication, one gene evolved for α-lactalbumin and one for lysozyme. Gene duplication provides the principal mechanism for increasing the size of the genome in the population and yet has utility in that the original function of the duplicated gene may be preserved. This is the case with the proteolytic enzymes where evolution has led to modification of enzymatic specificity without affecting mechanism [55]. However, it is quite possible that homologs exist which are structurally related but have acquired different functions. Such appears to be the case with nerve growth factor and insulin where it is proposed that nerve growth factor evolved from an ancestral proinsulin to form a protein with a new function, although structurally related to another protein [56]. The suggestion was made that the evolution of nerve growth factor from proinsulin was similar to the evolution of α-lactalbumin from lysozyme [56].

There is an abundance of evidence that is consistent with a high degree of structural homology between lysozyme and α-lactalbumin. However, there is a lack of direct evidence to suggest a functional relationship be-tween these two proteins. α-Lactalbumin does not have lysozyme activity nor does it interfere with the lysozyme reaction, and in a similar manner lysozyme does not have nor interfere with α-lactalbumin activity. A histidine residue in α-lactalbumin replaces the catalytically functional glutamic-35 residue in lysozyme, but carboxylmethylation of α-lactalbumin results in only small losses of activity of α-lactalbumin [57], suggesting that histidyl residues are not critical to the biological function of α-lactal-bumin. Aspartic-52 is involved in lysozyme activity, and this amino acid is retained as Asp-49 in α-lactalbumins [41]. Modification of the carboxyl groups in α-lactalbumin with 1-ethyl-3-(3-dimethylaminopropyl) carbodiim-ide and glycinamide caused loss of activity of α-lactalbumin, but the re-sults suggested that α-lactalbumin had no unique carboxyl groups as found in lysozyme [58]. Further work with a more discriminating carboxyl-modifying reagent is needed since the modification by the carbodiimide is not very specific. The evidence at present indicates that the active site in α-lactalbumin is probably different from the active site in lysozyme.

Further work is required to determine which amino acids in α-lactalbumin are associated with maintenance of activity of α-lactalbumin. Undoubtedly an x-ray crystallographic map of α-lactalbumin would be most useful.

V. β-(1 → 4) GALACTOSYLTRANSFERASE

A. General Properties

1. Distribution

The galactosyltransferase involved in the lactose synthetase reaction occurs as a particulate enzyme in mammary [21] and other tissues [59]. However, the enzyme is found in soluble form in skim milk and has been purified from bovine [60-63] and from human milk [64, 65]. The enzyme also exists in whole milk in a particulate form which is entrapped within the cream layer [21] and accounts for about 1% of the enzyme in whole milk. Presumably, the enzyme found soluble in milk comes from the enzymatic solubilization of the particulate form found in the tissue. The galactosyltransferase is a Golgi enzyme [66-68] and has been used as an enzymatic marker of this cellular organelle [69, 70].

2. Physical and Chemical Properties

The enzyme purified from bovine skim milk by conventional methods had a specific activity between 3-5 (μmol lactose/min/mg at 25° C) and had an apparent molecular weight of 70,000 to 75,000 on Sephadex G-100 [21]. Specific activities obtained by a similar isolation procedure were 3.5 to 4.7 when assayed at 37° C (1.5 to 2.0 at 25° C) [62]. The molecular weight by sedimentation equilibrium was 44,000, by sedimentation equilibrium in 6 M guanidine hydrochloride − 0.1 M mercaptoethanol was 48,000, and 50,000 by acrylamide gel electrophoresis in sodium dodecyl sulfate. Trayer and Hill [61] purified the galactosyltransferase from bovine skim milk by utilizing an affinity column whereby α-lactalbumin was covalently bound to Sepharose 4B. The specific activity of the final product was 14.1 at 37° C (5.8 at 25° C) and appeared to be homogenous under denaturing conditions. Electrophoresis on polyacrylamide gels in 6.25 M urea, pH 3.2, or in sodium dodecyl sulfate gave a single band. Molecular weights between 40,000 – 44,000 were obtained by sodium dodecyl sulfate gel electrophoresis, gel filtration, and sedimentation equilibrium. Sucrose density gradient experiments also gave a molecular weight of 42,000, which was consistent with a single polypeptide chain of molecular weight of about 42,000. The previous molecular weight determination of 75,000 on Sephadex G-100 columns appears to be too high [21] and is probably due to the fact that the galactosyl-transferase is a glycoprotein since a molecular weight of 44,000 is obtained with the same material on Bio-Gel P-200 columns and by sodium dodecyl sulfate gel electrophoresis [71]. The detailed amino acid composition is given in Table 2 and is based on the molecular weight of 42,000 and 12% carbohydrate. There are 330 amino acid residues, and the protein is characterized by a high amide and proline content but rather low amounts of the polar amino acids [61]. The enzyme was reported to contain about 2% sialic acids, 1.1% glucosamine, 1.1% galactosamine, and 8% neutral sugars consisting of galactose, mannose, and glucose, which is somewhat different from a later determination which reported 5% reducing sugars and 2 sialic acids per mole [62].

A detailed study on the carbohydrate composition of bovine milk galac-tosyltransferase is presented in Table 3 [72]. Further studies showed that the galactosyltransferase contained carbohydrate units linked to the poly-peptide chain through an O-glycosidic linkage involving galactosamine and serine and/or threonine. The presence of mannose in the glycoprotein and in isolated glycopeptide suggested the presence of carbohydrate units with the glycosylamine type of linkage involving N-acetylglucosamine and asparagine. Three major glycopeptide fractions were isolated from pronase digests of the enzyme, and each fraction appeared to contain each type of glycopeptide. All the glycopeptide fractions contained sialic acid. It is interesting to note that the galactosyltransferase contains relatively high amounts of aspartic acid, serine, and threonine, the amino acids (Table 2) involved in forming glycopeptide linkages. Complete removal of the sialic acid from the galactosyltransferase by neuraminidase had no effect on the specific activity of the enzyme.

A more recent purification procedure employs a variety of affinity columns including a UDP-hexanolamine agarose, N-acetylglucosamine hexanolamine agarose, and an α-lactalbumin agarose column [63]. Utilization of these affinity columns should ease the problem in purifica-tion of galactosyltransferase from milk and other sources. About 2 mg of purified galactosyltransferase may be isolated from skim milk at 40% yield from an original volume of 25.8 liters.

3. Multiple Forms

Earlier procedures for the purification of galactosyltransferase from bovine milk resulted in a protein with a molecular weight in the vicinity of 42,000 [61, 62]. As the purification procedures became more rapid, multiple forms of the enzyme were observed [62, 63, 71]. Klee and Klee [62] reported that the galactosyltransferase migrated as a broad zone of protein more or less resolved into discrete bands which had activity on

TABLE 2

Amino Acid Composition of Galactosyltransferase [a]

Amino acid	Residues/mole	Amino acid	Residues/mole
Lysine	18	Glycine	25
Histidine	11	Alanine	15
Arginine	17	Half-cystine	5
Aspartic	11	Valine	19
Asparagine	27	Methionine	8
Threonine	16	Isoleucine	16
Scrine	26	Leucine	27
Glutamic	10	Tyrosine	15
Glutamine	19	Phenylalanine	16
Proline	24	Tryptophan	5

[a] Based on a molecular weight of 42,000, less 12% for carbohydrate [61].

7.5% disc gels at pH 9.5 but displayed only a single band on sodium dodecyl sulfate disc gels. Magee et al. [71] reported two major bands in polyacrylamide gels using a continuous buffer of 50 mM phosphate, pH 8.0, and two major bands on sodium dodecyl sulfate gel electrophoresis which gave molecular weights of 45,000 and 58,000. Removal of the sialic acid did not alter the number of bands. Barker et al. [63] observed in most cases two bands in sodium dodecyl sulfate gels having molecular weights of 54,000 and 43,000, and in some cases a form of 49,000 mol wt was observed.

The origin of the multiple forms of the galactosyltransferase isolated from bovine milk has been further investigated [73, 74]. Three major possibilities for the observed multiple forms are heterogeneity in the amino acid composition, variation in the carbohydrate content, or proteolytic degradation of a larger form of the enzyme. The latter appears to be the principal cause for the observed forms on sodium dodecyl sulfate gels. The two forms found in bovine skim milk may be partially separated on Bio-Gel P-200, and both forms contain carbohydrate in the same molar ratios. Both forms, 58,000 and 42,000 mol wt, have similar catalytic

TABLE 3

Carbohydrate Composition of Bovine Milk Galactosyltransferase

Monosaccharide	Mol/100 mg	g/100 g of protein	Residues/molecule[a]
Hexoses	30.8	4.98	12.6
Mannose	13.4	2.17	5.9
Galactose	17.4	2.81	7.7
Hexosamines	26.0	5.32	11.5
Glucosamine	11.8	2.39	5.2
Galactosamine	14.2	2.93	6.3
Fucose	3.4	0.52	1.5
Sialic Acid	10.0	3.02	4.4
TOTAL	70.2	13.8	31

[a] Based on a molecular weight of 44,000 [72].

properties, and the apparent K_m's are similar for UDP-galactose, N-acetylglucosamine, and α-lactalbumin [75]. Both forms are inhibited to the same extent by sulfhydryl inhibitors and have the same heat denaturation curves. Both forms bind equally well to an α-lactalbumin-Sepharose column in the presence of either N-acetylglucosamine or UDP-galactose and cannot be differentiated by affinity columns. A variety of conventional protein purification techniques have failed to resolve completely these forms although molecular sieve columns are partially successful.

Extended storage of partially purified and purified samples of galactosyltransferase often led to a moderate loss of enzymatic activity and a decrease in the amount of the high molecular weight form with a concomitant increase in the lower molecular weight form. In addition, enzyme isolated from various milk preparations had different ratios of the two molecular weight forms. Milk contains a protease [76] and this gave credence to the idea that proteolysis may have occurred on storage. Indeed, highly purified galactosyltransferase, which was subjected to careful trypsin hydrolysis resulted in a first-order activity loss, but with very mild trypsin treatment at $0°$C there was a conversion of the higher molecular weight form to the lower form [74]. During the initial phase of treatment of 8 mg/ml of galactosyltransferase with 6 μg trypsin per ml at $0°$C, the relative amounts of high molecular weight to low molecular weight material changed from 4 to 1 in the controls to 1 to 3 though only

10% of the original activity was lost, which indicates that the 42,000 form
was active. This form was retained on the α-lactalbumin affinity column.
During the trypsinolysis, no carbohydrates were released from the galactosyltransferase, but with low molecular weight peptides were released.
Extended treatment with trypsin gives rise to small amounts of a 38,000
mol wt form, but it has little if any activity. There is evidence for a milk
protease [76] which has trypsin-like activity, and this enzyme may be
identical to blood plasmin [77]. The protease isolated from milk, trypsin,
and plasmin all gave similar results in converting the larger form to the
smaller form. Plasmin completely converted the 55,000 mol wt form to
the 42,000 mol wt form with only 35% loss in enzymatic activity (Fig. 2).
Galactosyltransferase (3 mg/ml) was incubated with 0.2 mg/ml plasmin
in 1 mM 2-mercaptoethanol-10 mM tris-Cl, pH 7.8, for 24 hours at
25°C. A control contained no plasmin. Samples were removed at intervals
for sodium dodecyl sulfate gel electrophoresis while other samples were
assayed periodically using N-acetylglucosamine as the acceptor substrate.
At the end of the incubation period an equal volume of buffered soybean
trypsin inhibitor (2 mg/ml)-bovine serum albumin (20 mg/ml)-ovalbumin
(20 mg/ml)-Blue Dextran (5 mg/ml) was added to each, and a portion of
each was separately applied to a Bio-Gel P-200 column equilibrated with
200 mM ammonium sulfate-2 mM MgCl$_2$-1 mM EDTA-1 mM 2-mercapto-
ethanol-20 mM tris-Cl, pH 7.3 [75]. Fractions were assayed with N-
acetylglucosamine as the substrate, activity being presented in milliunits/
ml, and the column void volume determined by absorbence at 6250 Å. In
both cases the Blue Dextran peak eluted at fraction number 32.1; the control activity peak was at 61.2, while the plasmin treated peak was at 64.4.
Gel scans of a control, including molecular weight graduations in thous-
ands, and the plasmin treated protein are shown in the upper right corner
[74]. In addition, some of the milk protease can copurify with the galactosyltransferase and cannot be removed by affinity columns. In fact, all
purified samples of enzyme examined in this laboratory contained protease.
Incubation at room temperature of the purified galactosyltransferase results in the conversion of the high molecular weight form to the low
molecular weight form, but this conversion can be largely prevented by
trypsin inhibitors. Plasmin has the property of binding to proteins, and
this may be the case with the galactosyltransferase isolated from milk.
It is of interest to note that the soluble galactosyltransferase found in
bovine blood has a molecular weight of about 45,000 and may also be a
degradation product of plasmin.

Palmiter [78] has reported enzymatic activities of galactosyltransferase associated with varying molecular weights in an ammonium sulfate
fraction prepared from mouse mammary tissue as determined on
Sephadex G-100. One form eluted at the exclusion volume, the major
form eluted between 100,000 and 130,000, and another form appeared at

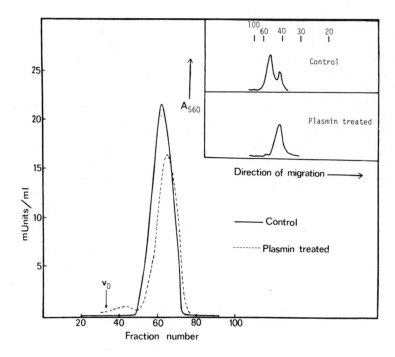

FIG. 2. Plasmin Conversion of Large Molecular Weight Form to Small Form

29,000 mol wt. These data should be interpreted with some caution since the mouse galactosyltransferase is probably a glycoprotein which may give anomalous migration on Sephadex. Furthermore, it is difficult to truly solubilize the enzyme from tissues.

The enzyme purified from human milk is less well characterized but appears to have properties similar to those of the bovine milk enzyme. Andrews [64] purified the enzyme from human milk, four to seven days post partum, using an α-lactalbumin-Sepharose column as the last step in the purification procedure and obtained an enzyme preparation in 36% yield with a specific activity of 5.7 at 25° C which is comparable to the bovine milk enzyme. A 40,000 to 42,000 mol wt was obtained from chromatography on Bio-Gel P-200, which is similar to the lower molecular weight form of the bovine enzyme. No evidence was obtained for a higher molecular weight form of the human milk enzyme. To find a higher molecular weight form, fresh milk would undoubtedly be required in conjunction with a rapid purification scheme.

The galactosyltransferase has also been purified from Japanese human milk [65]. The purified enzyme had a specific activity of 0.75 and appeared as a single band on paper electrophoresis at pH 9.5. The amino acid analysis showed a high percentage of proline and acidic amino acids, which is similar to the bovine milk enzyme. The enzyme is a glycoprotein and contained 1.10% sialic acid, 5.03% galactosamine, 2.47% glucosamine, 2.36% galactose, 0.87% mannose, and 0.87% fucose. Qualitatively, the carbohydrates are similar to those found in the bovine milk enzyme (Table 3), but quantitatively there is less mannose and sialic acid but larger amounts of galactosamine. The molecular weight as determined on Sephadex G-100 chromatography was 75,000 and was 72,000 by electrophoresis at pH 9.5 in 0.2% sodium dodecyl sulfate. These molecular weights do not correspond to 40,000 to 42,000 obtained by Andrews [64] for Caucasian human milk.

B. Catalytic Properties

1. Reactions Catalyzed

The general reaction catalyzed by the galactosyltransferase is as follows:

$$\text{UDP-galactose + acceptor} \xrightarrow{\text{Mn}^{2+}} \text{gal-acceptor + UDP} \qquad (3)$$

The major galactosyl acceptors are GlcNAc and terminal β-($1 \rightarrow 4$) glycosides of GlcNAc including the carbohydrate side-chain of glycoproteins. Glucose is a good acceptor only in the presence of α-lactalbumin, and under these conditions lactose is formed at meaningful rates. This latter reaction has been termed lactose synthetase, a term useful for describing this reaction. The principal function of the galactosyltransferase in most tissues is its involvement in glycoprotein biosynthesis except in the lactating mammary gland where the enzyme is also involved with the biosynthesis of lactose.

2. Substrate Specificity

a. Galactosyl Donor. Babad and Hassid [9] examined the galactosyl donor specificity of the bovine milk galactosyltransferase and found that dUDP-D-galactose was 80% as effective as UDP-D-galactose. No formation of lactose was observed when ADP-D-galactose, TDP-D-galactose, CDP-D-galactose, or GDP-D-galactose was used as a substrate.

b. Galactosyl Acceptor. Babad and Hassid [9] reported that α-D-glucose-1-phosphate, α-D-galactose-1-phosphate, L-glucose, D-xylose, maltose, and α-methyl-D-glucoside were not galactosyl acceptors. D-Glucosamine was 0.3% as effective as D-glucose, and cellobiose was 7.5%

as effective as glucose; N-acetyl-D-glucosamine was 25% as effective as glucose and the product of the latter reaction was identified as N-acetyllactosamine. The early studies by Babad and Hassid [9] were carried out with the galactosyltransferase in the presence of α-lactalbumin.

Brew et al. [10] showed that the galactosyltransferase could transfer galactose to N-acetylglucosamine in the absence of α-lactalbumin and that α-lactalbumin inhibited this reaction. This inhibition was examined in more detail by Schanbacher and Ebner [79], who showed that it was hyperbolic (intercept) with respect to N-acetylglucosamine. Hill et al. [16] also showed that orosomucoid (galactose removed) was a galactosyl acceptor. This reaction was partially inhibited by α-lactalbumin.

The galactosyltransferase acceptor specificity of the enzyme from bovine milk has been examined in detail [79]. In the absence of α-lactalbumin, the best substrates are GlcNAc and its β-(1→4) glucosides, $(GlcNAc)_2$, $(GlcNAc)_3$, $(GlcNAc)_4$, and ovalbumin with apparent K_m's in the low mM range. To date, the best substrate is p-nitrophenyl-β-N-acetylglucosamine, which has a K_m of 0.30 mM in the absence of α-lactalbumin. Other β-glycosides are poorer acceptors. These include cellobiose, cellobiulose, β-methylglucose, β-indoxylglucose, glucosylmannose, and gentiobiose. Alpha glycosides such as α-methylglucose and maltose were not active in the absence of α-lactalbumin but were slightly active in the presence of α-lactalbumin. Glucosamine and N-acetylmannosamine were poor acceptors; N-acetylmuramic acid was fair and 2-deoxy-D-glucose was an acceptor only in the presence of α-lactalbumin. Mannose, fucose, melibiose, and UDP-GlcNAc are not acceptors even in the presence of α-lactalbumin. Glycoproteins with carbohydrate side-chains of the type terminating in sialic acid-β-galactosyl-(1→4)-β-N-acetylglucosamine-glycoprotein upon removal of sialic acid by neuraminidase and galactose by β-galactosidase are good acceptors. These include orosomucoid, fetuin, and desialyzed bovine submaxillary mucin [23] and thyroglobulin glycopeptides freed of sialic acid and galactose [80].

α-Lactalbumin inhibits markedly the transfer of galactose to GlcNAc (15 μg/ml gave 50% inhibition) whereas it is a much less effective inhibitor of β-(1→4) glycosides such as $(GlcNAc)_3$ (1 mg/ml inhibits 30%) and ovalbumin, which has terminating N-acetylglucosaminyl residues.

The general conclusions resulting from the specificity studies are summarized in Figure 3, in which X and Y are different constituents. The best acceptors are β-(1→4) glycosides in which X = H and Y = N-acetyl, but this acceptor is inhibited markedly by α-lactalbumin. When X is not H, then α-lactalbumin is a less effective inhibitor but the compound is still a good substrate. When X = H and Y = OH (glucose), α-lactalbumin is required for the compound to be a good galactosyl acceptor; when X is not H (e.g., CH_3), then the compound is only a fair substrate. The apparent K_m's for good acceptors are in the low mM range.

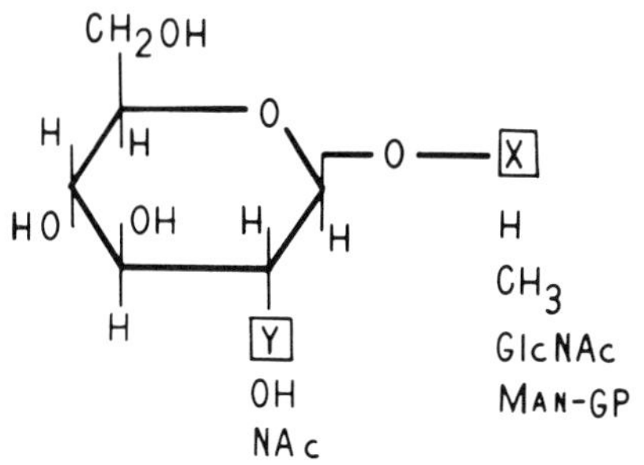

FIG. 3. Acceptor for β-(1 → 4) Galactosyltransferase

3. Metal Requirement

Babad and Hassid [9] reported that in crude bovine enzymic preparations Mg^{2+} was 25% as effective as Mn^{2+}, which has an optimum concentration of 13.3 mM at pH 7.5. Ca^{2+} could replace Mg^{2+} but Co^{2+}, Na^+, K^+,
or NH_4^+ had little stimulating activity. However, studies with more
highly purified enzyme [81] showed that no activity was observed in the
absence of Mn^{2+} or in the presence of Mg^{2+} or Ca^{2+}. Concentrations of
Mn^{2+} above 4 mM were inhibitory. It was possible that Mg^{2+} or Ca^{2+}
could displace Mn^{2+} from other proteins in the crude enzymatic preparation [9] and therefore would appear to activate the enzyme.

4. Inhibitors

Relatively few studies have been carried out on galactosyltransferase
to determine which amino acids constitute the active site. Babad and
Hassid [9] observed inhibition by p-chloromercuribenzene sulfonate which
could be abolished by adding 2-mercaptoethanol. These results suggest
the presence of a critical thiol residue. Kitchen and Andrews [82] observed that p-hydroxymercuribenzoate inhibited the galactosyltransferase
from human milk; the inhibition was reduced by including 1 mM UDP-
galactose and abolished if both UDP-galactose and Mn^{2+} were included.
Alone and in various combinations neither Mn^{2+}, glucose, N-acetylglucosamine, nor α-lactalbumin afforded any protection. Inactivated enzyme
was able to bind to an α-lactalbumin-Sepharose column. The native enzyme
reacted slowly with about 1 mol of 5,5'dithiobis-(2-nitrobenzoate) (Ellman's
reagent) per mole of enzyme. Their results suggest the presence of a
partly shielded thiol group which is involved with Mn^{2+} in the binding of
UDP-galactose since neither N-acetylglucosamine nor α-lactalbumin
afforded any protection.

Recent studies in this laboratory with the galactosyltransferase from bovine milk have shown that this enzyme is inhibited by p-chloromercuribenzoate in a manner similar to the human milk enzyme. The inhibition is prevented by the presence of Mn^{2+} and UDP-galactose but not by N-acetylglucosamine or glucose. N-ethylmaleimide also inhibits the enzyme, and the inhibition is prevented by Mn^{2+} and UDP-galactose, Mn^{2+} and UDP, Mn^{2+} and UTP, and only partially by Mn^{2+} and UMP. The protection with UDP, UTP, and UMP in the presence of Mn^{2+} occurs since they react with the same form of enzyme as UDP-galactose and Mn^{2+}. The rate of inactivation was slower and to a lesser extent with N-ethylmaleimide than with p-chloromercuribenzoate. Inhibition by $25 \mu M$ p-chloromercuribenzoate was 75% complete in 30 min in the absence of substrates but required 2 hr in 10 mM N-ethylmaleimide to obtain the same degree of inhibition.

C. Reaction Kinetics

The galactosyltransferase from bovine milk transfers galactose from UDP-galactose to GlcNAc. This reaction is inhibited by α-lactalbumin but in the presence of α-lactalbumin the enzyme will transfer galactose to glucose to form lactose [26]. On the basis of these observations, α-lactalbumin was called a specifier protein in that it changes the galactosyl acceptor specificity of the galactosyltransferase and distinguishes it from other regulatory proteins. However, it was subsequently shown that the galactosyltransferase catalyzed the transfer of galactose to glucose to form lactose in the absence of α-lactalbumin provided the glucose concentration was high ($K_m = 1.4$ M) and that the maximum velocity of the reaction was the same in the absence or presence of α-lactalbumin [60]. These results showed that the active site of the enzyme is on the galactosyltransferase, and the function of α-lactalbumin was to lower the apparent K_m of glucose to the low mM region. Similar results were obtained by Klee and Klee [83] with the bovine milk enzyme and by Andrews [84] with the human milk enzyme. In the assay for lactose synthetase, an optimum concentration of α-lactalbumin is required, and higher concentrations are inhibitory [60]. Efforts to detect an α-lactalbumin-galactosyltransferase complex under conditions of maximum product formation have been unsuccessful [61, 85], which is supportive evidence for the absence of a long-lived kinetic complex.

1. Steady-State Kinetics

The steady-state kinetic analysis of the galactosyltransferase reaction was undertaken to determine the functional role of α-lactalbumin in the lactose synthetase reaction [81, 86, 87]. The general approach was to study: (1) the galactosyltransferase reaction with N-acetylglucosamine as

the galactosyl acceptor in the absence of α-lactalbumin [81]. (2) the galactosyltransferase reaction with glucose as the galactosyl acceptor at a
fixed concentration of α-lactalbumin [86], and (3) the effects of α-lactalbumin on the reaction with either glucose or N-acetylglucosamine as the
galactosyl acceptor [87].

The equilibrium of the reaction is essentially in the direction of product
formation; no reversibility of the reaction could be demonstrated. These
results ruled out the use of product inhibition analysis to assist in determining the order of addition of substrates and release of products. UDP-
glucose was an inhibitory analog of UDP-galactose, but no suitable inhibitory
analog of N-acetylglucosamine was found from the compounds tested, which
were: D-sorbitol, glycerol, sucrose, inositol, glucose, N-acetylgalacto-
samine, N-acetylmannosamine, galactosamine, mannosamine, glucuronic
acid, and glucose-6-phosphate. It was also observed that N-acetylglucosamine caused substrate inhibition at high concentrations and that the
inhibition was noncompetitive with respect to Mn^{2+} and UDP-galactose. It
was postulated that at high concentrations the substrate formed inactive
dead-end complexes on both the substrate addition side and product release
side of the reaction mechanism. The steady-state kinetic results based on
initial velocity, dead-end inhibition, and substrate inhibition studies were
consistent with an equilibrium-ordered type of mechanism. Mn^{2+} adds
first to the free enzyme under conditions of thermodynamic equilibrium and
does not dissociate at each turn of the catalytic cycle. This is followed by
an ordered addition of UDP-galactose and GlcNAc followed by the ordered
release of products, N-acetyllactosamine and UDP.

The initial rate equation for the proposed mechanism in the absence of
α-lactalbumin and N-acetylglucosamine as the galactosyl acceptor can be
expressed as

$$v = \frac{VMAB}{K_{im}K_{ia}K_b + K_{ia}K_bM + K_{im}K_aB + K_aMB + K_bMA + MAB} \quad (4)$$

where V is the maximum velocity, M, A, and B represent Mn^{2+}, UDP-
galactose, and N-acetylglucosamine, respectively. K_{im} represents the
dissociation constant of M with free enzyme, and K_{ia} is the dissociation
constant of A with EM. K_a is the Michaelis constant for A and K_b is the
Michaelis constant for B [81].

Similar kinetic studies [86] were carried out with glucose as the
galactosyl acceptor and with the concentration of α-lactalbumin fixed at
4 μM, which is at a nonsaturating level. Substrate inhibition was observed
with glucose. UDP-glucose was a good inhibitor of the reaction. It was observed that a number of pentoses at relatively high concentrations (0.4 M)
inhibited the reaction: D-arabinose (13%); L-xylose (16%); D-xylose (36%);
L-sorbose (45%); L-arabinose (54%); D-ribose (59%); and 2-deoxyglucose

(68%). The inhibition by L-arabinose was competitive with respect to glucose and was used for dead-end inhibition analysis. The data from initial velocity and dead-end inhibition analysis provided support for the view that in the presence of glucose and α-lactalbumin, the galactosyltransferase had an ordered addition of substrates and release of products. It appeared that this sequence was not influenced by the galactosyl acceptor or by the presence of α-lactalbumin.

The previous studies [81, 86] showed that neither α-lactalbumin nor the nature of the galactosyl acceptor affected the order of addition of substrates or release of products. The kinetic effects of α-lactalbumin on the galactosyltransferase reaction when glucose or N-acetylglucosamine was the galactosyl acceptor were investigated to provide an explanation for the role of α-lactalbumin in the lactose synthetase reaction [87]. Previous studies had shown that N-acetylglucosamine [81] or glucose [86] at high concentrations would cause substrate inhibition. It was further observed that α-lactalbumin increased the extent of substrate inhibition observed either with glucose or N-acetylglucosamine, and it was concluded that α-lactalbumin in the presence of high concentrations of substrates formed additional dead-end complexes involving α-lactalbumin and the substrate dead-end complex [87]. However, α-lactalbumin does not appear to increase inhibition observed with L-arabinose when glucose is the substrate. The proposal that α-lactalbumin forms additional dead-end complexes with the substrate provides an explanation for the observation that high levels of α-lactalbumin inhibit the enzymatic assay for the galactosyltransferase.

A summary of the kinetic constants for the substrates of the galactosyltransferase in the absence and presence of α-lactalbumin are presented in Table 4. The true inhibition constant of UDP-glucose was 0.089 ± 0.006 mM in the absence of α-lactalbumin and 0.061 ± 0.002 mM in the presence of 4μM α-lactalbumin. The true inhibition constant for L-arabinose in the presence of $4\,\mu$M α-lactalbumin was 0.27 to 0.34 M.

A kinetic mechanism based on the steady-state kinetic analysis is illustrated in Figure 4 where UDP-gal represents UDP-galactose, α-LA is α-lactalbumin, and CHO represents a carbohydrate reactant which is either a substrate or an inhibitory substrate analog [87]. The main features of the mechanism are: (1) Mn^{2+}, UDP-galactose, and galactosyl acceptor (at nonsubstrate inhibiting concentrations) add to the enzyme in an ordered manner, and the resulting complex may then combine with α-lactalbumin, (2) Mn^{2+} does not leave the enzyme at each turn of the catalytic cycle whereas α-lactalbumin does leave at each turn of the cycle. The products dissociate in the order: galactosyl acceptor and UDP, (3) At higher concentrations the galactosyl acceptor can add randomly to free enzyme, enzyme-manganese (not shown), and enzyme-manganese-UDP-galactose, but an active complex is produced only if there has been prior

TABLE 4

Kinetic Constants for Substrates of the Galactosyltransferase Reaction

Variable Substrate	Kinetic Constant	None[a] Weighted Mean Value, mM	α-Lactalbumin	
			Fixed at 4μM[b] Weighted Mean Value, mM	Variable[c] Apparent Constant, mM
Mn^{2+}	$K_{(Mn^{2+})}$	1.35 ± 0.14^{d}	1.24 ± 0.02	0.20 ± 0.04
N-acetylglucosamine	$K_{(GlcNAc)}$	5.8 ± 0.06		
Glucose	$K_{(Glc)}$		5.3 ± 0.05	2.5 ± 0.6
α-Lactalbumin	$K_{(\alpha-LA)}$–Glc varied			$0.73 \pm 0.15\,\mu M$
	$K_{(\alpha-LA)}$–UDPGal varied			$2.4 \pm 0.2\,M$
	$K_{(\alpha-LA)}$–Mn^{2+} varied			$3.3 \pm 0.5\,\mu M$

[a] N-acetylglucosamine was the galactosylacceptor [81].

[b] Glucose was the galactosylacceptor [86].

[c] Glucose was the galactosylacceptor, and α-lactalbumin was varied from 0 to 5μM [87].

[d] Standard error.

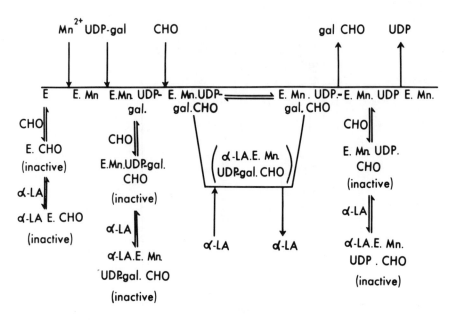

FIG. 4. Kinetic Scheme of the Galactosyltransferase Reaction

combination of Mn^{2+} and UDP-galactose. Otherwise, inactive dead-end complexes are formed which inhibit the overall reaction, (4) The effect of α-lactalbumin on inhibiting the reaction when N-acetylglucosamine is the galactosyl acceptor, activating the reaction when glucose is the galactosyl acceptor, increasing substrate inhibition by N-acetylglucosamine and glucose, and enhancing the dead-end inhibition by a substrate analog is essentially due to its ability to combine with a complex containing carbohydrate and cause displacement of already established equilibria.

For example, when N-acetylglucosamine is the substrate, the reaction proceeds well in the absence of α-lactalbumin [81]. When the concentration of N-acetylglucosamine is nonsaturating, the addition of α-lactalbumin forces more of the reaction flux via the branched pathway. If this pathway is slower than the linear pathway which occurs in the absence of α-lactalbumin, then inhibition of the reaction is predicted, which is consistent with the experimental findings. Similar arguments may be presented to explain how α-lactalbumin may act as an activator at concentrations of N-acetylglucosamine well below the K_m value. α-Lactalbumin can effectively increase the concentration of N-acetylglucosamine since it can reduce the apparent K_m of the substrate and cause part of the reaction flux to occur along the branched pathway. When the glucose concentration is 1 to 4 M, reaction occurs along the linear pathway in the absence of α-lactalbumin.

The addition of α-lactalbumin reduces the K_m for glucose to the mM region, and virtually all the reaction proceeds along the branched pathway [87].

Additional dead-end complexes containing α-lactalbumin were postulated to explain the enhancement of substrate inhibition by glucose and N-acetyl-glucosamine by α-lactalbumin. The addition of α-lactalbumin would increase the steady-state concentration of the total dead-end complexes, and thereby reduce the overall reaction rate by reducing the amount of enzyme available for product formation.

From this kinetic study [81, 86, 87] it would appear that α-lactalbumin is best classified as a modifier protein rather than as a specifier protein, as previously proposed [26], since the galactosyltransferase can by itself catalyze the formation of lactose in the absence of α-lactalbumin (high glucose). It would also appear that α-lactalbumin lowers the K_m for glucose so that glucose becomes an effective substrate. The fundamental role of α-lactalbumin appears to differ from the action of small organic molecules which can affect the binding of substrate and/or the maximum velocity of an enzyme by binding at a site distinct from the catalytic site. Under conditions of catalysis, α-lactalbumin requires the presence of Mn^{2+} and the two substrates on the enzyme prior to binding with the enzyme.

Kitchen and Andrews [88] investigated the effect of temperature and bovine α-lactalbumin on the rate of galactosyltransferase isolated from human milk. α-Lactalbumin activates the galactosyltransferase from bovine milk at low GlcNAc concentrations and inhibits at high concentrations [26, 83, 87] whereas human galactosyltransferase is inhibited at all GlcNAc concentrations at 25° C [84]. As the temperature is increased from 25° to 37° C, the human galactosyltransferase resembles the bovine enzyme in behavior. This has been attributed to a temperature effect. There is virtually no temperature effect (25–40° C) on the rate of reaction at 1 mM GlcNAc in the absence of α-lactalbumin, but in the presence of α-lactalbumin the enzyme shows a more normal temperature response so that the activating effect of α-lactalbumin at low GlcNAc is more apparent than real.

2. Detection of Reactant Complexes by Affinity Chromatography

The mechanism described in Figure 4 provides an explanation of the failure to detect an enzyme-Mn-UDP-galactose-glucose-α-lactalbumin complex under conditions of maximum product formation since it is postulated that α-lactalbumin dissociates rapidly from the enzyme at each turn of the catalytic cycle. However, it does predict the formation of complexes under conditions of dead-end inhibition, e.g., high substrate and/or high α-lactalbumin concentrations and, indeed, evidence for the presence of such complexes has been reported [61, 63, 64, 89].

Andrews [64] has purified the galactosyltransferase from human milk using the affinity chromatography technique whereby α-lactalbumin was chemically coupled to Sepharose 6B which had been previously activated with CNBr. He observed that the galactosyltransferase was retained on such a column in the presence of 3 mM N-acetylglucosamine and released from the column when N-acetylglucosamine was removed from the eluting buffer. Trayer and Hill [61] have used a similar procedure of affinity chromatography to purify the galactosyltransferase from bovine milk. The buffer contained Mg^{2+} and glucose and upon removal of glucose, the galactosyltransferase was eluted from the column. The reason the technique worked is that the Sepharose-α-lactalbumin columns contained a high concentration of α-lactalbumin and in the presence of carbohydrate substrates, dead-end complexes are formed by displacing established equilibria. In both the above cases no product can form because of the lack of Mn^{2+} and UDP-galactose in the reaction mixture. Upon removal of the substrate from the buffer, the complex dissociates and the galactosyltransferase is eluted in the buffer.

Recently, affinity columns prepared from derivatives of UDP and N-acetylglucosamine have been used as efficient methods for the purification of galactosyltransferase from bovine milk. Combinations of affinity columns should greatly facilitate the purification of the galactosyltransferase from a variety of sources.

It became apparent that the presence of reactant complexes predicted for the mechanism outlined in Figure 4 could be detected by affinity chromatography experiments using Sepharose-α-lactalbumin columns. The results of such studies [89] provided direct evidence for the formation of dead-end complexes involving α-lactalbumin which were predicted from the data obtained from the steady-state kinetic investigations [81, 86, 87].

No complexes were detected with free enzyme or E-Mn^{2+} but complexes were formed with free enzyme and/or E-Mn^{2+} with the substrates, glucose, N-acetylglucosamine, and UDP-galactose. The addition of Mn^{2+} to substrates or substrate analogs causes tighter retardation of the galactosyltransferase but is not essential for complex formation. All these complexes are not kinetically significant; they arise as a consequence of the high concentration of α-lactalbumin bound to the Sepharose. They do appear to be important in explaining the possible point of inhibition of lactose formation by high concentrations of α-lactalbumin [60]. Under identical conditions, glucose fails to retard the galactosyltransferase on a Sepharose-α-lactalbumin column at 20-25 mM; N-acetylglucosamine fails at 10-15 mM; and UDP-galactose fails at 2-3 μM. These results suggest that the inhibition of lactose formation by α-lactalbumin may be principally due to the formation of E-Mn-UDP-galactose-α-LA and E-UDP-galactose-α-LA complexes. The extent of retardation is dependent on the total and relative concentrations of Sepharose-α-lactalbumin and substrate or substrate analog. The product, lactose, does not cause retardation of the galactosyltransferase, which is consistent with the lack of inhibition exhibited by lactose on the reaction.

D. Relationship to Other Galactosyltransferases

There is a variety of galactosyl linkages found in biological materials and they appear in glycoproteins, mucins, mucopolysaccharides, collagen, cartilage, and blood group substances. Representative compounds having various galactosyl linkages are presented in Table 5. The first type is catalyzed by the galactosyltransferase reviewed in this chapter. That is, galactose is transferred to Glc or GlcNAc with the formation of a $\beta(1 \rightarrow 4)$ linkage. The enzymatic relatedness between this and other galactosyltransferases has not been studied in any detail, and most of the work has been with crude preparations. In general, all the transferases are particulate, require UDP-galactose as the galactosyl donor, and most require Mn^{2+}. A question not resolved at the present time is whether these reactions are catalyzed by separate and distinct enzymes or whether some enzymes can catalyze more than one type of reaction. The problem will be resolved when the galactosyl acceptor specificity of these enzymes can be adequately tested. It will be of interest to determine the size and which groups on the acceptor are essential for specificity.

There is some evidence to indicate that there may be a distinct galactosyltransferase for each linkage formed. For example, Helting [90] was able to demonstrate in particles prepared from embryonic chick cartilage that a galactosyltransferase which formed N-acetyllactosamine was distinct from two other galactosyltransferases involved in the synthesis of cartilage. Previous work had established that the galactosyltransferases involved in cartilage biosynthesis were catalyzed by two independent reactions [91]. These studies provide some support for the view that there may be a separate galactosyltransferase for a specific type of linkage and acceptor.

The enzymatic basis for blood groups in man also supports the view that glycosyltransferases specified by separate genes determine the blood groups [92]. The B gene transfers galactose from UDP-galactose to form

$$\alpha\text{-Gal-}(1 \rightarrow 3)\text{-}\beta\text{-Gal-}(1 \rightarrow 3)\text{-}\beta\text{-GlcNAc-R} \atop \underset{\alpha\text{-Fuc}}{\overset{\mid}{\underset{\uparrow}{\overset{2}{\frown}}}} \tag{5}$$

TABLE 5

Representative Galactosyl Linkages

Type	Compounds	Structure
1. β-Gal-(1→4)-	Lactose N-acetyllactosamine Ovalbumin Fetuin(-NANA, -Gal)	β-Gal-(1→4)-Glc β-Gal-(1→4)-GlcNAc β-Gal-(1→4)-β-GlcNAc- β-Gal-(1→4)-β-GlcNAc-
2. β-Gal-(1→4)-β-	Lactosylceramide	β-Gal-(1→4)-β-Glc-(1→1)-ceramide
3. β-Gal-(1→4)-β-xyl	Cartilage	β-Gal-(1→4)-β-xyl-protein
4. β-Gal-(1→3)-β-GlcNAc	Lac ter-ceramide	β-Gal-(1→3)-β-GlcNAc-(1→3)- β-Gal-(1→4)-β-Glc-ceramide
5. β-Gal-(1→3)-β-Gal-	Proteoglycan	β-GlcUA-(1→3)-β-Gal-(1→3)- β-Gal(1→4)-β-xyl-protein
6. β-Gal-hydroxylysine	Collagen	2-α-Glc-β-Gal-(1→2)-β-hydroxylysine
7. α-Gal-(1→4)-β-Gal	Globoside	β-GalNAc-(1→3)-α-Gal-(1→4)- β-Gal-(1→4)-β-Glc-ceramide
8. α-Gal-(1→3)-β-Gal- (1→2) α-Fuc	Blood group B	α-Gal-(1→3)-β-Gal-(1→3)-β- GlcNAc- (1→2) α-Fuc
9. β-Gal-(1→4)-GalNAc	Blood group substances	α-Fuc -β-Gal-(1→4)-GalNAc-Ser or Thr

but will not transfer galactose to β-Gal-(1 → 3)-β-GlcNAc-R, indicating that the enzyme recognizes fucose as well as galactose.

Examination of Table 5 shows that there are a number of distinct galactosyl linkages, each possibly being catalyzed by a separate enzyme. Purification of these enzymes should be possible by judicious use of affinity columns. An example of the effective use of affinity columns has been with the purification of a xylosyltransferase from embryonic chick cartilage [93, 94] using an affinity column prepared by coupling a Smith-degraded proteoglycan to Sepharose. The highly purified xylosyltransferase has a molecular weight of 90,000 to 95,000 and consists of four subunits. The product of this reaction, xyl-Ser-proteoglycan-Sepharose, was used to partially purify a galactosyltransferase which transfers galactose to the xylose residue [94] in the proteoglycan.

VI. INTERACTION BETWEEN α-LACTALBUMIN AND GALACTOSYLTRANSFERASE

A. Physical Interactions

The steady-state mechanism outlined in Figure 4 predicts several α-lactalbumin-galactosyltransferase complexes. Under conditions of maximum product formation, which implies minimum dead-end complexes, no interaction between α-lactalbumin and galactosyltransferase could be detected by a number of physical means including density gradient, gel filtration, and fluorescence experiments [85]. However, the mechanism does predict the formation of dead-end complexes, and some of these have been detected by affinity chromatographic methods in which α-lactalbumin was coupled covalently to Sepharose (Sec. V, C, 2). Several reports provide direct evidence for a physical interaction between the galactosyltransferase and α-lactalbumin. Andrews [64] determined from chromatography on Bio-Gel P-200 that the molecular weight of human milk galactosyltransferase was 40,000 to 42,000. This did not change in the presence of bovine α-lactalbumin (100μg/ml) but in the presence of bovine α-lactalbumin (100 or 200 μg/ml) and GlcNAc (3 mM or 6 mM) the molecular weight was about 60,000 indicative of a 1:1 complex of galactosyltransferase-α-lactalbumin and GlcNAc. These results are consistent with the mechanism in Figure 4 and represent dead-end complexes formed in the presence of a carbohydrate-containing substrate. Klee and Klee [62] have obtained similar results with galactosyltransferase isolated from bovine milk and bovine α-lactalbumin by using band sedimentation techniques. Complex formation depended upon Mn^{2+} and N-acetylglucosamine. Data from a Scatchard plot indicated that the complex contained 1 mol of each protein; the binding constant was close to 10^5. Similar results have been obtained by Ivatt and Rosemeyer [95]. They report an association constant of $1 \times 10^5 (1/\text{mol})^2$ for a complex of galactosyltransferase-GlcNAc-α-lactalbumin and an association constant of 5×10^5 $(1/\text{mol})^2$ for a complex of galactosyltransferase-Glc-α-lactalbumin. The available data clearly show that α-lactalbumin and galactosyltransferase in the presence of a galactosyl acceptor can form a 1:1 complex even though this complex is not catalytically active.

The question arises whether α-lactalbumin binds at or very near the catalytic site on the galactosyltransferase or at a site distant from the catalytic site. To date there is no direct evidence to resolve this question. The plots obtained in the steady-state analysis were all linear and as such provided no evidence for cooperative interactions. If the binding site for α-lactalbumin on the galactosyltransferase were distinct from the catalytic site, it would be anticipated that chemically modified derivatives of α-lactalbumin would act differently with the substrates Glc and GlcNAc. It would be expected that such derivatives would not stimulate the reaction when glucose was the galactosyl acceptor and not inhibit when GlcNAc was the acceptor. To date such experiments have been negative with acetylated and nitro-derivatives of α-lactalbumin. The present data, though not firmly based, would favor the idea that α-lactalbumin was binding at or near the catalytic site of the galactosyltransferase.

B.   Chemical Modification of α-Lactalbumin

Some progress has been made regarding the nature of the amino acids in α-lactalbumin required for activity in modifying the galactosyltransferase so that glucose becomes an efficient substrate or conversely inhibiting when N-acetylglucosamine is the substrate. Unfortunately, both activities have not been measured in all the studies reported. The facts that a model for α-lactalbumin is based on the coordinates for lysozyme [31] and that the x-ray crystallography of α-lactalbumin is under active investigation have stimulated research in this area both with respect to activity of α-lactalbumin with galactosyltransferase and its structural relationship to lysozyme.

Modification of the carboxyl groups in α-lactalbumin with 1-ethyl-3-(3-dimethylaminopropyl) carbodiimide and glycinamide resulted in modification of 20 carboxyl groups on the average and resulted in total loss of activity of α-lactalbumin to form lactose with galactosyltransferase and glucose and to inhibit the formation of N-acetyllactosamine. The carboxyl-modified α-lactalbumin did not inhibit the activity of native α-lactalbumin [58]. The reagent used in this study modified all the carboxyl groups in α-lactalbumin; hence, it was not possible to determine if a critical carboxyl group was essential for the activity of α-lactalbumin.

Carboxymethylation of α-lactalbumin with iodoacetate resulted in the carboxymethylation of a single methionine and three histidines over a period of 12 days [57]. After one day, Methionine 90 (in α-lactalbumin) was carboxymethylated but this resulted in no change in the apparent K_m of α-lactalbumin or V_{max}. Carboxymethylation of Methionine 90 and Histidine 68 resulted in a 1.6-fold increase in the apparent K_m and a 20% decrease in V_{max}; with Methionine 90, and Histidine 68, 32, and 107 carboxymethylated, the apparent K_m for α-lactalbumin increased 2.6-fold and the V_{max} decreased 44%. These results were obtained by measuring lactose formation and clearly show that modification of Methionine 90 resulted in no loss of activity of α-lactalbumin. Progressive modification of the histidines resulted in an increase in apparent K_m of α-lactalbumin and decrease in V_{max} but did not result in total loss of activity.

Three of the four tryptophan residues of α-lactalbumin are alkylated after reaction with 2-hydroxy-5-nitrobenzyl bromide (HNB) and these have been identified as Trp 26, Trp 104 and Trp 118 with Trp 60 not being reactive [96]. Further studies have shown that about 50% of the activity of α-lactalbumin as measured by lactose synthesis is lost when 1 mol of HNB is incorporated, and no activity is present when 2.5 mol of HNB are incorporated per mole of α-lactalbumin [97]. The activity measured was attributed to native α-lactalbumin, and further studies using monolabeled α-lactalbumin isomers which could be partially separated on DEAE cellulose indicated that they were inactive except for HNB-(Trp 118)-α-lactalbumin, which might possess very low activity. Though no gross conformational changes were detected, the monolabeled derivatives were more susceptible to trypsin than native α-lactalbumin.

α-Lactalbumin treated with maleic anhydride, specific for amino groups, resulted in the incorporation of 13.9 maleyl groups per mole of protein [98]. No activity loss was detected until more than four maleyl groups were incorporated. After this there was a slow linear loss of activity until 12 groups were incorporated, but 80% of the activity remained. Fully maleylated α-lactalbumin was active, but the apparent K_m increased 50%. However, when 2,4,6-trinitrobenzensulfonic acid was used to modify the amino groups, there was a rapid loss of activity and virtually none was left when 8 mol of trinitrophenyl per mole of α-lactalbumin was incorporated. The results were interpreted that the large bulky trinitrophenyl group introduced inactivation due to steric hindrance. The incorporation of 1 mol of dansyl chloride per mole of α-lactalbumin does not appreciably affect its activity [43]. Recent work in this laboratory [99] has shown that acetylation of α-lactalbumin with N-acetylimidazole results in acetylation of nine amino groups and four tyrosines, which results in a nearly complete loss of activity. However, de-O-acetylation with hydroxylamine results in a complete regain of activity due to removal of the acyl groups from the tyrosines even though the amino groups remain acetylated.

Modification of the tyrosyls of α-lactalbumin by chemical iodination or nitration with tetranitromethane resulted in loss of activity of α-lactalbumin [100]. This loss of activity closely paralleled the loss of the four tyrosines in α-lactalbumin by either iodination or nitration. Nitration also resulted in a loss of tryptophan though the rate of loss was slower than for tyrosine. As indicated above, treatment of α-lactalbumin with N-acetylimidazole results in loss of activity of α-lactalbumin but the activity can be fully restored by de-O-acetylation of the tyrosyls by hydroxylamine [99] indicating that modification of a tyrosyl(s) results in loss of activity. The tyrosyls on α-lactalbumin are essentially exposed to the above reagents; hence, it has not been possible to determine which tyrosyl may be involved in the activity of α-lactalbumin. Recent studies of iodination of α-lactalbumin by lactoperoxidase shows that all the activity is lost, again indicating tyrosyl involvement [101]. However, in the presence of galactosyltransferase and substrates there is protection of α-lactalbumin to iodination by lactoperoxidase. These results should allow for future experiments to determine which tyrosine is essential for the activity of α-lactalbumin.

The area in α-lactalbumins of known sequences (bovine, human, and guinea pig) which has the longest region of identical sequences is between Gly-100 and Lys-108 [32, 33]. This area includes Trp-104, His-107, Lys-108, and Tyr-103 (Table 1) and may be the region important in interaction with the galactosyltransferase. The partial amino acid sequence of kangaroo α-lactalbumin has a tyrosine in the 3-position whereas it is in the 18-position in the bovine, human, and guinea pig. The tyrosine at 36 in the bovine, human, and guinea pig is absent in the kangaroo. Thus, it would appear that Tyr-50 and Tyr-103 are the prime candidates for the tyrosine to be involved in the activity of α-lactalbumin. The most likely candidate is Tyr-103 since it is in the region of high homology. It is also apparent from the results of the chemical modification experiments that several amino acids are essential for α-lactalbumin activity. It is most likely that they constitute a binding domain which is located on the top surface region of the proposed cleft in α-lactalbumin as based on the lysozyme model.

VII. BIOLOGICAL SIGNIFICANCE

In most tissues, the principal physiological role of the galactosyltransferase involved in lactose biosynthesis is to transfer galactose to a terminal N-acetylglucosamine residue of the carbohydrate side-chain of a glycoprotein to form a product such as β-Gal-(1 $\rightarrow$ 4)-β-GlcNAc-glycoprotein.

The enzyme is widely distributed [6, 16, 59, 102], is a particulate enzyme in most tissues, and has been used as an enzymic marker of the

Golgi apparatus [68-70]. The enzyme readily transfers galactose to proteins such as ovalbumin and proteins which have been treated with neuraminidase and β-galactosidase to remove terminal sialic acid and galactose,
e.g., fetuin. The β-Gal-(1→4)-β-GlcNAc-linkage is the most prominent
linkage found in glycoproteins. The enzyme will transfer galactose to N-
acetylglucosamine, but this reaction is of little biological importance since
there is little if any free N-acetylglucosamine found in biological systems.
The enzyme is found in a soluble form in milk, porcine serum [103], rat
serum [104, 105], and bovine serum [105] and presumably becomes
solubilized by enzymatic degradation of tissues.

The situation in the mammary gland is somewhat unique in that both
lactose and glycoproteins are synthesized at the same time. The major
proteins synthesized by the mammary gland are the caseins, and some of
these are glycoproteins. The results of the kinetic analysis and substrate
specificity studies indicate that α-lactalbumin does not appreciably inhibit
the transfer of galactose to glycoproteins even though it inhibits markedly
the transfer of galactose to N-acetylglucosamine. Thus, the enzyme can
carry out both reactions at the same time in the presence of α-lactalbumin
and allow for meaningful rates of synthesis of both lactose and glycoproteins.

Proposals have been made that lactose and milk proteins are exported
into milk by the way of the Golgi system [106]. It would also be predicted
from the kinetic studies that the amount of lactose synthesized and found in
milk should be proportional to the amount of α-lactalbumin found in milk
since the rate of synthesis of lactose depends on the concentration of
α-lactalbumin. Indeed, there is a good correlation between the lactose and
α-lactalbumin contents of milk [107]. An extreme case occurs with the
fur seal in which there is only a very low concentration of lactose and a
corresponding low level of α-lactalbumin [108]. There is also additional
evidence to show that α-lactalbumin and lactose are cosecreted in that both
are detected in the serum of lactating rats and bovines [105].

A schematic model relating lactose and glycoprotein biosynthesis [105]
is presented in Figure 5 and is an extension of an earlier model proposed
by Kuhn [18] based on the ideas expressed by Brew [17, 109]. The key
to the figure is as follows: GAL-T, is the galactosyltransferase bound in
the Golgi membrane; Mn, manganese; Lac, lactose ●——■ ; α-LA, α-
lactalbumin; ● , galactose; GP is a glycoprotein with a free N-acetylglucosaminyl residue linked β to another sugar in the carbohydrate side chain;
●——GP, galactosyl glycoprotein. In essence, the model proposes that
there is vectorial synthesis of lactose as well as the established pathway
for the incorporation of carbohydrates into the carbohydrate side-chain of
glycoproteins [110]. The galactosyltransferase is located in the Golgi
membranes in association with Mn^{2+}, and the substrates are assumed to

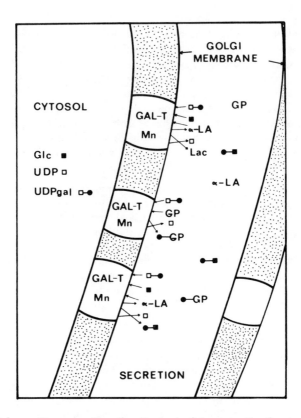

FIG. 5. Scheme Representing the Control of Lactose Synthesis and Galactosyl Transfer to Glycoproteins

be available to the enzyme. During lactation, due to the influence of the
appropriate hormones, α-lactalbumin is synthesized on the ribosomes and
subsequently passes into the Golgi system where it can come in direct contact with the galactosyltransferase and substrates to form lactose. α-Lactalbumin is released from the galactosyltransferase at each turn of the
catalytic cycle and hence can interact in a vectorial sense with more than
one molecule of galactosyltransferase as it passes through the Golgi. Both
lactose and α-lactalbumin are excreted into milk, and the amount of lactose
formed is dependent on the relative concentrations of substrates and α-
lactalbumin if it is assumed that the galactosyltransferase is not rate limiting. The model predicts that the concentration of lactose found in milk
would be directly related to the concentration of α-lactalbumin. This appears to be the case [107]. The galactosyltransferase can also transfer
galactose to the appropriate carbohydrate side-chain of glycoproteins (GP).

The rate of transfer of galactose to glycoproteins or glucose is dependent on the relative concentration of α-lactalbumin, glycoprotein, UDP-galactose, and glucose. The model can also provide an explanation for the formation of glyco-α-lactalbumin [111, 112] which makes up about 7% of the total α-lactalbumin found in bovine milk. Glyco-α-lactalbumin contains galactose; if the linkage of galactose were to an N-acetylglucosamine residue, this galactosyltransferase could effect the transfer of galactose. If galactose were linked to another residue, a different galactosyltransferase would be involved.

The control of lactose biosynthesis appears to be directly related to the concentration and rate of flow of α-lactalbumin through the Golgi system. This type of control appears to be effective in regulation of lactose synthesis. In a sense it acts as a switching mechanism which is directly under hormonal control. The mammary gland in essence has utilized an existing enzyme, a galactosyltransferase, for the biosynthesis of lactose, and its rate of synthesis is regulated by α-lactalbumin. It is not clear at present why the mammary gland utilizes this unique biochemical control mechanism, but this is undoubtedly related to the overall hormonal control of the mammary gland and may represent a control system whereby a cytoplasmic component can regulate membrane function.

Glycosyltransferases have been postulated to function in intercellular adhesion in the sense that all opposing surfaces contain complementary molecules which in this case would be glycosyltransferases and the complex carbohydrates found on cell surfaces [113]. The idea is that a glycosyltransferase-acceptor may play the role in the adhesion process, and in essence these would represent dead-end complexes (Fig. 4). There are some reports [114, 115] that would support the above hypothesis which indicate that glycosyltransferases may be located on the cell surface of certain cell types.

ACKNOWLEDGEMENTS

These investigations were supported in part by grants AM 10764 from N. I. H. and GB 24291 from N. S. F. Kurt E. Ebner is the recipient of a Career Development Award, GM 42396.

ADDENDUM

Khatra et al. [116] have examined in detail the steady-state kinetics of human galactosyltransferase isolated from colostrum and have concluded that the reaction mechanism differs from the enzyme isolated from bovine milk. Powell and Brew [117] have isolated the galactosyltransferase from bovine colostrum and obtained a protein with a molecular weight of 51,000. This form could be converted by trypsin to a protein of 41,000 molecular weight. The kinetic properties of these two forms appeared to be somewhat different which coupled with assaying under different conditions may offer an explanation for observed differences in the kinetic mechanisms.

Geren and Ebner [118] have shown by circular dichroism studies that galactosyltransferase from bovine skim milk has a major positive change in the mean residue ellipticity in the 2650 and 2750-2900 Å region upon the addition of UDP-galactose to galactosyltransferase-Mn^{2+}, suggesting that a conformational change has occurred under these conditions.

Osborne and Steiner [119] have recently studied the interaction of α-lactalbumin with galactosyltransferase isolated from bovine skim milk and concluded that the modifier action of α-lactalbumin is pH dependent and that dead-end complex formation is enhanced by substrates.

Earlier studies by Kirschbaum and Bosmann [120-122] reported that folic acid and lysolecithin enhanced the activities of a number of glycosyltransferases, including galactosyltransferase. However, recent studies by Geren and Ebner [123] have shown that folic acid does not activate purified galactosyltransferase but, in crude systems, folic acid inhibits a nucleotide pyrophosphatase which rapidly cleaves the substrate, UDP-galactose. 5'AMP is an effective inhibitor of this enzyme and is useful in assay mixtures when assaying galactosyltransferase in crude systems to prevent destruction of substrate.

Aronson et al. [124] reported that a plasma membrane component responsible for the binding of asialoglycoproteins was UDP-galactose-glycoprotein galactosyltransferase. Hudgin and Ashwell [125] have isolated a hepatic binding protein with high specificity for ^{125}I-asialo-orosomucoid. This binding protein was devoid of glycosyltransferase activity, including galactosyltransferase. α-Lactalbumin inhibits the binding of orosomucoid to the purified binding protein, and it would appear that glycosyltransferases are not involved in binding asialoglycoproteins.

REFERENCES

1. R. Jenness, in Milk Proteins, Vol. 1 (H. A. McKenzie, ed.), Academic Press, New York, 1970, pp. 17-40.
2. W. G. Gordon and W. F. Semmett, J. Am. Chem. Soc., 75, 328 (1953).
3. W. G. Gordon, in Milk Proteins, Vol. 2 (H. A. McKenzie, ed.), Academic Press, New York, 1971, pp. 332-361.
4. U. Brodbeck and K. E. Ebner, J. Biol. Chem., 241, 762 (1966).

5. K. E. Ebner, W. L. Denton, and U. Brodbeck, Biochem. Biophys. Res. Commun., 24, 232 (1966).
6. U. Brodbeck, W. L. Denton, N. Tanahashi, and K. E. Ebner, J. Biol. Chem., 242, 1391 (1967).
7. W. M. Watkins and W. Z. Hassid, J. Biol. Chem., 237, 1432 (1962).
8. H. Babad and W. Z. Hassid, J. Biol. Chem., 239, PC946 (1964).
9. H. Babad and W. Z. Hassid, J. Biol. Chem., 241, 2672 (1966).
10. K. Brew, T. C. Vanaman, and R. L. Hill, Proc. Natl. Acad. Sci., U. S., 59, 491 (1968).
11. K. E. Ebner and U. Brodbeck, J. Dairy Sci., 51, 317 (1968).
12. K. E. Ebner, Accounts of Chem. Res., 3, 41 (1970).
13. K. E. Ebner, J. Dairy Sci., 54, 1229 (1971).
14. K. E. Ebner, R. Mawal, and J. F. Morrison, in Biochemistry of the Glycosidic Linkage (R. Piras and H. G. Pontis, eds.), P. A. A. B. S. Symp. 2, Academic Press, New York, 1972, pp. 267-279.
15. K. E. Ebner, in The Enzymes, Vol. 9 (P. D. Boyer, ed.), Academic Press, New York, 1973, p. 363.
16. R. L. Hill, K. Brew, T. C. Vanaman, I. P. Trayer, and P. Mattock, Brookhaven Symp. in Biol., 21, 139 (1968).
17. K. Brew, in Essays in Biochemistry (P. N. Campbell and F. Dickens, eds.), Vol. 6, London, 1970, p. 93.
18. N. J. Kuhn, in Lactation (I. R. Falconer, ed.), Butterworths, London, 1971, p. 161.
19. J. E. Gander, W. E. Petersen, and P. D. Boyer, Arch. Biochem. and Biophys., 60, 259 (1956).
20. J. E. Gander, W. E. Petersen, and P. D. Boyer, Arch. Biochem. and Biophys., 69, 85 (1957).
21. U. Brodbeck and K. E. Ebner, J. Biol. Chem., 241, 5526 (1966).
22. N. Tanahashi, U. Brodbeck, and K. E. Ebner, Biochim. Biophys. Acta, 154, 247 (1968).
23. E. J. McGuire, G. W. Jourdian, D. M. Carlson, and S. Roseman, J. Biol. Chem., 240, PC4112 (1965).
24. K. T. Yasunobu and P. E. Wilcox, J. Biol. Chem., 231, 309 (1958).
25. K. Brew and P. N. Campbell, Biochem. J., 102, 258 (1957).
26. K. Brew, T. C. Vanaman, and R. L. Hill, J. Biol. Chem., 242, 3747 (1967).
27. R. L. Hill, K. Brew, T. C. Vanaman, I. P. Trayer, and P. Mattock, Brookhaven Symp. in Biol., 1, 139 (1968).
28. K. Brew, F. J. Castellino, T. C. Vanaman, and R. L. Hill, J. Biol. Chem., 245, 4570 (1970).
29. K. Brew and R. L. Hill, J. Biol. Chem., 245, 4559 (1970).
30. T. C. Vanaman, K. Brew, and R. L. Hill, J. Biol. Chem., 245, 4583 (1970).
31. W. J. Browne, A. C. T. North, D. C. Phillips, K. Brew, T. C. Vanaman, and R. L. Hill, J. Mol. Biol., 42, 65 (1969).

32. J. B. C. Findlay and K. Brew, Eur. J. Biochem., 27, 65 (1972).
33. K. Brew, Eur. J. Biochem., 27, 341 (1972).
34. K. Brew, H. M. Steinman, and R. L. Hill, J. Biol. Chem., 248, 4739 (1973).
35. P. N. Lewis and H. A. Scheraga, Arch. Biochem. Biophys., 144, 584 (1971).
36. R. Aschaffenburg, R. E. Fenna, B. O. Handford, and D. C. Phillips, J. Mol. Biol., 67, 525 (1972).
37. R. Aschaffenburg, R. E. Fenna, and D. C. Phillips, J. Mol. Biol., 67, 529 (1972).
38. W. R. Krigbaum and F. R. Kugler, Biochemistry, 9, 1216 (1970).
39. E. K. Achter and I. D. A. Swan, Biochemistry, 10, 2976 (1971).
40. J. H. Bradbury and N. L. R. King, Austr. J. of Chem., 24, 1703 (1971).
41. D. A. Cowburn, K. Brew, and W. B. Gratzer, Biochemistry, 11, 1228 (1972).
42. P. B. Sommers, M. J. Kronman, and K. Brew, Biochem. Biophys. Res. Commun., 52, 98 (1973).
43. A. B. Rawitch, Arch. Biochem. Biophys., 151, 22 (1972).
44. A. O. Barel, M. Turneer, and M. Dolmans, Eur. J. Biochem., 30, 26 (1972).
45. R. A. Bradshaw and D. A. Dernanleau, Biochemistry, 9, 3310 (1970).
46. F. M. Robbins and L. G. Holmes, J. Biol. Chem., 247, 3062 (1972).
47. J. A. Magnuson and N. S. Magnuson, Biochem. Biophys. Res. Commun., 45, 1513 (1971).
48. M. Z. Atassi, A. F. S. A. Habeeb, and L. Rystedt, Biochim. Biophys. Acta, 200, 184 (1970).
49. K. S. Iyer and W. A. Klee, J. Biol. Chem., 248, 707 (1973).
50. A. D. Strosberg, C. Nihoul-Deconinck, and L. Kanarek, Nature, 227, 1241 (1970).
51. A. Faure and P. Jolles, FEBS Letters, 10, 237 (1970).
52. R. Arnon and E. Maron, J. Mol. Biol., 51, 703 (1970).
53. R. Arnon and E. Maron, J. Mol. Biol., 61, 225 (1971).
54. E. M. Prager and A. C. Wilson, J. Biol. Chem., 246, 5976 (1971).
55. H. Neurath, K. A. Walsh, and W. P. Winter, Science, 158, 1638 (1967).
56. W. A. Frazier, R. H. Angeletti, and R. A. Bradshaw, Science, 176, 482 (1972).
57. F. J. Castellino and R. L. Hill, J. Biol. Chem., 245, 417 (1970).
58. T. Y. Lin, Biochemistry, 9, 984 (1970).
59. D. K. Fitzgerald, L. M. McKenzie, and K. E. Ebner, Biochim. Biophys. Acta, 235, 425 (1971).
60. D. K. Fitzgerald, U. Brodbeck, I. Kiyosawa, R. Mawal, B. Colvin, and K. E. Ebner, J. Biol. Chem., 245, 2103 (1970).

61. I. P. Trayer and R. L. Hill, J. Biol. Chem., 246, 6666 (1971).
62. W. A. Klee and C. B. Klee, J. Biol. Chem., 247, 2336 (1972).
63. R. Barker, K. W. Olsen, J. H. Shaper, and R. L. Hill, J. Biol. Chem., 247, 7135 (1972).
64. P. Andrews, FEBS Letters, 9, 297 (1970).
65. T. Nagasawa, K. Kiyosawa, and N. Tanahashi, J. Dairy Sci., 54, 835 (1971).
66. R. G. Coffey and F. J. Reithel, Biochem. J., 109, 169 (1968).
67. R. G. Coffey and F. J. Reithel, Biochem. J., 109, 177 (1968).
68. T. W. Keenan, D. J. Morre, and R. D. Cheetham, Nature, 228, 1105 (1970).
69. B. Fleischer, S. Fleischer, and H. Ozawa, J. Cell Biol., 43, 59 (1969).
70. D. E. Leelavathi, L. W. Estes, D. S. Feingold, and B. Lombardi, Biochim. Biophys. Acta, 211, 124 (1970).
71. S. C. Magee, R. Mawal, and K. E. Ebner, Fed. Proc., 31, 499 (1972).
72. E. D. Lehman, B. G. Hudson, and K. E. Ebner, submitted.
73. S. C. Magee and K. E. Ebner, Fed. Proc., 32, 541 (1973).
74. S. C. Magee, R. Mawal, and K. E. Ebner, J. Biol. Chem., 249, 6992 (1974).
75. S. C. Magee, R. Mawal, and K. E. Ebner, Biochemistry, 13, 99 (1974).
76. J. R. Dulley, J. Dairy Research, 39, 1 (1972).
77. S. Kaminogawa, H. Mizobuchi, and K. Yamauchi, Agr. Biol. Chem., 12, 2163 (1972).
78. R. D. Palmiter, Biochim. Biophys. Acta., 178, 35 (1969).
79. F. L. Schanbacher and K. E. Ebner, J. Biol. Chem., 245, 5057 (1970).
80. M. J. Spiro and R. G. Spiro, J. Biol. Chem., 243, 6529 (1968).
81. J. F. Morrison and K. E. Ebner, J. Biol. Chem., 246, 3977 (1971).
82. B. J. Kitchen and P. Andrews, Biochem. J., 130, 80P (1972).
83. W. A. Klee and C. B. Klee, Biochem. Biophys. Res. Commun., 39, 833 (1970).
84. P. Andrews, Biochem. J., 111, 14P (1969).
85. F. L. Schanbacher and K. E. Ebner, Biochim. Biophys. Acta, 229, 226 (1971).
86. J. F. Morrison and K. E. Ebner, J. Biol. Chem., 246, 3985 (1971).
87. J. F. Morrison and K. E. Ebner, J. Biol. Chem., 246, 3992 (1971).
88. B. J. Kitchen and P. Andrews, FEBS Letters, 26, 333 (1972).
89. R. Mawal, J. F. Morrison, and K. E. Ebner, J. Biol. Chem., 246, 7106 (1971).
90. T. Helting, Biochim. Biophys. Acta, 227, 42 (1971).
91. T. Helting and L. Roden, J. Biol. Chem., 244, 2790 (1969).
92. V. Ginsburg, Adv. in Enzymol., 36, 131 (1972).

93. J. R. Baker, L. Roden, and A. C. Stoolmiller, J. Biol. Chem., 247, 3838 (1972).

94. L. Roden, H. B. Schwarz, and A. Dorfman, Abst. 66 Biol. Div., 166th. Amer. Chem. Soc. Meeting, Chicago, 1973.

95. R. J. Ivatt and M. A. Rosemeyer, FEBS Letters, 28, 195 (1972).

96. T. E. Barman, Biochim. Biophys. Acta, 258, 297 (1972).

97. T. E. Barman and W. Bagshaw, Biochim. Biophys. Acta, 278, 491 (1972).

98. B. J. Kitchen and T. E. Barman, Biochim. Biophys. Acta, 298, 861 (1973).

99. C. R. Merriman and K. E. Ebner, in preparation.

100. W. L. Denton and K. E. Ebner, J. Biol. Chem., 243, 4053 (1971).

101. D. K. Fitzgerald, L. Walker, and K. E. Ebner, unpublished data.

102. T. Helting, J. Biol. Chem., 246, 815 (1971).

103. R. L. Hudgin and H. Schachter, Can. J. Biochem., 99, 838 (1971).

104. R. R. Wagner and M. A. Cynkin, Biochem. Biophys. Res. Commun., 45, 57 (1971).

105. K. E. Ebner and L. M. McKenzie, Biochem. Biophys. Res. Commun., 49, 1624 (1972).

106. J. L. Linzell and M. Peaker, Physiol. Revs., 51, 564 (1971).

107. J. M. Ley and R. Jenness, Arch. Biochem. Biophys., 138, 464 (1970).

108. D. V. Schmidt and K. E. Ebner, Biochim. Biophys. Acta, 243, 273 (1971).

109. K. Brew, Nature, 222, 671 (1969).

110. H. Schachter, I. Jabbal, R. L. Hudgin, L. Pinteric, E. J. McGuire, and S. Roseman, J. Biol. Chem., 245, 1090 (1970).

111. T. E. Barman, Biochim. Biophys. Acta, 214, 242 (1970).

112. E. J. Hindle and J. V. Wheelock, Biochem. J., 119, 14P (1970).

113. S. Roseman, Chem. Phys. Lipids, 5, 270 (1970).

114. S. Roth, E. J. McGuire, and S. Roseman, J. Cell. Biol., 51, 536 (1971).

115. H. B. Bosmann, Biochem. Biophys. Res. Commun., 48, 523 (1972).

116. B. S. Khatra, D. G. Herries, and K. Brew, Eur. J. Biochem., 44, 537 (1974).

117. J. T. Powell and K. Brew, Eur. J. Biochem., 48, 217 (1974).

118. C. R. Geren and K. E. Ebner, Biochemistry, in press.

119. J. C. Osborne and R. F. Steiner, Arch. Biochem. and Biophys., 165, 615 (1974).

120. B. B. Kirschbaum and H. B. Bosmann, Biochem. Biophys. Res. Commun., 50, 510 (1973).

121. B. B. Kirschbaum and H. B. Bosmann, Biochim et Biophys Acta, 320, 416 (1973).

122. B. B. Kirschbaum and H. B. Bosmann, FEBS Letters, 34, 129 (1973).

123. L. M. Geren and K. E. Ebner, <u>Biochem. Biophys. Res. Commun.</u>, <u>59</u>, 14 (1974).

124. N. N. Aronson, L. Y. Tan, and B. P. Peters, <u>Biochem. Biophys. Res. Commun.</u>, <u>53</u>, 112 (1973).

125. R. L. Hudgin and G. Ashwell, <u>J. Biol. Chem.</u>, <u>249</u>, 7369 (1974).

Chapter 5

ACETYL COENZYME A CARBOXYLASE

M. Daniel Lane
S. Efthimios Polakis

Joel Moss*

Department of Physiological Chemistry
The Johns Hopkins University
School of Medicine
Baltimore, Maryland

Department of Medicine
The Johns Hopkins University
School of Medicine
Baltimore, Maryland

I. INTRODUCTION — 182

II. THE ACETYL CoA CARBOXYLATION REACTION — 183
   A. The Biotinyl Prosthetic Group — 184
   B. Partial Reactions — 187

III. THE ESCHERICHIA COLI ACETYL CoA CARBOXYLASE
SYSTEM — 189

IV. ANIMAL ACETYL CoA CARBOXYLASES — 193
   A. Polymeric Filamentous Form — 193
   B. Paracrystalline Form — 199
   C. Protomeric Form and Subunit Structure — 201
   D. Dependence of Catalytic Activity on the Protomer ⇌
   Polymer Equilibrium — 204
   E. Citrate Activation Mechanism — 212

V. YEAST AND PLANT ACETYL CoA CARBOXYLASES — 214

*Presently affiliated with the Laboratory of Cellular Metabolism,
Heart and Lung Institute, National Institutes of Health, Bethesda, Maryland.

I. INTRODUCTION

Structural considerations at several levels (i.e., at the active site, subunit, protomer, and oligomer level) are of particular importance in understanding catalysis and regulation of acetyl CoA carboxylase. The role of
biotin in the carboxylation reaction is firmly established [1, 2]. In its
capacity as mobile CO_2 carrier, the covalently bound biotinyl prosthetic
group of the carboxylase functionally links active sites located on different
subunits, each of which carries out part of the complex carboxylation
process. Moreover, it is through effects on the interplay between the subunits of acetyl CoA carboxylase, mediated by the prosthetic group, that
regulation is exerted [2-6]. In the case of the animal carboxylases, allosteric activation and inhibition are associated with dramatic changes in the
polymeric state of the enzyme [2, 3].

Acetyl CoA carboxylase was first recognized as one of two enzyme
fractions, prepared from liver extracts, that were required for fatty acid
synthesis from acetyl CoA [7-22]. The observation that fatty acid synthesis was bicarbonate-dependent [12-14, 23] led to the discovery [11, 16,
24] that one fraction catalyzed the ATP-dependent carboxylation of acetyl
CoA [reaction (1)].

$$\text{acetyl CoA} + HCO_3^- + ATP \underset{\substack{\text{acetyl CoA} \\ \text{carboxylase}}}{\overset{Me^{2+}}{\rightleftharpoons}} \text{malonyl CoA} + ADP + P_i \qquad (1)$$

$$1 \text{ acetyl CoA} + 7 \text{ malonyl CoA} + 14 \text{ NADPH} + 14 \text{ H}^+ + H_2O \xrightarrow{\quad\quad}$$

$$\underset{\text{synthetase}}{\overset{\text{fatty acid}}{}}$$

$$\text{palmitic acid} + 7 \text{ CO}_2 + 8 \text{ CoA} + 14 \text{ NADP}^+ \qquad (2)$$

Subsequently, it was found that the second enzyme fraction, which contained
the fatty acid synthetase multienzyme complex, utilizes malonyl CoA generated by the carboxylase for condensation with the aliphatic acyl CoA being reductively elongated [reaction (2)] [25-27].

One unique property of the acetyl CoA carboxylases from animal tissues
is their requirement for a tricarboxylic acid activator, notably citrate or
isocitrate. This activating effect was noted first in 1952 by Brady and
Gurin [28], who observed that the synthesis of fatty acids from acetate
catalyzed by pigeon liver extracts was markedly stimulated by tricarboxylic
acids. Further investigations revealed that citrate or isocitrate activated
the fatty acid synthesizing systems from avian liver, rat liver, adipose
tissue, and mammary gland as well [4, 8, 9, 29-39]. When the components (acetyl CoA carboxylase and fatty acid synthetase) of the system had

been resolved and the chemical steps elucidated [11, 12, 14, 16, 19-22, 25], the acetyl CoA carboxylase-catalyzed reaction was shown to be the locus of citrate activation [39-47]. Vagelos et al. [48, 49] then made the interesting finding that citrate activation of the carboxylase was accompanied by an increase in its sedimentation velocity on sucrose density gradients. This increased sedimentation velocity was found in Lane's laboratory [50-52] to result from the polymerization of inactive protomer to yield catalytically active filamentous structures. The polymerization phenomenon has subsequently been shown to be characteristic of acetyl CoA carboxylases isolated from many animal species and diverse organ systems [50, 52-59]. Compelling evidence has now accumulated which indicates that the carboxylase-catalyzed step controls fatty acid synthesis in animal cells through an allosteric regulatory mechanism (citrate activation and feedback inhibition by long-chain fatty acyl CoA derivatives) [2].

Unlike its counterpart in animal systems, E. coli acetyl CoA carboxylase is not subject to regulation by citrate [60, 61]. However, recent evidence [62, 63] indicates that fatty acid synthesis in E. coli is under stringent control through the gene product of the rel (RC) locus. Polakis et al. [62] have shown that the mediator of stringent control (p)ppGpp* specifically blocks acetyl CoA carboxylase, hence the rate of lipid synthesis. It appears, therefore, that lipid synthesis is coordinated with the rate of protein synthesis, hence growth, through the action of (p)ppGpp.

II. THE ACETYL CoA CARBOXYLATION REACTION

The overall reactions catalyzed by the biotin-dependent carboxylases can be partitioned into two discrete steps (Fig. 1) [1]. The initial event — common to all biotin-dependent carboxylations — is the formation of a carboxybiotinyl enzyme intermediate by nucleophilic attack of the ureido ring of biotin at an electrophilic center of the carboxyl donor, bicarbonate (step 1, Fig. 1). In the second step, carboxyl transfer from enzyme-biotin-CO_2^- to an appropriate acceptor substrate occurs, the nature of this acceptor being dependent on the specific enzyme involved. With acyl CoA, α-keto acid or amido carboxylases transfer to the position adjacent to the carbonyl group occurs (steps 2, 3, or 4, respectively, Fig. 1), giving rise to a malonyl CoA derivative, a β-keto acid, or N-carboxyurea, respectively [1]. Exceptions to this are β-methylcrotonyl CoA and geranyl CoA carboxylases in which the electrophilic site, hence the site of carboxylation, is displaced to the γ-position by conjugation [1].

---

*(p)ppGpp refers collectively to ppGpp (guanosine 3'-diphosphate, 5'-diphosphate) and pppGpp (guanosine 3'-diphosphate, 5'-triphosphate).

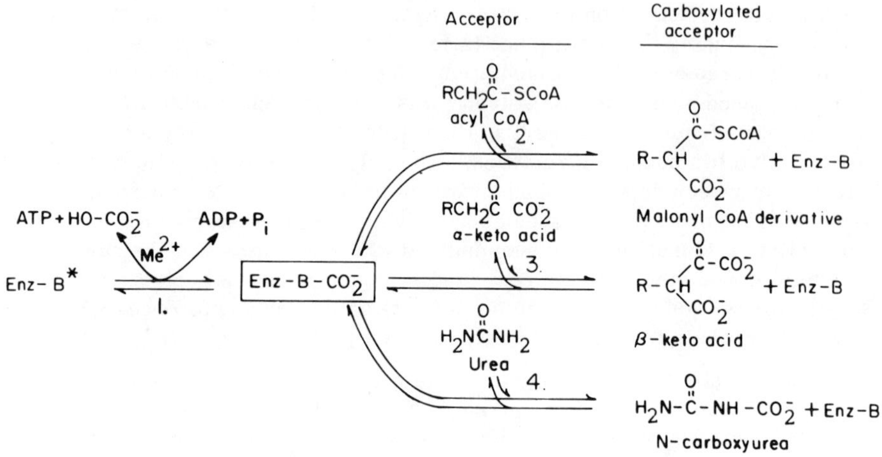

*Enz–B = enzyme-biotin

FIG. 1.   Central role of enzyme-biotin-CO_2^- (Enz–B–CO_2^-) in reactions catalyzed by the biotin-dependent carboxylases. (From Moss and Lane [1].)

A.   The Biotinyl Prosthetic Group

Intensive efforts in several laboratories [1] led to the elucidation of the correct structure of biotin by du Vigneaud and coworkers [64, 65], which was subsequently confirmed by chemical synthesis [66-70] and x-ray crystallography [71-73]. Only the (+)-biotin isomer (Fig. 2) is enzymatically active [1, 2], the active isomer having the ureido and tetrahydrothiophene rings fused cis and the aliphatic side chain possessing a cis configuration with respect to the ureido ring [71]. The absolute stereochemistry of (+)-biotin was unequivocally established by x-ray crystallographic analysis of a derivative of the active isomer (bis-p-bromoanilide of 1'-N-carboxybiotin) [72, 73].

X-ray crystallographic analysis of (+)-biotin [71] and of the same carboxybiotin derivative described previously [72] revealed that the bicyclic ring system has a boat-like configuration (Fig. 3). The planar ureido ring projects upward at an angle of 62.0° with respect to a plane (plane I) formed by the four carbon atoms of the tetrahydrothiophene ring.

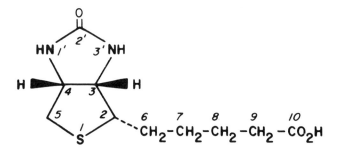

FIG. 2. Absolute configuration of (+)-biotin (2'-keto-3,4-imidazolido-2-tetrahydrothiophene-n-valeric acid).

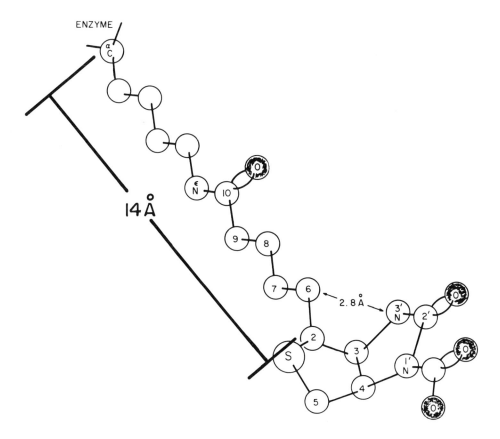

FIG. 3. Diagrammatic representation of ε-N-(1'-N-carboxy-(+)-biotinyl)-L-lysyl enzyme (correct absolute configuration). (From Moss and Lane [1].)

Another plane comprising S-1, C-2, and C-5 tilts upward at an angle of
37.6° with respect to plane I [72]. Because of the cis orientation of the
ureido ring with respect to the aliphatic side chain, C_6 resides approximately 2.8 Å from the 3'-N- of the ureido ring [74]. Repulsion between
the 3'-N and C-6 positions is thought to cause the greater-than-anticipated
C_3-C_2-C_6 angle of 119° [74].

The terminal step in the synthesis of biotin enzymes is the covalent attachment of the prosthetic group to the apoenzyme to form an active holoenzyme [1]. The biotinyl moiety is attached via an amide linkage through
its side-chain carboxyl group to a lysyl ε-amino group of the apoenzyme
(Fig. 3) [75-85]. Measurements of ε-N-(+)-biotinyl-L-lysine reveal that
the maximal distance from the α-carbon of lysine to C-2 of the tetrahydrothiophene ring is 14 Å [51]. Thus, although covalent attachment firmly
anchors the prosthetic group to the apoprotein, the long side-chain gives
the bicyclic ring sufficient mobility to permit translocation between remote
catalytic sites (see Sec. II, B and III).

It was first noted in Lynen's laboratory [86] that biotin per se was the
enzymatic site of carboxylation since bacterial β-methylcrotonyl CoA
carboxylase could catalyze the carboxylation of free d-biotin in the absence
of its natural CoA-thioester substrate. Presumably, free d-biotin can
gain access to the carboxylation site of biotin enzymes because the prosthetic group is attached to the apoenzyme through its long (14 Å) side-chain
and can oscillate in and out of the carboxylation site. The carboxylated
free biotin product was found to be relatively unstable, particularly at acid
pH, and therefore was converted to its dimethyl ester with diazomethane
to stabilize it during isolation and characterization [1]. This derivative was
found to be identical with the isomer of the dimethyl ester of N-carboxy-
biotin formed predominantly by chemical synthesis [87, 88], where acyla-
tion of free biotin methyl ester with methyl chloroformate yields a mixture
of 93.5% 1'-N- and 6.5% 3'-N-methoxycarbonyl d-biotin methyl ester.
The 1'-N rather than the 3'-N position was favored as the site of carboxylation, since x-ray crystallographic studies of Traub [71, 74] indicated steric
hindrance of the 3'-N position by C-6 of the aliphatic side-chain (Fig. 3).
Subsequently x-ray crystallography confirmed the fact that the methoxy-
carbonyl group of the methylated carboxylation product was located at the
1'-N position [72]. Similarly, the labile carboxyl group of enzyme-
biotin-^{14}CO$_2^-$, was stabilized by methylation with diazomethane and the
methylated protein subjected to digestion with proteolytic enzymes. The
principal radioactive product in these proteolytic digests was identified as
ε-N- (1'-N-methoxycarbonyl-d-biotinyl)-L-lysine [1]. Experiments of
this type with acetyl CoA carboxylase [83, 84, 89] and several other biotin
enzymes [76, 77, 80, 81] led to the conclusion that the site of carboxylation
is the 1'-N-ureido N of biotin (Fig. 3). More recently, Bruice and Hegarty
[90] correctly pointed out that this structural assignment was equivocal;

had carboxylation occurred on the ureido O atom, as in Scheme (1), methylation would have led to a product which, on chemical grounds, would be expected to rearrange to the more thermodynamically stable 1'-N substituted biotin derivative.

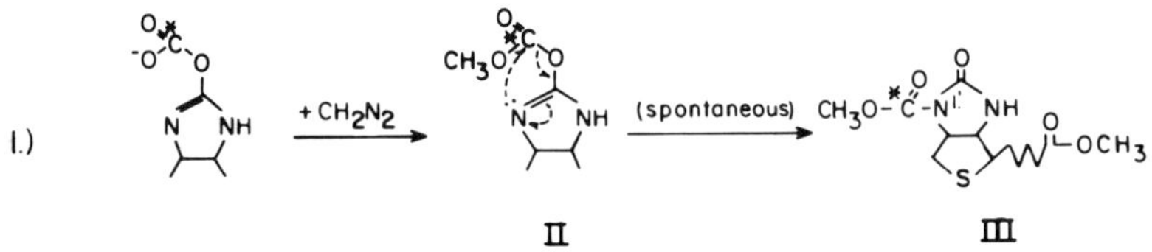

I.)

II   III

SCHEME 1

Recent investigations in Lane's laboratory [91, 92] show definitively that the 1'-N position of biotin is the carboxylation site. "Carboxy-d-biotins" generated enzymatically by the biotin carboxylase or carboxyl transferase components of E. coli acetyl CoA carboxylase system have stability properties indistinguishable from that of chemically synthesized 1'-N-carboxy-d-biotin; first-order decarboxylation rates over the pH range 5 to 9 are identical for the enzymatic products and the authentic 1'-N carboxybiotin derivative. Moreover, chemically synthesized 1'-N-carboxy-d-biotinol is an efficient carboxyl donor substrate for transcarboxylation to acetyl CoA catalyzed by the carboxyl transferase component of E. coli acetyl CoA carboxylase [Scheme (2)] [91, 92].

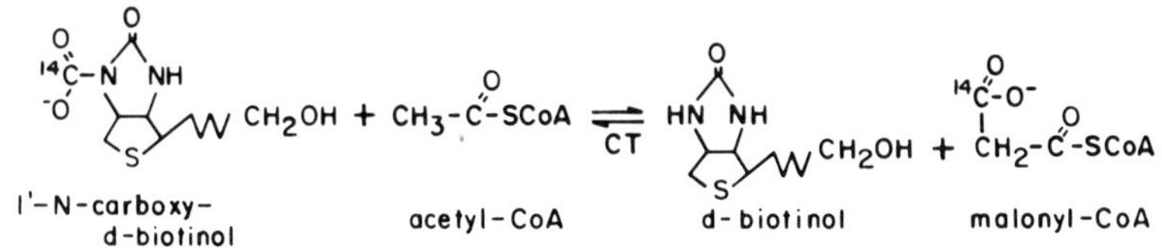

1'-N-carboxy-   acetyl-CoA   d-biotinol   malonyl-CoA
d-biotinol

SCHEME 2

These results and the fact that authentic 1'-N-carboxy-d-biotin also acts as substrate for biotin carboxylase-catalyzed net formation of ATP from ADP and P_i [92] indicate that the 1'-N position of biotin is the enzymatic site of carboxylation.

B. Partial Reactions

The carboxylation of acetyl CoA [reaction (5)] catalyzed by enzymes from a variety of animal, microbial, and plant sources can be subdivided into two fundamental half-reactions [reactions (3) and (4)].

$$\text{end-biotin} + HCO_3^- + ATP \xrightleftharpoons{Me^{2+}} \text{enz-biotin-}CO_2^- + ADP + P_i \quad (3)$$

$$\text{enz-biotin-}CO_2^- + CH_3\text{-}\overset{\overset{\text{O}}{\|}}{C}\text{-SCoA} \rightleftharpoons \text{enz-biotin} + H_2C \overset{\overset{\overset{\text{O}}{\|}}{C}\text{-SCoA}}{\underset{CO_2^-}{\diagdown}} \quad (4)$$

$$\text{net: } HCO_3^- + ATP + \text{acetyl CoA} \xrightleftharpoons{Me^{2+}} \text{malonyl CoA} + ADP + P_i$$

$$(5)$$

Evidence for this minimal reaction sequence has been obtained by isotopic exchange, stoichiometric enzyme carboxylation-transcarboxylation, and model reaction studies both with the intact stable multisubunit carboxylases from animal tissues [47, 51, 52, 83, 93-98] and with the resolved enzymatically active subunit components of the E. coli carboxylase system [91, 92, 99-106]. For example, the participation of the first step [reaction (3)] in the overall carboxylation process is supported by carboxylase-catalyzed P_i-, HCO_3^--, and divalent cation-dependent ATP-[^{14}C] ADP exchange [95, 104]. A reciprocal ATP-$^{32}P_i$ exchange, also catalyzed by the enzyme, requires ADP, HCO_3^-, and divalent cation [47, 94, 98, 104]. The occurrence of the second half-reaction (reaction 4) involving carboxyl transfer from carboxybiotin to acetyl CoA is indicated by the demonstration [47, 94, 96, 97] that acetyl CoA carboxylase catalyzes malonyl CoA-[^{14}C] acetyl CoA exchange which is independent of the components of the first partial reaction [reaction (3)].

Compelling evidence (see Sec. III) indicates that these half-reactions take place at separate catalytic sites and are functionally linked by the biotinyl prosthetic group which acts as a mobile carboxyl carrier, as illustrated in Figure 4. Consistent with this, Numa and co-workers [107-110] have obtained kinetic evidence which indicates that the carboxylation of biotin occurs via an ordered ping-pong mechanism with ATP binding prior to HCO_3^-, and following carboxybiotin formation, P_i release subsequent to ADP. The reaction appears to proceed by a "bi bi uni uni ping-pong" mechanism as shown in reaction (6).

$$
\begin{array}{cccccccc}
\text{ATP} & HCO_3^- & \text{ADP} & P_i & & \text{acetyl} & \text{malonyl} & \\
 & & & & & \text{CoA} & \text{CoA} & \quad (6) \\
\downarrow & \downarrow & \uparrow & \uparrow & & \downarrow & \uparrow & \\
\hline
\text{enz-biotin} & & & & \text{enz-biotin-}CO_2^- & & & \text{enz-biotin}
\end{array}
$$

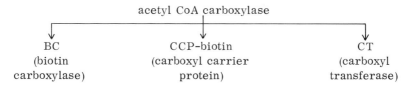

FIG. 4. Hypothetical scheme illustrating the translocation of 1'-N-carboxybiotinyl prosthetic group from the carboxylation site (I) to the transcarboxylation site (II) of acetyl CoA carboxylase.

It should be pointed out that nonclassical ping-pong kinetics have been observed for the reaction catalyzed by the related biotin-dependent enzyme, pyruvate carboxylase [111, 112].

III. THE ESCHERICHIA COLI ACETYL CoA CARBOXYLASE SYSTEM

The E. coli carboxylase system is readily resolved into three essential protein components (below) which retain their characteristic enzymatic activities.

acetyl CoA carboxylase

BC	CCP-biotin	CT
(biotin carboxylase)	(carboxyl carrier protein)	(carboxyl transferase)

In contrast, the animal acetyl CoA carboxylases have stable multisubunit structures and undergo purification without loss of structural integrity (see Sec. IV, A, C; Ref. 94 and Ref. 113).

As revealed by investigations in Vagelos' [105, 106] and Lane's [61, 99, 100] laboratories, the carboxylation of acetyl CoA [reaction (5)] requires the presence of biotin carboxylase, carboxyl carrier protein which contains the covalently bound biotin prosthetic group, and carboxyl transferase. Since each of these resolved proteins can catalyze (BC and CT) or participate in (CCP-biotin) its respective partial reaction, it has been possible to assign function to each subunit. Hence, the E. coli carboxylase is well suited for mechanistic studies since the catalytic centers involved in the overall process can be resolved and their respective partial reactions studied independently using prosthetic group models (i.e., free biotin derivatives as substrates).

All three components of the carboxylase system have been obtained in homogeneous form and their respective molecular properties determined (Table 1) [61, 100, 114, 115]. Biotin carboxylase, the catalytic element responsible for the first half-reaction [reaction (7)], has been obtained in crystalline form [116] as illustrated in Figure 5. Crystals were obtained in 10mM potassium phosphate, pH 7.0, containing 1 mM EDTA and 2 mM dithiothreitol at $4\,^{\circ}C$ as described previously [116].

$$\text{CCP-biotin} + \text{HCO}_3^- + \text{ATP} \underset{\text{BC}}{\overset{\text{Me}^{2+}}{\rightleftharpoons}} \text{CCP-biotin-CO}_2^- + \text{ADP} + P_i \quad (7)$$

$$\text{CCP-biotin-CO}_2^- + \text{acetyl CoA} \underset{\text{CT}}{\rightleftharpoons} \text{CCP-biotin} + \text{malonyl CoA} \quad (8)$$

This component has a molecular weight of about 100,000 ($s_{20, w} = 5.7S$) and is a dimer composed of apparently identical 51,000 dalton subunuts [61]. It has been demonstrated that catalytically active biotin carboxylase is free of biotin [61, 106]. The native form of carboxyl carrier protein which contains the covalently bound biotinyl prosthetic group exists as a 44,000 dalton dimer of identical 22,000 dalton subunits [115]; each subunit contains 1.0 covalently attached biotinyl group per molecule. It has now been established [115] that the previously reported 9,000 dalton carboxyl carrier protein, which has been crystallized [117], was derived from the "native" 22,000 dalton species by proteolysis during isolation. As with other biotin-dependent enzymes [1, 75-77, 79-82, 85], the side-chain carboxyl of the prosthetic group is covalently attached in amide linkage to a lysyl ε-amino group of the apoprotein [78, 83]. The enzyme component responsible for the second half-reaction [reaction (8)], i.e., carboxyl transferase, appears to have an $A_2 B_2$ tetrameric structure

TABLE 1

Properties of the Components of the E. coli
Acetyl CoA Carboxylase System

	Biotin carboxylase [a]	Carboxyl transferase [a]	Carboxyl carrier protein [b]
Molecular weight	98,000	130,000	44,000 (or higher order aggregates)
$s_{20,w}$	5.7	–	5.66^{c}
Subunit(s) weight A	51,000	35,000	22,000
B	–	30,000	–
Subunits/molecule	A_2	A_2B_2	A_2
Biotin content	none	none	1.0 per 22,000 daltons
Catalytic function:	catalyzes the carboxylation of CCP-biotin (reaction 7)	catalyzes carboxyl transfer from CCP-biotin- CO_2^- to acetyl CoA (reaction 8)	attached biotin acts as "mobile carboxyl carrier" between BC and CT
Inhibition by ppGpp[d]	No	Yes	–

[a] See Refs. 61, 99, and 100.

[b] See Refs. 114 and 115.

[c] This sedimentation velocity most likely corresponds to the tetramer [115].

[d] Ref. 62.

composed of subunits having molecular weights of 30,000 and 35,000 [100]. As with biotin carboxylase [61, 106], the carboxyl transferase [99, 100, 105] component has been found not to contain biotin. Regulation of the acetyl CoA carboxylase system is mediated through the action of an allosteric inhibitor, ppGpp, which binds specifically to the carboxyl transferase component [62].

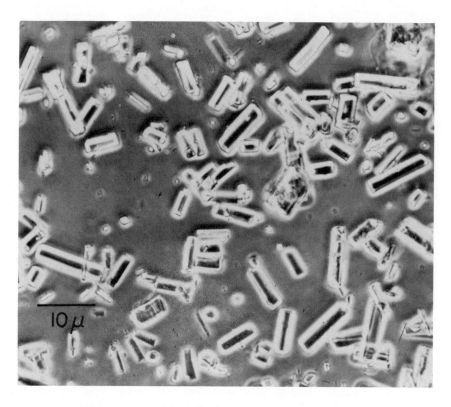

FIG. 5. Crystalline Biotin Carboxylase from <u>E. coli</u>

It has become evident that biotin carboxylase and carboxyl transferase
have distinct functions unique to the first [reaction (3)] and second [reaction
(4)] half-reactions, respectively, while the carboxyl carrier protein
serves a dual role, participating in both processes. Functions for biotin
carboxylase and carboxyl carrier protein in the first half-reaction [reaction (3)] have been indicated by the dependence of the ATP-[^{14}C] ADP and
ATP-^{32}P$_i$ exchanges upon these components but not on the carboxyl transferase [104]. Like all acyl CoA carboxylases, these exchange reactions
also require the presence of ATP, ADP, P$_i$, HCO$_3^-$, and Mg^{2+} [104].
Free biotin is unable to replace the biotin-containing carrier protein for
either exchange. These results plus the fact that the <u>E. coli</u> biotin
carboxylase, which is devoid of covalently bound biotin, also catalyzes the
ATP · Mg^{2+}-dependent carboxylation of free d-biotin [61, 100, 106] shows
that the active site for the first half-reaction [reaction (3)] is housed on
this subunit. Stoichiometric carboxylation of the biotinyl prosthetic group
of <u>E. coli</u> carboxyl carrier protein catalyzed by biotin carboxylase [104,
106] can be demonstrated in the presence of ATP, HCO$_3^-$, and divalent

metal ion. As with other biotin-dependent enzymes [76, 77, 80, 81], the site of carboxylation was found to be the 1'-N position of the biotin (see Fig. 3 and Sect. II, A) [91, 92, 101].

The participation of the carboxyl transferase and carboxyl carrier protein components in the second half-reaction [reaction (4)] is supported by: (1) the dependence of malonyl CoA-[^{14}C]acetyl CoA exchange upon both components, but not on biotin carboxylase [99, 101, 104, 105]; (2) stoichiometric transcarboxylation fron CCP-biotin-CO_2^- to acetyl CoA catalyzed by carboxyl transferase yielding malonyl CoA [104, 114]; and (3) the ability of carboxyl transferase to carry out transcarboxylation from malonyl CoA to free d-biotin derivatives in the absence of carboxyl carrier protein [92, 99, 104]. It should be pointed out that all the above-mentioned reactions are blocked by avidin [104], the specific biotin-binding protein from egg-white; avidin combines nearly irreversibly ($K_{D(biotin)} \cong 10^{-15}$ M) with free d-biotin derivatives including the biotinyl prosthetic group of acetyl CoA carboxylase [1, 118, 119].

As is apparent from the above discussion, neither catalytic component, i.e., biotin carboxylase [61, 106, 116] or carboxyl transferase [99, 105], contains covalently bound biotin; yet both catalyze model reactions with free biotin, which accounts for their respective catalytic roles in the half-reactions [reactions (7) and (8)]. Hence, both the biotin carboxylase and carboxyl transferase components must contain distinct binding sites for the bicyclic ring of the prosthetic group. Both sites exhibit a high degree of specificity for the biotinyl moiety, structural or stereochemical alterations to the naturally occurring bicyclic ring of d-biotin resulting in greatly reduced activity [61, 92, 99, 104]. The fact that the catalytic sites reside on different subunits, distinct from the carboxyl carrier protein, indicates that either the prosthetic group must oscillate between sites or the subunits bearing these sites must move with respect to the prosthetic group. Since the functional bicyclic ring of the prosthetic group resides at the end of a flexible 14 Å side-chain which anchors it to the apo-CCP [51, 120], the biotinyl group should be capable of oscillating between the remote carboxylation and transcarboxylation sites on biotin carboxylase and carboxyl transferase as visualized in Figure 6 [51, 99]. In Figure 6, BC refers to biotin carboxylase, CCP to carboxyl carrier protein, and CT to carboxyl transferase.

IV. ANIMAL ACETYL CoA CARBOXYLASES

A. Polymeric Filamentous Form

Acetyl CoA carboxylases purified from several animal tissues exist as enzymatically active polymeric filaments of high molecular weight [50, 55, 57, 113, 119] and have similar electron microscopic, hydrodynamic, and

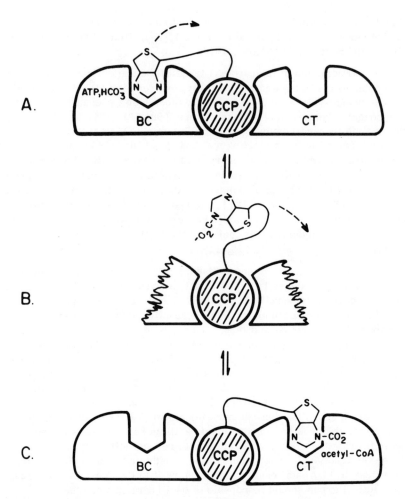

FIG. 6. Postulated scheme for intersubunit translocation of the Carboxylated biotin prosthetic group of E. coli acetyl CoA carboxylase. (From
Guchhait et al. [99].

catalytic properties (Table 2) [50-52, 55, 56, 59, 113, 119, 121-123].
Examination by electron microscopy of the homogeneous enzymes from
avian liver and bovine adipose tissue reveal that their filamentous structures are virtually indistinguishable (Fig. 7a and 7b; Ref. 55). Dilute
solutions (20 µg/ml) of avian liver [50] or bovine perirenal adipose tissue
[55, 113] carboxylase in 50 mM tris (Cl⁻) buffer containing 10 mM potassium citrate, 5 mM 2-mercaptoethanol and 0.1 mM EDTA at pH 7.5 were

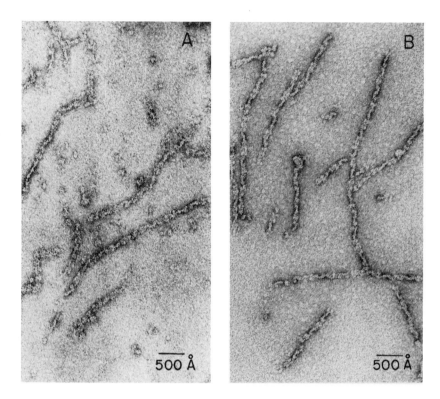

FIG. 7. Filamentous forms of avian liver (a) and bovine adipose tissue (b) acetyl CoA carboxylases in the presence of citrate.

applied to carbon support films. After staining with 4% aqueous uranyl acetate, the preparations were examined in the electron microscope. The filaments have a twisted appearance with indentations along the longitudinal axis suggestive of a helical structure [50, 55, 113]. The characteristic wide-narrow-wide pattern and a longitudinal periodicity of about 120-140 Å is best explained by the twisting of a flat helical filament composed of protomeric units, 130 Å in length [123, 124], as illustrated in Figure 8. The experimental conditions were as described for Figure 7. The true helical repeat from measurements on individual filaments and laterally aggregated filaments in paracrystalline form is about 1000 Å or every 8 protomers (see Sec. IV, B) [124, 125]; widths of the filaments range from 70-100 Å with lengths up to 5000 Å. Such a filamentous structure is compatible with the hydrodynamic properties (Table 2) of this form of the enzyme including a high intrinsic viscosity ($[\eta] = 83$ cm^{3}g^{-1}, avian enzyme; Ref. 119), a hypersharp sedimenting boundary in the analytical

TABLE 2

Molecular Properties of Animal Acetyl CoA Carboxylases

Enzyme	Molecular species	Molecular weight	$s_{20,w}$ analyt. ultracentr.	SDG	Biotin content (# of sites)	Refs.
Avian liver	Polymeric filaments	4-10 million	72 S [a]	47-50 S	Multiple	50-52, 55
	Protomer	410,000	13.1 S [a]	13-15 S	0.93	50-52

<table>
<tr><td></td><td colspan="2"># of sites per protomer</td></tr>
<tr><td>Covalent HCO$_3^-$ binding sites</td><td>1.0</td></tr>
<tr><td>Acetyl CoA binding sites</td><td>0.95</td></tr>
<tr><td>Citrate binding sites</td><td>1.1</td></tr>
<tr><td>-SH (Ellmans) groups</td><td>43-44</td></tr>
<tr><td>Half-cystine residues</td><td>44</td></tr>
</table>

Enzyme	Molecular species	Molecular weight	$s_{20,w}$ analyt. ultracentr.	SDG	Biotin content (# of sites)	Refs.
	Subunits[c]		4.3 S [b]			51, 147
	A	117,000			+	
	B	117,000			0	
	C	130,000			0	
	D	140,000			0	
Bovine adipose tissue	Polymeric filaments	Several million	68 S	47-50 S	Multiple	55, 113
	Protomer	560,000	14.4 S	13-15 S	2	113
Rat liver	Polymeric filaments	—	45 S	—	Multiple	122, 128
	Protomer[d]	215,000	—	—	1	122
	Subunits					122
	A	118,000	—	—	—	
	B	125,000	—	—	—	

ultracentrifuge (Fig. 9) [50, 122], and a marked concentration dependence of $s_{20,w}$ and $\eta_{sp/c}$ [119]. Furthermore, light scattering data [121] confirm the rod-like shape of the polymeric enzyme and indicate that the carboxylase possesses a molecular weight of 1.1×10^7 and a length of 3000 Å, in agreement with data obtained from sedimentation equilibrium [52, 55] and electron microscopy [55].

It can be calculated from the specific activity of the purified enzyme [94], the total carboxylase activity of avian liver (0.9 unit per gram wet weight; Ref. 94), and the fraction of total liver water in the cytosol [126], that the acetyl CoA carboxylase concentration is about 170 µg per gram wet weight of liver. Furthermore, using the value of Gates et al. [127] of 2×10^8 hepatocytes per gram of liver and the data of Gregolin et al. [51, 52], which indicate that an average carboxylase filament is composed of 20 protomeric units, a single avian liver cell would contain about 50,000 filaments.

Similar filamentous forms of the human [55] and rat [128] liver carboxylases have also been observed. In the case of the rat liver enzyme, the polymeric form also exhibits a high sedimentation velocity and hyper-sharp sedimenting boundary [122], in agreement with its being a fila-mentous asymmetric molecule, similar to its avian counterpart [121]. Besides possessing a similar polymeric structure, the animal acetyl CoA carboxylases exhibit a high degree of molecular relatedness, as judged by their immunological cross-reactivity [108, 129-133]. Virtually identical equivalence points are obtained for the interaction of rabbit antibody against avian liver enzyme with the carboxylases from avian liver, rat liver, and rat adipose tissue. Moreover, antibody to the avian liver enzyme exhibits good cross-reactivity as well with the human liver [134] and cultured skin fibroblast [113] carboxylases. Consistent with this pat-tern of relatedness, there is a striking similarity of the amino acid com-positions of avian [52] and rat [122] liver acetyl CoA carboxylases.

(Footnotes to Table 2.)

[a] $s_{20,w}^o$.

[b] In the presence of 0.1% sodium dodecyl sulfate (SDS).

[c] Although subunits A and B are not resolved on SDS gels, the biotin content of the protomer indicates that only one of these subunits contains biotin.

[d] Based only on biotin content.

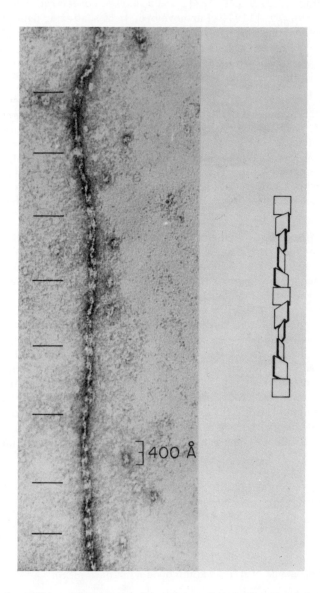

FIG. 8. Filamentous Form of Avian Liver Acetyl CoA Carboxylase

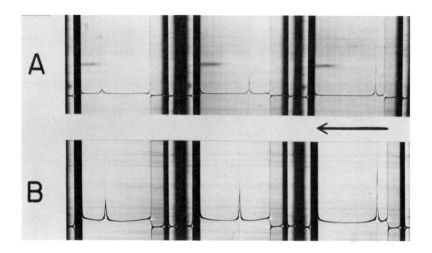

FIG. 9. Sedimentation patterns of avian liver acetyl CoA carboxylase.
(A) Enzyme in 0.05 M potassium phosphate, pH 7.0, 0.1 mM EDTA,
5 mM 2-mercaptoethanol; protein concentration, 1.08 mg/ml; centrifugation at 25,980 r.p.m. and 22°C; photographs taken at 9, 23, 51 min after
reaching speed; direction of centrifugation, right to left. (B) Enzyme in
0.05 M potassium phosphate, pH 7.0, 0.1 M ammonium sulfate, 0.1 mM
EDTA, 5 mM 2-mercaptoethanol; protein concentration, 1.70 mg/ml;
centrifugation at 25,980 r.p.m. and 20°C; photographs taken 20, 48, 68
min after reaching speed; direction of centrifugation, right to left. From
Gregolin et al. [50].

B. Paracrystalline Form

An interesting property of the filamentous form (Figs. 7, 8; Sec. IV,
A) of avian liver acetyl CoA carboxylase is its tendency to aggregate
laterally to form paracrystalline fibers [123-125]. This process occurs
slowly at room temperature and neutral pH in the presence of 20% saturated ammonium sulfate giving rise to structures suitable for analysis by
electron microscopy [123-125]. Illustrated in Figure 10 is an electron
micrograph of the paracrystalline form of the enzyme which shows considerable structural detail. Optical diffraction measurements made
on electron micrographs (Fig. 10, inset) reveal strong horizontal
periodicity with a spacing of about 130 Å; reflections along the
meridian of the diffraction pattern as far as the eighth layer line
indicate the presence of an eightfold screw axis in the vertical direction of the image [124, 125]. A schematic representation of the packing
of filaments into the paracrystalline aggregate as shown by the optically

FIG. 10. Electron micrograph of the paracrystalline form of avian liver
acetyl CoA carboxylase; stained with uranyl acetate. (From Leonard et al.
[125].)

filtered reconstructed image is presented in Figure 11a and 11b. The filaments are packed side-by-side, each being staggered by one-half of a
helical repeat with respect to its nearest neighbors. Interactions between
adjacent helical filaments appear to occur in a side-to-side and a back-to-
back fashion, but not in a face-to-face direction, which results in a zig-zag
perturbation of the individual filaments. In the region of the image where
side-to-side packing is evident, the filament shows two parallel chains. In
the side-on view of the filament, where back-to-back packing occurs, these
overlap in projection. Taking into account the various views which can be
seen in the filtered image, a model for one repeat unit of the filament,
presumably the protomer, can be constructed (Fig. 12) [125]. The dimensions of this model are consistent with the molecular weight of the protomer,
i.e., $\sim 450,000$, and the asymmetric shape of the protomer as indicated by
its relatively high intrinsic viscosity ($\eta = 11.5$; see Sec. IV, C).

C. Protomeric Form and Subunit Structure

The filamentous polymeric forms of the animal acetyl CoA carboxylases
can be dissociated into weight-homogeneous high molecular weight protomeric species which are themselves complex multisubunit structures
(Table 2) [51, 52, 55, 113]. Treatment of the avian liver and bovine
adipose tissue carboxylases with 0.5 M NaCl at pH 8-9 causes dissociation
to the protomer level, the protomeric species having molecular weights of
410,000 ($s_{20,w}^0 = 13.1$ S) [51, 52] and 560,000 ($s_{20,w}^0 = 14.7$ S) [113],
respectively. The carboxylase protomer from avian liver is a relatively
asymmetric structure as indicated by viscosity measurements ($[\eta] =
11.5 \text{ cm}^3\text{g}^{-1}$) and by the marked dependence of $s_{20,w}$ on enzyme concen-
tration [119]. Consistent with a protomer weight of 410,000 for the avian
liver carboxylase are the findings (Table 2) that the enzyme possesses 1.0
covalent bicarbonate-binding site, 0.95 acetyl CoA-binding sites ($K_D =
4 \mu M$), and 1.1 citrate binding sites ($K_D = 2.9 \mu M$) per 410,000 daltons
[51]. Moreover, the biotin content of the homogeneous carboxylase
(0.55 μg per mg of refractometrically determined protein [51, 52] corresponds to 1.0 biotinyl prosthetic group per 441,000 daltons or about one
per protomer molecular weight ($M_{SD} = 410,000 \pm 21,000$). On the basis of
the retention of the holoenzyme to avidin-Sepharose 4B affinity columns
under conditions (1 M guanidine·HCl) where the polymeric form is completely dissociated to the protomer level, Lane et al. [5] concluded that
each individual protomer contains one covalently bound biotinyl prosthetic
group. On the basis of its biotin content, Inoue and Lowenstein suggest
[122] that the protomeric form of the rat liver carboxylase is smaller,
i.e., about 215,000 daltons; however, direct molecular weight measurements will be necessary to validate this size.

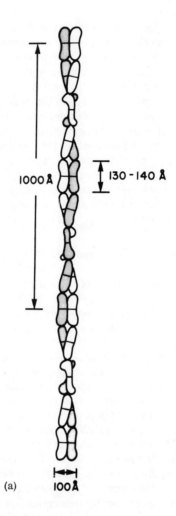

(a)

FIG. 11. Model of avian liver acetyl CoA carboxylase filament (a) and
paracrystalline form (b). (From Leonard et al. [125].)

The subunit structure of the protomeric form has been investigated in
detail only with the avian liver enzyme [5, 51, 52, 94, 119, 135]. Using
cross-linking analysis by the dimethylsuberimidate method [136, 137],
Guchait et al. [135] (Fig. 13) found that four species, a monomer
(117,000 daltons), dimer (234,000 daltons), trimer (350,000 daltons),
and tetramer (470,000 daltons), were formed. Analysis of the four
species resolved by dodecyl sulfate-acrylamide gel electrophoresis
(Fig. 13) revealed that each contained covalently bound biotin. Cross-

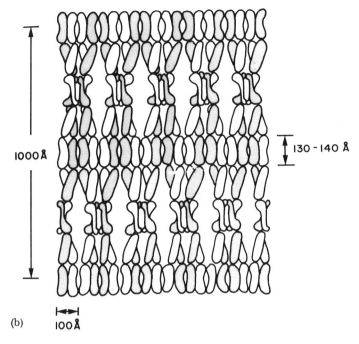

1000 Å

130 - 140 Å

(b)   100 Å

FIG. 11(b).

linking of subunits of the avian liver acetyl CoA carboxylase protomer (in
0.5 M NaCl) was carried out according to the method of Davies and Stark
[136]. For gel 1, 300 μg of carboxylase protomer per ml of 0.5 M NaCl
was reacted with dimethylsuberimidate (DMS, 5 mg/ml) at pH 8.3 and
37° C for 2 hrs; carboxylase for gels 2, 3, 4, and 5 were treated similarly
except that 0.5 mg/ml of DMS was used and the reaction was allowed to
proceed for 30, 60, 120, and 120 min, respectively. Ovalbumin for gel 8
was cross-linked for use as markers (monomer, 45,000; dimer, 90,000;
trimer, 135,000; tetramer, 180,000; pentamer, 225,000; hexamer,
270,000; heptamer, 315,000; and octamer, 360,000) by the method of Carpenter and Harrington [137]. All cross-linked proteins as well as rabbit
muscle myosin (gel 7) and acetyl CoA carboxylase (gel 6), which were not
cross-linked, were dissociated with 1% sodium dodecyl sulfate (SDS) for
3 hrs at 37° C and subjected to electrophoresis on 3.5% acrylamide gels
[136]; 30 μg of protein were applied to gels 1-4, 6, and 8 and 60 μg to
gels 5 and 7. While the subunits of acetyl CoA carboxylase can be resolved with other SDS-polyacrylamide gel electrophoresis systems (see
Sec. IV, C), only one band (gel 6) of molecular weight 115,000-120,000
is obtained with the system employed here. These results strongly indicate that the protomeric form of the carboxylase is a tetramer. Moreover,
the fact that the enzyme contains 1 mol of covalently bound biotin, has one

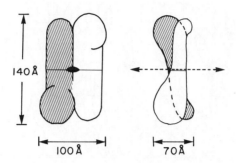

FIG. 12. Model of the "repeat unit" (protomer) of the avian liver carboxylase filament (see Figs. 8 and 11a). (From Leonard et al. [125].)

carboxylation site, and binds tightly 1 mol of citrate and 1 mol of acetyl CoA per 410,000 grams of protein, shows that the subunits are nonidentical and is consistent with a tetrameric structure. Although earlier subunit analyses [51] by standard SDS-acrylamide gel electrophoresis and by sedimentation equilibrium in 0.1% SDS suggested the presence of a single subunit weight class of 110,000-114,000 daltons, high resolution acrylamide gel electrophoresis in 6 M urea-0.1% sodium dodecyl sulfate resolved the subunits into three weight classes [135] approximating those indicated above. The approximate mass ratio of the three weight class species, i.e., 117,000, 130,000, and 140,000 daltons, appears to be 2:1:1, respectively, providing additional support for a tetrameric structure for the protomer. The covalently bound biotin prosthetic group is found to be associated only with the 117,000 dalton subunit(s) [135]. In comparison, Inoue and Lowenstein [122] have reported resolution of the rat liver protomer into subunits of two weight classes, 118,000 and 125,000 daltons. Taken together the structural studies with the avian liver enzyme indicate that the carboxylase filaments are composed of identical protomeric units, each of which contains a single biotin prosthetic group covalently bound to one of four nonidentical subunits (Fig. 14). On the basis of numbers and types of sites and the precedent established with the E. coli carboxylase system (see Sec. III), it seems likely that the biotin carboxylating site, carboxyl transfer site, citrate activator site, and biotinyl prosthetic group are housed on different subunits within the protomer.

D. Dependence of Catalytic Activity on the Protomer ⇌ Polymer Equilibrium

In 1963, Vagelos et al. [49] reported that activation of a crude adipose tissue acetyl CoA carboxylase preparation by citrate was accompanied by

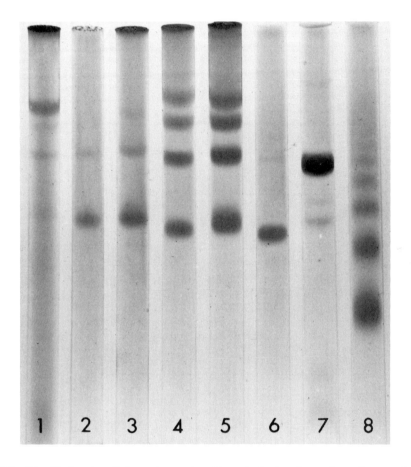

FIG. 13. Dodecyl sulfate-polyacrylamide gel electrophoresis patterns of dimethylsuberimidate cross-linked avian liver acetyl CoA carboxylase.

an increased sedimentation velocity in sucrose density gradients. This observation was subsequently confirmed by Numa et al. [84] with rat liver carboxylase and in Lane's laboratory [50, 52] with chicken liver carboxylase. Since the earlier investigations involved impure carboxylase preparations, it could not be concluded whether conformational changes per se, complex formation between carboxylase and another component in the preparation, or carboxylase aggregation per se was responsible for the sedimentation velocity effect. It was subsequently shown by Gregolin et al. [50, 52] that activation of homogeneous avian liver acetyl CoA carboxylase by tricarboxylic acid activator (Fig. 15) was accompanied by an increased sedimentation velocity of carboxylase protein per se in sucrose density

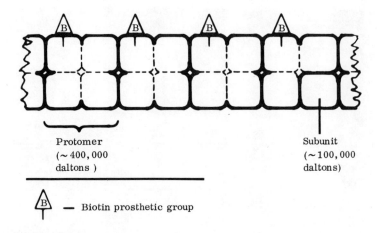

Protomer
($\sim 400,000$
daltons)

Subunit
($\sim 100,000$
daltons)

B — Biotin prosthetic group

FIG. 14. Model of the structure of the polymeric form of avian liver acetyl
CoA carboxylase. (From Lane et al. [5].)

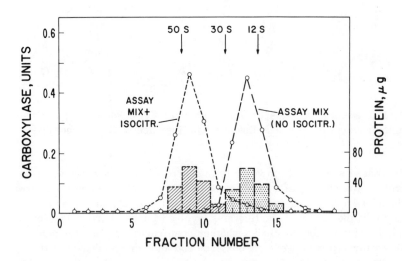

FIG. 15. The effect of tricarboxylic acid activator on the sedimentation
velocity of avian liver acetyl CoA carboxylase in assay reaction mixture.
(From Gregolin et al. [50].)

gradients containing the carboxylase assay reaction mixture components with or without activator; under conditions approximating those of the enzymatic assay, citrate or isocitrate induces a change in $s_{20,w}$ from 13-15 S to 47-50 S [3, 50-52, 55, 113] (Table 2). For Figure 15, sedimentation was carried out using 5-20% sucrose density gradients, total volume 4.5 ml. Sucrose density gradients contained the components of the carboxylase assay reaction mixture: 0.06 M tris (Cl^-), pH 7.5; 2 mM ATP; 8 mM $MgCl_2$; 0.01 M $KHCO_3$; 0.2 mM acetyl CoA; and 0.1 mM EDTA. 200 μg of homogeneous acetyl CoA carboxylase in 0.2 ml of 50 mM $K \cdot P_i$ buffer, pH 7.0, 0.1 mM EDTA were applied to each gradient. Gradients were centrifuged at 39,000 rpm (sw 39 Beckman-Spinco rotor) for 60 min at 25°C. 20 mM DL-isocitrate was added where indicated. o---o---o refers to carboxylase activity and bar graphs to carboxylase protein. Moreover, this change in sedimentation velocity corresponds to a transition from a protomeric to a polymeric filamentous form as judged by electron microscopy [50, 55, 113] (Fig. 16) and a large increase in intrinsic viscosity, i.e., from [η] of 11.3 for the protomer to 83 for the polymer [119].

The slowly sedimenting (13-15 S) species of the avian liver carboxylase found in assay reaction mixture without activator corresponds to the protomeric form, $s_{20,w}^0 = 13.1$ S, obtained by dissociating the polymer in 0.5 M NaCl, pH 8.0 [52]. This form has a molecular weight, determined by sedimentation equilibrium, of 410,000, which is in good agreement with the minimal molecular weight calculated from the biotin content of the enzyme and binding data [51, 52]. Thus it appears that acetyl CoA carboxylases from animal tissues exist in a catalytically inactive protomeric state and a catalytically active polymeric state — the position of the protomer-polymer equilibrium (Fig. 16) determining the level of catalytic activity. "True" activators, such as citrate and isocitrate, are able to shift the equilibrium toward the catalytically active state presumably by inducing a productive conformational change in the protomer, which favors polymerization [4, 50, 94, 113, 119]. Therefore, when allowed to depolymerize fully to the protomeric state in assay medium, prior to addition of citrate or isocitrate, these enzymes exhibit an absolute requirement for tricarboxylic acid activator [50, 51, 113, 119]. On the other hand, certain di- and tricarboxylic acids, including malonate, methylmalonate, malate, and tricarballylate — a closely related analogue of citrate and isocitrate — behave as "pseudo" activators under certain circumstances. These acids appear to activate when the assay is initiated with carboxylase in the active polymeric form but in fact merely retard the rate of transition of active polymer to inactive protomer and are unable to reverse this process [4, 50, 51, 94]. In addition to the pseudo activators, orthophosphate and a large number of other mono- and dibasic acids are without activity [51, 94]. Thus, the structural requirements for activation are quite specific.

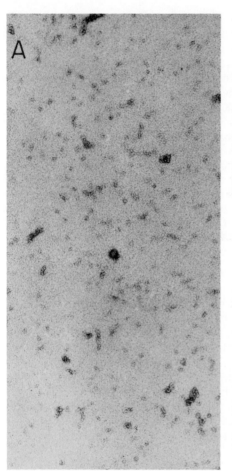 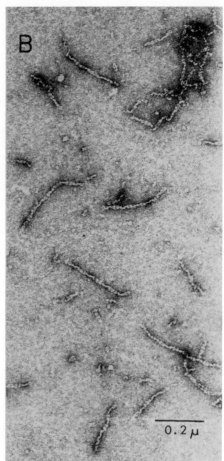

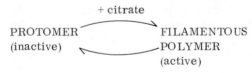

Equilibrium toward
protomer favored by:

$ATP \cdot Mg^{2+} + HCO_3^-$
malonyl CoA
fatty acyl CoA
high [NaCl]

Equilibrium toward
polymer favored by:

citrate, isocitrate
phosphate

FIG. 16. Protomer–polymer equilibrium (avian liver acetyl CoA
carboxylase).

It is significant that tricarballylate, P_i, and malonate, which are not activators, increase the sedimentation velocity of the enzyme in simple buffering media at neutral pH, whereas under carboxylase assay conditions only citrate and isocitrate are able to maintain the rapidly sedimenting polymeric form. Tricarballylate, P_i, and malonate are not able to do so [51]. This is due to the fact that the protomer-polymer equilibrium is shifted toward the protomer by converting the enzyme to its carboxylated form, enzyme-biotin-CO_2^- [51, 119]. Sucrose density gradient centrifugation conducted under assay conditions for the carboxylase-catalyzed forward reaction, ATP-$^{32}P_i$ and malonyl CoA-[^{14}C] acetyl CoA exchanges, all of which are activated by citrate or isocitrate, showed that there is a strict correlation between activation and the transition from protomeric to polymeric form [50-52, 113, 119]. Interestingly, under these conditions none of the other compounds, such as tricarballylate, P_i, or malonate (which are capable of maintaining the polymeric form in simple buffering media) could be substituted for citrate or isocitrate [51]. Moreover, none of these factors was capable of activating any of the above-mentioned reactions. It became apparent that some component(s) in the carboxylation, exchange, and decarboxylation reaction mixtures must promote disaggregation to the protomeric form. The causative agents were found to be those substrates and cofactors, or combination thereof, which were capable of carboxylating the enzyme to form "enzyme-biotin-CO_2^- " [51, 52, 119]. Hence, the introduction of a carboxyl function at the 1'-N-position of the biotinyl prosthetic group produces sufficient conformational strain such that only true activators — i.e., citrate and isocitrate, but not tricarballylate, P_i, or malonate — are able to constrain the enzyme and prevent its depolymerization [51]. Malonyl CoA, which can carboxylate the enzyme, is a competitive inhibitor of the overall reaction, $K_i = \sim 10^{-5}$ M, with respect to citrate [52, 138]. This competitive kinetic relationship apparently results from the opposing effects of citrate and malonyl CoA on the protomer-polymer equilibrium (Fig. 16). This effect has been verified by direct electron microscopic investigation as well. Avian liver acetyl CoA carboxylase in 50 mM potassium phosphate buffer, pH 7.0, exists in the polymeric filamentous form; however, on the addition of as little as 10 μM malonyl CoA, which carboxylates the enzyme, there is an instantaneous (within seconds) depolymerization to yield protomeric forms having the typical protomer dimensions of 50-130 Å [51].

It is evident that conformational strain induced by carboxylation of the enzyme may be utilized for the electrophilic activation of the N-carboxyl group of the biotinyl prosthetic group (a notoriously poor electrophile) by changing its bond angle or length as a result of strain or distortion induced by conformational changes at the active site. Thus, although carboxylation at the 1'-N-position of the biotinyl moiety induces sufficient conformational strain to cause depolymerization, it is evident that the binding of its allosteric activator, citrate, prevents this dissociation presumably by constraining the enzyme in the active conformation compatible with the polymeric state. Under conditions where the enzyme is not carboxylated, the K_D for citrate is $3 \mu M$; however, under assay conditions — that is, where the enzyme is carboxylated — the activator constant for citrate, K_A, is 3 mM [51]. It is suggested that the one thousandfold difference between K_D and K_A may reflect the binding energy needed for structural constraint to maintain the active conformation of enzyme-biotin-CO_2^-. In this connection, Edwards and Lane [139] have observed that avidin, which binds free- or enzyme-bound biotin with remarkable affinity ($K_{D, \text{ free biotin}}$ = 10^{-15} M) [118], enhances the decarboxylation of enzyme-biotin-CO_2^-; the $t_{1/2}$ of enzyme-biotin-CO_2^- at $2 °C$, pH 7.5, of about 200 min is reduced to < 1 min when treated with avidin. It is suggested that this effect may result from the strain induced in the carboxyureido system owing to the extraordinarily tight binding of the carboxybiotinyl group by avidin.

A number of other factors have been found to promote depolymerization, hence reversible inactivation, of the enzyme. These include elevated salt concentration (e.g., 0.5 M sodium chloride), elevated pH, and a number of inhibitors of potential physiological importance, such as long-chain acyl CoA derivatives [52, 84, 140].

Although the position of the protomer-polymer equilibrium appears to determine the level of acetyl CoA carboxylase activity, it has been uncertain whether activation is a consequence of polymerization per se. Whereas citrate-induced polymerization and activation are complete within seconds, the methods that have been used to assess changes in polymeric state (sedimentation velocity, viscosity, and electron microscopy) are too slow to be of use in following the kinetics of polymerization. The depolymerization process, however, is several orders of magnitude slower and has been studied in relation to the loss of catalytic activity [119]. By taking advantage of the slow rate of depolymerization of the filamentous form of the avian liver carboxylase in assay reaction mixture without activator at low temperature and high enzyme concentration, Moss and Lane [119] have obtained kinetic evidence for the coupling of the loss of activity to the polymer → protomer transition. As illustrated in Figure 17, the viscosity of the polymeric form of the enzyme added to assay reaction mix without activator undergoes first-order decay ($t_{1/2}$ = 9 1/2 min) at a rate equal to that of activity decay ($t_{1/2}$ = 10 min) indicating tight coupling of the polymer → protomer transition with loss of catalytic activity. The initial value at t = O reflects only polymer, while the $\eta_{sp/c}$ obtained following extensive decay is due exclusively to protomer. The extrapolated value of $\eta_{sp/c}$ at t = O agrees well with the intrinsic viscosity ($[\eta]$ = 83) of the polymeric form [119]. Addition of citrate following inactivation (Fig. 17) causes an immediate rise in $\eta_{sp/c}$, as well as reactivation of the inactive enzyme; although some denaturation occurs, it is evident that the relative activity regain approximates that of relative $\eta_{sp/c}$ regain induced by

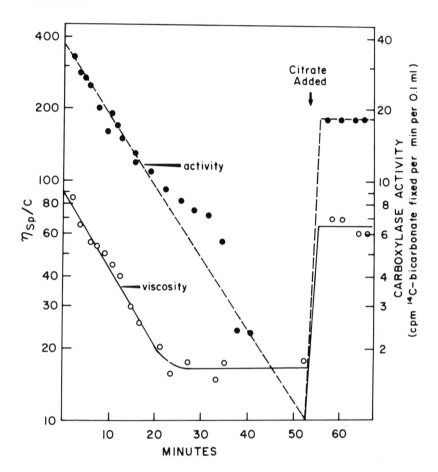

FIG. 17. Kinetic correlation of reversible viscosity and activity decay of polymeric acetyl CoA carboxylase. The carboxylation reaction was initiated by the addition of 194 μg of homogeneous acetyl CoA carboxylase in its polymeric state (in 50 mM potassium phosphate, pH 7.0) per ml of carboxylase assay reaction mixture (minus citrate) of the same composition as in Fig. 8. Viscosity measurements were made directly on the reaction mixture and 0.1 ml aliquots were taken for the determination of $[^{14}C]$ bicarbonate fixation rates at the time indicated (times recorded for $\eta_{sp/c}$ represent the midpoints of the determinations). The reaction mixture was held at 2°C during all operations. After 52 min potassium citrate (final concentration, 10 mM) was added to the reaction mixture after which additional viscosity and carboxylase activity measurements were made. From Moss and Lane [119].

citrate addition. On the basis of the kinetics of viscosity and activity changes, Moss and Lane [119] suggest that the depolymerization mechanism involves a "one shot" or concerted disruption of the filament structure with no, or short-lived intermediate oligomeric species. This model visualizes the occurrence of an initial break in the filament followed by a series of far more rapid cleavages which quickly lead to complete depolymerization. The absence of a rapid drop in viscosity prior to significant decay of activity tends to negate a random cleavage mechanism or depolymerization in which smaller active oligomeric intermediates accumulate. Interestingly, the presence of avidin in the assay reaction mixture during decay does not alter the $t_{1/2}$ for activity or viscosity decay. Avidin is known [93, 119] to instantaneously inactivate the protomeric form of the carboxylase by binding irreversibly to the biotinyl prosthetic group; in contrast, the polymeric form in the presence of citrate is completely resistant to avidin [119]. Thus, during activity decay associated with depolymerization only that fraction of carboxylase activity remaining (i.e., that fraction still in the polymeric state) can be protected by citrate from further erosion of activity and inactivation by avidin [119]. The fact that activity and viscosity decay, the kinetics of disappearance of the avidin-resistant polymeric form, and the rate of appearance of the avidin-sensitive protomeric species are closely correlated strongly indicates tight coupling between the loss of catalytic activity and the polymer → protomer transition. Thus, a strong case is made for a causal relationship between the activation of animal acetyl CoA carboxylases and a polymerization-associated conformational change(s) induced by citrate. As will be discussed in the next section, the inaccessibility of the biotin prosthetic group in the citrate-activated polymeric form of the enzyme indicates an activator-induced conformational change in the vicinity of the active site (i.e., near the biotinyl prosthetic group).

E. Citrate Activation Mechanism

As is evident from its inaccessibility to avidin in the citrate-activated carboxylase, the biotin prosthetic group must become shielded by neighboring groups due to conformational changes at the active site(s) (see preceding Sec. IV, D). This and several other lines of evidence suggest that citrate activation is mediated through effects on the biotinyl group. Since both half-reactions are citrate-activated [47, 52, 94, 95], it appears that a substituent at the active site of the carboxylase, common to both half-reactions, is the focal point of the citrate-induced conformational changes. The facts that citrate is an activator of the V_{max} type [2, 94, 113] and that isotopic exchange and model reactions which characterize each half-reaction [reactions (3) and (4)] involve different substrates show that this common participant is not a substrate [47, 53, 93-95]. The most obvious substituent common to both partial reactions is the covalently bound biotin prosthetic group.

In view of the apparent changes in conformation near the biotin prosthetic group, which accompany citrate activation, the possibility has been considered that the reactivity of the biotinyl- or N-carboxybiotinyl group might also be altered. Studies on the effect of tricarboxylic acid activator on malonyl CoA and enzyme-biotin-CO_2^- decarboxylation, both of which involve decarboxylation of the 1'-N-carboxybiotinyl prosthetic group, indicate that the activator greatly enhances the reactivity of this group [3, 5, 93, 141].

Further insight was gained in experiments in which "free" d-biotin served as a model for the covalently bound prosthetic group, i.e., in the ATP-dependent carboxylation of "free" biotin to yield 1'-N-carboxybiotin [a model for reaction (3)] and in carboxyl transfer from malonyl CoA to "free" biotin to form 1'-N-carboxybiotin [a model for reaction (4)] [5, 120, 141]. The high degree of specificity for d-biotin or closely related derivatives in both of these reactions indicates that binding and carboxylation of the free biotin species occurs at the same specific site normally occupied by the bicyclic ring of the biotinyl prosthetic group. As shown in Figure 3, the functional bicyclic ring resides at the end of a flexible 14 Å side chain, which anchors it to the apoprotein; in addition, secondary binding sites must be required to precisely orient the ureido ring system with respect to the substrates with which it must react. This is directly analogous to the E. coli acetyl CoA carboxylase system just described (Sec. III and Fig. 6).

One possible mechanism for the citrate effect, illustrated in Figure 18, is that the carboxybiotin prosthetic group may be brought into closer proximity to substrate binding sites by activator-induced conformational

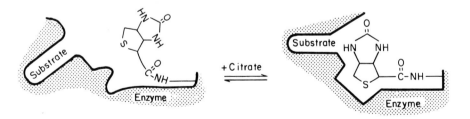

FIG. 18. Postulated scheme for the effect of tricarboxylic acid activator on the reorientation of the active site(s) of acetyl CoA carboxylase.(From Lane et al. [5].)

changes [1, 3]. Were this the case it would be expected that free d-biotin
and its closely related analogues would probably not compete favorably
with the bicyclic ring of the prosthetic group for the carboxylation site or
the carboxyl transfer site in the presence of citrate. Consistent with this
prediction is the fact that the K_m values for free d-biotin, d-homobiotin,
and biocytin in the ATP-dependent carboxylation reaction at the biotin
carboxylase site are markedly increased by citrate [5, 120]. A similar
situation seems to prevail at the carboxyl transferase site in that citrate
nearly completely blocks carboxyl transfer from malonyl CoA to "free"
d-biotin and biocytin [141]. On the basis of these findings, Lane and coworkers [1, 3-5, 141] have proposed that regulation of the animal acetyl
CoA carboxylases by citrate activation is mediated through a conformational
change(s) which perfects the orientation of the biotin prosthetic group with
respect to the biotin carboxylase and carboxyl transfer sites.

V. YEAST AND PLANT ACETYL CoA CARBOXYLASES

An avidin-sensitive acetyl CoA carboxylase [96] has been purified from
yeast, the specific activity of the preparation [142] being comparable to
that of the animal carboxylases [94, 113]. Structurally the protein appears
to be quite distinct from its animal counterparts; it is a high molecular
weight enzyme ($\sim$ 600,000) with an $s_{20, w}$ of 15.5 S [142], and is a
tetrameric structure composed of apparently identical 6 S subunits. This
contrasts markedly with the avian and rat liver carboxylases in which the
basic protomeric unit is composed of nonidentical subunits, as well as with
the E. coli system where nonidentical subunits have been purified and
physical and functional separation obtained. The enzyme has been isolated
in forms responsive [143] and unresponsive to citrate [144]. Rasmussen
and Klein [143] have noted activation effects by fructose-1, 6-diphosphate,
α-glycerophosphate, and α-ketobutyrate as well as by tricarboxylic acids.
In contrast to the animal carboxylases, activation is not accompanied by
a change in sedimentation behavior [143].

Acetyl CoA carboxylases of plant origin have been obtained from lettuce
[145] and spinach chloroplasts [146] and wheat germ [97, 98]. The
wheat germ enzyme has a structure of intermediate stability relative to the
readily dissociable E. coli carboxylase system (see Sec. III) and the
animal carboxylases which maintain their structural integrity during
purification. Wheat germ acetyl CoA carboxylase is dissociable during
purification into two enzymatically active components with sedimentation
coefficients of 7.3 S and 9.4 S, biotin being associated with the latter
species. Presumptive evidence suggests that the 7.3 S component corresponds to the carboxyl transferase subunit of E. coli acetyl CoA carboxylase,
while the 9.4 S species has characteristics similar to the E. coli biotin

carboxylase-carboxyl carrier protein complex [98]. The enzyme from chloroplasts is structurally similar to the E. coli enzyme in that it can be easily dissociated into enzymatically active subunits [146]. In broken chloroplasts, the component containing biotin is particle-bound, being localized in the thylakoids, while the biotin carboxylase and carboxyl transferase activities are found in the stroma [146]. Neither the wheat germ nor the chloroplast enzyme is activated by citrate.

ACKNOWLEDGEMENT

The authors wish to thank Mrs. Norma Mitchell for her expert and tireless assistance in the preparation of this manuscript. The unpublished investigations conducted in the authors' laboratory were supported by research grants AM-14574 and AM-14575, United States Public Health Service and a research grant from the American Heart Association, Inc.

REFERENCES

1. Joel Moss and M. Daniel Lane, Adv. in Enzymol., 35, 321 (1971).
2. M. D. Lane, J. Moss, and S. E. Polakis, Current Topics Cell. Regula., 8, 139 (1974).
3. M. D. Lane, J. Moss, E. Ryder and E. Stoll, Adv. Enz. Regulation, 9, 237 (1971).
4. M. D. Lane and J. Moss, in Metabolic Regulation (Henry J. Vogel, ed.), Vol. 5, Academic Press, New York, 1971, p. 23.
5. M. D. Lane, J. Edwards, E. Stoll and J. Moss, Vitamins Hormones, 28, 345 (1970).
6. M. D. Lane, R. B. Guchhait, S. E. Polakis, and J. Moss, in Molecular Basis of Biological Activity (K. Gaede, B. L. Horecker, and W. J. Whelan, eds.), Academic Press, New York, 1972, p. 103.
7. E. B. Titchener and D. M. Gibson, Fed. Proc., 16, 262 (1957).
8. J. W. Porter, S. J. Wakil, A. Tietz, M. I. Jacob, and D. M. Gibson, Biochim. Biophys. Acta, 25, 35 (1957).
9. J. W. Porter, and A. Tietz, Biochim. Biophys. Acta, 25, 41 (1957).
10. S. J. Wakil, E. B. Titchener and D. M. Gibson, Biochim. Biophys. Acta, 29, 225 (1958).
11. S. J. Wakil, J. Am. Chem. Soc., 80, 6465 (1958).
12. D. M. Gibson, E. B. Titchener and S. J. Wakil, Biochim. Biophys. Acta, 30, 376 (1958).
13. E. B. Titchener, D. M. Gibson, and S. J. Wakil, Fed. Proc., 17, 322 (1958).

14. D. M. Gibson, E. B. Titchener, and S. J. Wakil, J. Am. Chem. Soc., 80, 2908 (1958).

15. S. J. Wakil, E. B. Titchener, and D. M. Gibson, Biochim. Biophys. Acta, 34, 227 (1959).

16. J. V. Formica and R. O. Brady, J. Am. Chem. Soc., 81, 752 (1959).

17. S. J. Wakil and D. M. Gibson, Biochim. Biophys. Acta, 41, 122 (1960).

18. S. J. Wakil, J. Lipid Res., 2, 1 (1961).

19. S. J. Wakil, 5th Intl. Congr. Biochem., Moscow, Symposium VII, 1 (1961).

20. I. B. Fritz, Physiol. Rev., 41, 52 (1961).

21. F. Lynen, Federation Proc., 20, 941 (1961).

22. P. R. Vagelos, Ann. Rev. Biochem., 33, 139 (1964).

23. C. L. Squires, P. K. Stumpf, and C. Schmid, Plant Physiol., 33, 365 (1958).

24. F. Lynen, J. Cell. Comp. Physiol., 54 (Suppl. 1), 33 (1959).

25. R. O. Brady, Proc. Natl. Acad. Sci., U. S., 44, 993 (1958).

26. S. J. Wakil and J. Ganguly, J. Am. Chem. Soc., 81, 2597 (1959).

27. J. Ganguly, Biochim. Biophys. Acta, 40, 110 (1960).

28. R. O. Brady and S. Gurin, J. Biol. Chem., 199, 421 (1952).

29. G. Popják, and A. Tietz, Biochem. J., 60, 147 (1955).

30. R. O. Brady, A.-M. Mamoon, and E. R. Stadtman, J. Biol. Chem., 222, 795 (1956).

31. A. Tietz, Biochim. Biophys. Acta, 25, 303 (1957).

32. S. J. Wakil, J. W. Porter, and D. M. Gibson, Biochim. Biophys. Acta, 24, 453 (1957).

33. F. Dituri, W. N. Shaw, J. V. B. Warms, and S. Gurin, J. Biol. Chem., 226, 407 (1957).

34. R. G. Langdon, J. Biol. Chem., 226, 615 (1957).

35. S. Abraham, K. J. Matthes, and I. L. Chaikoff, J. Biol. Chem., 235, 2551 (1960).

36. R. Dils and G. Popják, Biochem. J., 80, 47 P (1961).

37. S. Abraham, K. J. Matthes, and I. L. Chaikoff, Biochim. Biophys. Acta, 49, 268 (1961).

38. R. Dils and G. Popják, Biochem. J., 83, 41 (1962).

39. D. B. Martin and P. R. Vagelos, J. Biol. Chem., 237, 1787 (1962).

40. S. Abraham, E. Lorch, and I. L. Chaikoff, Biochem. Biophys. Res. Commun., 7, 190 (1960).

41. R. G. Kallen and J. M. Lowenstein, Arch. Biochem. Biophys., 96, 188 (1962).

42. D. B. Martin and P. R. Vagelos, Biochem. Biophys. Res. Commun., 7, 101 (1962).

43. M. Waite, Fed. Proc., 21, 287 (1962).

44. M. Matsuhashi, S. Matsuhashi, S. Numa, and F. Lynen, Fed. Proc., 21, 288 (1962).

45. D. B. Martin and P. R. Vagelos, Fed. Proc., 21, 289 (1962).
46. M. Waite and S. J. Wakil, J. Biol. Chem., 237, 2750 (1962).
47. M. Matsuhashi, S. Matsuhashi, and F. Lynen, Biochem. Z., 340, 263 (1964).
48. P. R. Vagelos, A. W. Alberts, and D. B. Martin, Biochem. Biophys. Res. Communs., 8, 4 (1962).
49. P. R. Vagelos, A. W. Alberts, and D. B. Martin, J. Biol. Chem., 238, 533 (1963).
50. C. Gregolin, E. Ryder, A. K. Kleinschmidt, R. C. Warner, and M. D. Lane, Proc. Natl. Acad. Sci., U. S., 56, 148 (1966).
51. C. Gregolin, E. Ryder, R. C. Warner, A. K. Kleinschmidt, H.-C. Chang, and M. D. Lane, J. Biol. Chem., 243, 4236 (1968).
52. C. Gregolin, E. Ryder, R. C. Warner, A. K. Kleinschmidt, and M. D. Lane, Proc. Natl. Acad. Sci., U. S., 56, 1751 (1966).
53. T. Goto, E. Ringelmann, B. Riedel, and S. Numa, Life Sciences, 6, 785 (1967).
54. S. Numa, T. Goto, E. Ringelmann, and B. Riedel, Eur. J. Biochem., 3, 124 (1967).
55. A. K. Kleinschmidt, J. Moss, and M. D. Lane, Science, 166, 1276 (1969).
56. A. L. Miller and H. R. Levy, J. Biol. Chem., 244, 2334 (1969).
57. J. Moss, M. Yamagishi, A. K. Kleinschmidt, and M. D. Lane, Fed. Proc., 28, 1548 (1969).
58. E. Shrago, T. Spennetta, and E. Gordon, J. Biol. Chem., 244, 2761 (1969).
59. S. Numa, E. Ringelmann, and B. Riedel, Biochem. Biophys. Res. Commun., 24, 750 (1966).
60. A. W. Alberts and P. R. Vagelos, Proc. Natl. Acad. Sci., U. S., 59, 561 (1968).
61. P. Dimroth, R. B. Guchhait, E. Stoll, and M. D. Lane, Proc. Natl. Acad. Sci., U. S., 67, 1353 (1970).
62. S. E. Polakis, R. B. Guchhait, E. Zwergel, M. D. Lane, and T. G. Cooper, J. Biol. Chem., 249, 6657 (1974).
63. Y. Sokawa, E. Nakao, and Y. Kaziro, Biochem. Biophys. Res. Commun., 33, 108 (1968).
64. D. B. Melville, A. W. Moyer, K. Hofmann, and V. du Vigneaud, J. Biol. Chem., 146, 487 (1942).
65. V. du Vigneaud, D. B. Melville, K. Folkers, D. E. Wolf, R. Mozingo, J. C. Keresztesy, and S. A. Harris, J. Biol. Chem., 146, 475 (1942).
66. S. A. Harris, D. E. Wolf, O. R. Mozingo, and K. Folkers, Science, 97, 1447 (1943).
67. S. A. Harris, D. E. Wolf, R. Mozingo, R. C. Anderson, G. E. Arth, N. R. Easton, D. Heyl, A. N. Wilson, and K. Folkers, J. Am. Chem. Soc., 66, 1756 (1944).

68. S. A. Harris, N. R. Easton, D. Heyl, A. N. Wilson, and K. Folkers, J. Am. Chem. Soc., 66, 1757 (1944).

69. S. A. Harris, R. Mozingo, D. E. Wolf, A. N. Wilson, G. E. Arth, and K. Folkers, J. Am. Chem. Soc., 66, 1800 (1944).

70. S. A. Harris, D. E. Wolf, R. Mozingo, G. E. Arth, R. C. Anderson, N. R. Easton, and K. Folkers, J. Am. Chem. Soc., 67, 2096 (1945).

71. W. Traub, Nature, 178, 649 (1956).

72. C. Bonnemere, J. A. Hamilton, L. K. Steinrauf, and J. Knappe, Biochemistry, 4, 240 (1965).

73. J. Trotter, and J. A. Hamilton, Biochemistry, 5, 713 (1966).

74. W. Traub, Science, 129, 210 (1959).

75. D. P. Kosow, and M. D. Lane, Biochem. Biophys. Res. Commun., 7, 439 (1962).

76. J. Knappe, K. Biederbick, and W. Brümmer, Angew. Chem., 74, 432 (1962).

77. M. D. Lane and F. Lynen, Proc. Natl. Acad. Sci., U. S., 49, 379 (1963).

78. M. Waite and S. J. Wakil, J. Biol. Chem., 238, 81 (1963).

79. H. G. Wood, H. Lochmüller, and F. Lynen, Fed. Proc., 22, 537 (1963).

80. J. Knappe, B. Wenger, and U. Wiegand, Biochem. Z., 337, 232 (1963).

81. H. G. Wood, H. Lochmüller, C. Riepertinger, and F. Lynen, Biochem. Z., 337, 247 (1963).

82. S. H. G. Allen, B. E. Jacobson, and R. Stjernholm, Arch. Biochem. Biophys., 105, 494 (1964).

83. S. Numa, E. Ringelmann, and F. Lynen, Biochem. Z., 340, 228 (1964).

84. S. Numa, W. M. Bortz, and F. Lynen, Advan. Enzyme Regulation, 3, 407 (1965).

85. Y. Kaziro, A. Grossman, and S. Ochoa, J. Biol. Chem., 240, 64 (1965).

86. F. Lynen, J. Knappe, E. Lorch, G. Jütting, and E. Ringelmann, Angew. Chem., 71, 481 (1959).

87. J. Knappe, E. Ringelmann, and F. Lynen, Biochem. Z., 335, 168 (1961).

88. F. Lynen, J. Knappe, E. Lorch, G. Jütting, E. Ringelmann, and J.-P. LaChance, Biochem. Z., 335, 123 (1961).

89. M. Waite and S. J. Wakil, J. Biol. Chem., 241, 1909 (1966).

90. T. C. Bruice and A. F. Hegarty, Proc. Natl. Acad. Sci., U. S., 65, 805 (1970).

91. R. B. Guchhait, S. E. Polakis, M. I. Siegel, and C. Fenselau, Fed. Proc., 31, 419 (1972).

92. R. B. Guchhait, S. E. Polakis, D. Hollis, C. Fenselau, and M. D. Lane, J. Biol. Chem., 249, 6646 (1974).

93. E. Ryder, C. Gregolin, H.-C. Chang, and M. D. Lane, Proc. Natl. Acad. Sci., U. S., 57, 1455 (1967).

94. C. Gregolin, E. Ryder, and M. D. Lane, J. Biol. Chem., 243, 4227 (1968).

95. C. Gregolin, E. Stoll, and M. D. Lane, unpublished results.

96. M. Matsuhashi, S. Matsuhashi, S. Numa, and F. Lynen, Biochem. Z., 340, 243 (1964).

97. M. D. Hatch and P. K. Stumpf, J. Biol. Chem., 236, 2879 (1961).

98. P. F. Heinstein and P. K. Stumpf, J. Biol. Chem., 244, 5374 (1969).

99. R. B. Guchhait, J. Moss, W. Sokolski, and M. D. Lane, Proc. Natl. Acad. Sci., U. S., 68, 653 (1971).

100. R. B. Guchhait, S. E. Polakis, P. Dimroth, E. Stoll, J. Moss, E. Zwergel, and M. D. Lane, J. Biol. Chem., 249, 6633 (1974).

101. R. B. Guchhait, W. Sokolski, J. Moss, S. E. Polakis, and M. D. Lane, Fed. Proc., 30, 162 (1971).

102. S. E. Polakis and R. B. Guchhait, Fed. Proc., 31, 895 (1972).

103. S. E. Polakis, R. B. Guchhait, and M. D. Lane, J. Biol. Chem., 247, 1335 (1972).

104. S. E. Polakis, R. B. Guchhait, E. Zwergel, M. D. Lane, and T. G. Cooper, J. Biol. Chem., 249, 6657 (1974).

105. A. W. Alberts, S. G. Gordon, and P. R. Vagelos, Proc. Natl. Acad. Sci., U. S., 68, 1259 (1971).

106. A. W. Alberts, A. M. Nervi, and P. R. Vagelos, Proc. Natl. Acad. Sci., U. S., 63, 1319 (1969).

107. T. Hashimoto, N. Iritani, S. Nakanishi, and S. Numa, Proc. Japanese Conf. on Biochem. Lipids Meeting, Idaho, July, 1970, p. 21.

108. S. Numa, S. Nakanishi, T. Hashimoto, N. Iritani, and T. Okazaki, Vitamins Hormones, 28, 213 (1970).

109. T. Hashimoto, H. Isano, N. Iritani, and S. Numa, Eur. J. Biochem., 24, 128 (1971).

110. T. Hashimoto and S. Numa, Eur. J. Biochem., 18, 319 (1971).

111. W. R. McClure, H. A. Lardy, M. Wagner, and W. W. Cleland, J. Biol. Chem., 246, 3579 (1971).

112. R. E. Barden, C.-H. Fung, M. F. Utter, and M. C. Scrutton, J. Biol. Chem., 247, 1323 (1972).

113. J. Moss, M. Yamagishi, A. K. Kleinschmidt, and M. D. Lane, Biochemistry, 11, 3779 (1972).

114. R. R. Fall, A. M. Nervi, A. W. Alberts, and P. R. Vagelos, Proc. Natl. Acad. Sci., U. S., 68, 1512 (1971).

115. R. R. Fall and P. R. Vagelos, J. Biol. Chem., 247, 8005 (1972).

116. P. Dimroth, R. B. Guchhait, and M. D. Lane, Z. Physiol. Chem.,
     352, 351 (1971).
117. A. M. Nervi, A. W. Alberts, and P. R. Vagelos, Arch. Biochem.
     Biophys., 143, 401 (1971).
118. N. M. Green, Biochem. J., 89, 585 (1963).
119. J. Moss and M. D. Lane, J. Biol. Chem., 247, 4944 (1972).
120. E. Stoll, E. Ryder, J. B. Edwards, and M. D. Lane, Proc. Natl.
     Acad. Sci., U. S., 60, 986 (1968).
121. G. Henninger and S. Numa, Hoppe-Seyler's Z. Physiol. Chem., 353,
     S. 459 (1972).
122. H. Inoue and J. M. Lowenstein, J. Biol. Chem., 247, 4825 (1972).
123. A. K. Kleinschmidt and M. D. Lane, Biophys. J., 9, 106 Abstr. (1969).
124. A. K. Kleinschmidt, K. R. Leonard, M. Pendergast, R. Guchhait,
     and M. D. Lane, Abstract of the 9th Internat. Congr. of Biochem.,
     Stockholm, 2 k 43, 1973, p. 86.
125. K. Leonard, A. K. Klineschmidt, R. B. Guchhait, and M. D. Lane,
     in preparation.
126. A. L. Lehninger, in The Mitochondrion, W. A. Benjamin, New
     York, 1965, p. 32.
127. G. A. Gates, K. Henley, H. M. Pollard, E. Schmidt, and F. W.
     Schmidt, J. Lab. Clin. Med., 57, 182 (1961).
128. S. Numa, personal communication.
129. P. W. Majerus, R. Jacobs, M. B. Smith, and H. P. Morris,
     J. Biol. Chem., 243, 3588 (1968).
130. P. W. Majerus and E. Kilburn, J. Biol. Chem., 244, 6254 (1969).
131. S. Nakanishi and S. Numa, Eur. J. Biochem., 16, 161 (1970).
132. S. Nakanishi and S. Numa, Proc. Natl. Acad. Sci., U. S., 68, 2288
     (1971).
133. R. A. Jacobs, W. S. Sly, and P. W. Majerus, J. Biol. Chem.,
     248, 1268 (1973).
134. R. B. Guchhait, J. Moss, and M. D. Lane, in preparation.
135. R. B. Guchhait, E. Zwergel, and M. D. Lane, J. Biol. Chem.,
     249, 4776 (1974).
136. G. Davies and G. R. Stark, Proc. Natl. Acad. Sci., U. S., 66,
     651 (1970).
137. F. H. Carpenter and K. T. Harrington, J. Biol. Chem., 247, 5580 (1972).
138. H.-C. Chang, I. Seidman, G. Teebor and M. D. Lane, Biochem.
     Biophys. Res. Commun., 28, 682 (1967).
139. J. B. Edwards and M. D. Lane, unpublished data.
140. S. Numa, E. Ringelmann, and F. Lynen, Biochem. Z., 343, 243
     (1965).
141. J. Moss and M. D. Lane, J. Biol. Chem., 247, 4952 (1972).
142. M. Sumper and C. Riepertinger, Eur. J. Biochem., 29, 237 (1972).

143. R. K. Rasmussen and H. P. Klein, Biochem. Biophys. Res. Commun., 28, 415 (1967).

144. M. Matsuhashi, Methods in Enzymology, XIV, 3 (1969).

145. D. Burton and P. K. Stumpf, Arch. Biochem. Biophys., 117, 604 (1966).

146. C. G. Kannangara and P. K. Stumpf, Arch. Biochem. Biophys., 152, 83 (1972).

147. R. B. Guchhait and M. D. Lane, unpublished experiments.

Chapter 6

TRYPTOPHAN SYNTHETASE

Irving P. Crawford

Microbiology Department
Scripps Clinic and Research Foundation
La Jolla, California

I. INTRODUCTION 224
   A. Subunit Structure 224
   B. Enzymatic Activity 225
   C. Side Reactions 226

II. PHYSICAL AND IMMUNOLOGICAL PROPERTIES OF THE ENTERIC BACTERIAL ENZYME AND ITS SUBUNITS 227

III. MODIFICATION OF SPECIFIC AMINO ACID RESIDUES OF THE ENTERIC BACTERIAL ENZYME 230
   A. Mutational Modification 230
   B. Chemical Modification 233

IV. REACTION MECHANISM 234

V. REGULATION OF ENZYME SYNTHESIS 235

VI. EVOLUTIONARY RELATIONSHIPS 241
   A. Chromosomal Organization 241
   B. Amino Acid Sequence and Structure 249
   C. Evolutionary Trends 256

VII. RECENT DEVELOPMENTS 258

I. INTRODUCTION

A. Subunit Structure

Tryptophan synthetase [EC 4.2.1.20, L-serine hydrolyase (adding indole) and EC 4.1.2.8, indole-3-glycerolphosphate D-glyceraldehyde-3-phosphate-lyase], as befits one of the first known multicomponent enzymes, has been the subject of many reviews, some quite recent [1-4]. Ref. 1 surveys most of the studies with this enzyme up to the end of 1970, focusing on its use as a gene-enzyme system and the concept of its active site obtained through genetic and biochemical approaches. Margolin [2] discusses the regulation of all the enzymes of the tryptophan synthetic pathway in bacteria. Imamoto [3] covers transcriptional and translational studies using the enteric bacterial trp genes, and Kirschner and Wiskocil [4] describe recent enzymological and kinetic investigations with tryptophan synthetase. Topics treated extensively in these reviews will be covered only briefly here.

In all bacteria studied to date, tryptophan synthetase has an $\alpha_2\beta_2$ subunit structure. Its four constituent polypeptide chains are not covalently linked to each other, and even in neutral buffers of moderate ionic strength there is some dissociation, proceeding according to Eq. (1).

$$\alpha_2\beta_2 \rightleftharpoons \alpha + \alpha\beta_2 \rightleftharpoons 2\alpha + \beta_2 \tag{1}$$

The dissociated subunits shown on the right in Eq. (1) were easily obtained as separate proteins from the Escherichia coli enzyme by virtue of the fact that the α-chain (originally called the A component of the enzyme) remains in solution in acetate buffers at pH 4.5, whereas the β_2 subunit (the B component) is denatured. Conversely, the β_2 subunit survives heating above 55°C, while α-chains do not. Isolated α and β_2 subunits prepared in this way can recombine to form apparently normal enzyme. The $\alpha_2\beta_2$ complex binds two molecules of its cofactor, pyridoxal-5'-phosphate (PLP). Dissociated β_2 subunits retain these cofactor molecules. The β_2 dimer dissociates spontaneously to some extent [Equation (2)] ; this dissociation is augmented when the protein is resolved from its cofactor, brought to alkaline pH, or subjected to dissociating agents such as urea or sodium dodecyl sulfate.

$$\beta_2 \rightleftharpoons 2\beta \tag{2}$$

Purified α-chains from three enteric bacteria, E. coli, Salmonella typhimurium, and Enterobacter aerogenes, all have molecular weights near 29,000. Their complete covalent structure is known [5-7]. Enteric bacterial β-chains have a molecular weight of about 45,000 [8]; only a portion of the covalent structure of the E. coli β-chain has been deduced so far [9, 10]. Tryptophan synthetase from both Pseudomonas putida [11] and Bacillus subtilis [12, 13] is similar in size and subunit structure to the enteric bacterial enzyme.

The subunit structure of fungal tryptophan synthetases is not as well established as that of the bacterial ones and probably differs significantly from it. This subject will be considered more fully in Sec. VI, B.

B. Enzymatic Activity

Reaction (3) is the raison d'etre for tryptophan synthetase in those prokaryotic and eukaryotic cells possessing the enzyme. In its performance indole-3-glycerolphosphate (InGP) and L-serine are converted to L-tryptophan and D-glyceraldehyde-3-phosphate (G-3-P).

$$\text{InGP} + \text{L-serine} \xrightarrow{\text{PLP}} \text{L-tryptophan} + \text{G-3-P} \tag{3}$$

In bacteria only those forms of the enzyme containing both α and β chains and PLP can catalyze Reaction (3).

If presented with the proper substrates, the $\alpha_2\beta_2$ holoenzyme will catalyze two subreactions, reaction (4) (sometimes termed the A reaction) and reaction (5) (the B reaction). Reactions (4) and (5) performed in succession quite obviously add up to reaction (3).

$$\text{InGP} \rightleftarrows \text{Indole} + \text{G-3-P} \tag{4}$$

$$\text{Indole} + \text{L-serine} \xrightarrow{\text{PLP}} \text{L-tryptophan} + \text{H}_2\text{O} \tag{5}$$

Purified E. coli α-chains have a weak but definite ability to catalyze reaction (4), being about 1% as active alone as when saturated with β_2 subunits. Similarly, the β_2 subunit with its bound PLP displays some activity in reaction (5). Here the activity of the subunit depends strongly on the ionic environment [14], ranging from 3% of an equivalent amount of complex in the absence of monovalent inorganic cations to over 60% in strong ammonium ion-containing solutions. Refs. 1 and 4 may be consulted for detailed discussions of the mechanism whereby association of the subunits mutually increases their intrinsic catalytic efficiency in the A and B subreactions. Suffice it to say here that although direct evidence is still scanty, opinion now favors the hypothesis that a conformational change occurs as each subunit enters the complex, and this rearrangement provokes the activity increases observed.

When catalyzed by the $\alpha_2\beta_2$ complex, reaction (3) apparently proceeds through a concerted performance of the two half-reactions, (4) and (5), each portion having its catalytic site on a different subunit. It has been

known for a long time that indole produced from InGP is not released from
the enzyme complex, provided L-serine is present. Therefore, Creighton
[15], among others, has proposed that in the $\alpha_2\beta_2$ complex the active site
for the A reaction on the α subunit exists in juxtaposition to the active site
for the B reaction on the β_2 subunit. Each PLP molecule on the β_2 subunit
acts independently of the other [16], so there must be two such composite
active sites on the complex. As reaction (4) is reversible, there is a
topological requirement for four indole-binding sites on the molecule, two
on α subunits and two on the β_2 subunit. It is noteworthy that as early as
1962 DeMoss [17] was led to the conclusion that Neurospora crassa
tryptophan synthetase has two kinds of indole-binding sites, one for reaction (5) and the other for the reversal of reaction (4). This fungal enzyme
has not been dissociated into catalytically active subunits as the bacterial
one has, but just as with the bacterial enzyme, indole is not released from
the protein during the performance of reaction (3) [18].

The current view [4] of the enzymology of E. coli tryptophan synthetase
can be summarized as follows. Identical reaction mechanisms are used
for the catalysis of reaction (4) by the α subunit or the $\alpha_2\beta_2$ complex.
Similarly, the mechanism for reaction (5) is the same whether it is catalyzed by the β_2 subunit or the $\alpha_2\beta_2$ complex. The indole-binding sites
which exist on the α and β_2 subunits are retained in the α_2 and β_2 complex.
The increase in catalytic efficiency in reactions (4) and (5) exhibited by the
subunits when they enter the complex is associated with an increase in
affinity for the substrates InGP, indole, and serine and reflects a conformational change in each subunit. The failure to observe indole as a free intermediate in reaction (3) suggests that indole is transferred rapidly and ef-
ficiently ("channeled") from the site of its release from InGP to the site of
its combination with serine, so it is highly likely that the preexisting sites
for the A and B reactions on the individual subunits are in extremely close
proximity in the complex, i.e., form a "composite active center"[4, 15].

Although there are some differences between the N. crassa and E. coli
enzymes in reaction rates [15, 17], there are no compelling reasons to
postulate that the mechanism of catalysis, or even the geometry of the
active site itself, differs in the fungal type of enzyme, which does not dissociate into active subunits, from that of the bacterial and plant types, which
do.

C. Side Reactions

Several side reactions have been described for bacterial tryptophan
synthetase. Held and Smith [19] showed that 7-methyl-InGP, formed from
3-methylanthranilate, can be converted both to 7-methylindole and to 7-
methyltryptophan by resting E. coli cells containing normal tryptophan
synthetase. Indoles methylated in the 4-, 5- or 6-positions can also be

converted to the corresponding methyltryptophans both in E. coli [20] and N. crassa [21]. This failure to discern substituents on the benzene portion of the indole ring may have regulatory consequences for the cell under laboratory conditions [19] but is of minor interest in elucidating reaction mechanisms.

Some more remarkable side reactions of the $\alpha_2\beta_2$ complex and the β_2 subunit are shown in Table 1. These all involve substrates reacting with PLP on the enzyme, and include β-elimination reactions with 3- or 4-carbon amino acids [22, 23] and β-replacement reactions adding indole or thiols to these compounds [23, 24]. Although at first glance reactions (6) through (16) in Table 1 present a bewildering variety of substrates and products, all are variations of a single process. Each reaction starts with a β-substituted 3- or 4-carbon α-amino acid. This substrate presumably attaches by its amino group to PLP in the active site through a pyridoxylidine (Schiff base) linkage. Subsequent events are triggered by the elimination of the β-substituent, with the transitory appearance of an enzyme-bound intermediate having a double bond between the α and β carbons (enzyme-bound α-aminoacrylate in the case of 3-carbon substrates). With either the native complex or the isolated β_2 subunit, indole [reactions (8-10)] or, if present in high concentrations, thiols [reactions (6-7)] can be added across this double bond to form β-alanyl or S-cysteinyl derivatives. With the β_2 subunit, however, in the absence of an adding substrate the unsaturated PLP-bound intermediate can hydrolyze and be released as an α-keto acid plus ammonia [reactions (11-15)]. When the adding substrate is a thiol such as mercaptoethanol or dithioerythritol, the β_2 subunit can also catalyze a dismutation reaction in which the amino group of serine remains attached to the cofactor [reaction (16)] [24].

The major point of interest in these side reactions is that they demonstrate yet a third role played by the α subunit in the native complex. Not only does it contribute the active site for InGP binding and the performance of subreaction (4), and not only does it increase the affinity for substrates and the V_{max} at the active site for subreaction (5) on the β_2 subunit, it also increases the specificity of the β_2 subunit's active site by abolishing those side reactions leading to deamination [reaction (11)] and transamination [reaction (16)] of the natural substrate, L-serine. Whether this effect is reciprocal, with the β-chains also enhancing the specificity (in addition to the activity) of the α-chain active site in reaction (4), is not known but is an area open to investigation.

II. PHYSICAL AND IMMUNOLOGICAL PROPERTIES OF THE ENTERIC BACTERIAL ENZYME AND ITS SUBUNITS

Sucrose density gradient centrifugation, the method originally employed in studying the dissociation of the $\alpha_2\beta_2$ complex of E. coli [26], has

TABLE 1

Side Reactions Catalyzed by $\underline{E.\ coli}$ Tryptophan Synthetase or Its β_2 Subunit

Reaction	Reaction	Active form $\alpha_2\beta_2$	β_2	Ref.
A. β-Replacement				
L-Serine + $CH_3SH \rightarrow$ S-methyl-L-cysteine + H_2O	(6)	+	+	23
L-Serine + $HOCH_2CH_2SH \rightarrow$ S-hydroxyethyl-L-cysteine + H_2O	(7)	+	+	24, 25
S-Methyl-L-cysteine + indole $\rightarrow$ L-tryptophan + CH_3SH	(8)	+	+	23
O-Methyl-L-serine + indole $\rightarrow$ L-tryptophan + CH_3OH	(9)	+	+	23
β-Chloro-L-alanine + indole $\rightarrow$ L-tryptophan + HCl	(10)	+	+	23
B. β-Elimination (deamination, etc.)				
L-Serine $\rightarrow$ pyruvate + NH_3	(11)	–	+	22
L-Threonine $\rightarrow \alpha$-ketobutyrate + NH_3	(12)	–	+	22
S-Methyl-L-cysteine + $H_2O \rightarrow$ pyruvate + CH_3SH + NH_3	(13)	–	+	23
O-Methyl-L-serine + $H_2O \rightarrow$ pyruvate + CH_3OH + NH_3	(14)	–	+	23
β-Chloro-L-alanine + $H_2O \rightarrow$ pyruvate + HCl + NH_3	(15)	–	+	23
C. Transamination				
L-Serine + $HOCH_2CH_2SH$ + PLP[a] $\rightarrow$ S-hydroxyethylmercapto-pyruvate + PMP[a]	(16)	–	+	24

a Abbreviations: PLP, pyridoxal-5'-phosphate; PMP, pyridoxamine-5'-phosphate.

been used recently in studies of the S. typhimurium enzyme [27]. As would be predicted from their structural [6, 7] and immunological [28, 29] similarities, enzymes from these two enteric organisms are nearly identical in their sedimentation and dissociation behavior. One new observation is that at 5° C, even in gradients containing both PLP and L-serine, both enzymes are more dissociated when centrifuged at 50,000 rpm than at 39,000 rpm. Increasing the rotor temperature to 20° C under these conditions restores the expected amount of $\alpha_2 \beta_2$ complex even at the higher speeds. The authors interpret these results to mean that pressure effects drive the dissociation process [Eq. (1)] to the right at high speeds, and that the major forces tending to hold the subunits together are hydrophobic ones, which increase in strength as the temperature increases.

Adachi and Miles [30] have purified the β_2 subunit of the E. coli enzyme by a new and more gentle technique avoiding the drastic heat steps in the original [8] procedure. Adachi and Miles describe the crystallization of β_2 subunits from both heated and unheated preparations; these two crystalline preparations are identical in their crystal appearance, specific activities, mobilities on SDS gels, and spectral characteristics. The protein crystallizes as the apoenzyme from ammonium sulfate solutions and as the holoenzyme from potassium phosphate solutions. This work, coupled with the previously reported preliminary x-ray diffraction study of crystalline α subunit and one of its mutants [31] and the likelihood that crystals of the $\alpha_2 \beta_2$ subunit can also be formed [32, 33] opens up the prospect of detailed crystallographic studies of the enzyme and its subunits.

The dissociation of the β_2 subunit into monomers [Eq. (2)] has been studied immunologically [34], with the conclusion that the monomer maintains many of its antigenic sites after dissociation, making it cross-react rather strongly with the dimer. This result contrasts rather vividly with certain other dimeric proteins such as E. coli alkaline phosphatase [35] and imidazolylacetolphosphate aminotransferase [36], whose antigenic structures change drastically on monomerization. Additional immunological data concerning comparative studies with the α and β_2 subunits of the enzyme will be described in Sec. VI, B.

The state of the histidine residues of the dissociated α subunit of the E. coli enzyme has been investigated by observing the nuclear magnetic resonance of enzyme in which the histidine C_2 (ring) positions had been enriched in ^{13}C and labeled with deuterium [37]. The authors demonstrate that all four histidine residues are highly immobilized within the protein matrix.

III. MODIFICATION OF SPECIFIC AMINO ACID RESIDUES
OF THE ENTERIC BACTERIAL ENZYME

A. Mutational Modification

The nature and location of nine sites for inactivating missense changes in the α chain and one similar site in the β chain have been reviewed [1, 38]. Table 2 summarizes much of this information along with the location of two recently positioned ochre (nonsense) mutations in the α chain [39]. These nonsense mutants, at positions 15 and 243 in the 268 residue α chain, are the genetic markers nearest the ends of the trpA gene. Their approximate positions were known from recombinational data; the precise location was determined by sequence analysis of the α chain of a spontaneous revertant ($lys_{15} \rightarrow tyr_{15}$) and a suppressed amber strain ($gln_{243} \rightarrow$ ochre $\rightarrow$ amber $\rightarrow tyr_{243} su3^+$) [39].

The missense mutant trpA218, a phe $\rightarrow$ leu change at position 22, was originally isolated in a strain whose α chain carried a gly $\rightarrow$ ser change at position 211. Evidence exists that trpA218, which is completely auxotrophic in this background, is a prototroph in the wild type (gly_{211}) background [38]. This elegant demonstration that a so-called neutral change (the ser_{211} α-chain appears to be as active as wild type, Ref. 40) may have unexpected results elsewhere in the protein suggests a proximity of residues 22 and 211 in the three-dimensional structure of the enzyme. It was shown earlier that auxotrophy due to the trpA46 change at residue 211 is partially alleviated by trpA446 at residue 175 and that trpA187 at 213 (which is auxotrophic only on the condition that gly_{211} has been changed to val_{211}) is similarly ameliorated by trpA487 at residue 177; these results were used to suggest that residues 175 and 177 lie close to 211 and 213 in the three-dimensional structure [41]. Possibly the structural relationships uncovered by studies of mutants will be confirmed by x-ray crystallographic results in the near future.

Although the number of missense mutants in the trpA gene examined has grown to over 100, with many repeated occurrences of previously characterized changes and several instances of two or three alternate replacements at a given residue (Table 2), yet the number of sites for inactivating missense changes remains restricted to nine, with two of those being conditional on differences from the wild type sequence elsewhere in the molecule. Therefore, it is apparent that the number of possibilities for this kind of change really is quite limited. Obviously there may be many other sites where a change may result in a diminished but not abolished enzymatic activity. These would escape selection by the requirement for complete auxotrophy.

By use of the potent bacterial frameshift mutagen, ICR-191A [45], it has become possible to extend the study of amino acid substitutions to

TABLE 2

Mutational Substitutions in the E. coli Tryptophan
Synthetase α and β Chains

Mutant designation	Location	Substitution	Phenotype	Ref.
trpA38	α_{15}	Lys → ochre	Chain termination	39
trpA218	α_{22}	Phe → Leu	Inactive α subunit[a]	39
trpA3, 11 or 33	α_{49}	Glu → Val, Gln or Met	Inactive α subunit	42
trpA1	α_{49}	Glu → amber	Chain termination[b]	42
trpA446	α_{175}	Tyr → Cys	Inactive α subunit[c]	44
trpA487	α_{177}	Leu → Arg	Inactive α subunit[d]	44
trpA223	α_{183}	Thr → Ile	Inactive α subunit	40
trpA23 or 46	α_{211}	Gly → Arg or Glu	Inactive α subunit[e]	40, 43
trpA187	α_{213}	Gly → Val	Inactive α subunit[f]	38
trpA78 or 58	α_{234}	Gly → Cys or Asp	Inactive α subunit	40
trpA169	α_{235}	Ser → Leu	Inactive α subunit	44
trpA96	α_{243}	Gln → ochre	Chain termination	39
trpB244	$\beta_{\omega-15}$	Lys → Asn	Inactive β_2 subunit	10

[a] Inactive only when Ser replaces Gly at position α_{211}.

[b] By suppressor selection the protein with Gln at position α_{49} has also been obtained from trpA1.

[c] Reactivates and serves as a second site revertant for trpA46.

[d] Reactivates and serves as a second site revertant for trpA187.

[e] By suppressor selection the protein with an Asp residue at position α_{211} has also been obtained and found to be enzymatically inactive. Val, Ile, Asn, and Thr residues at this position are partially active, and Ala and Ser are as active as wild type.

[f] This mutation is completely inactive only when Val occupies position α_{211}.

residues that are not the sites of primary auxotrophic mutations. Thus Figure 1 demonstrates that one residue proximal and five distal to position α_{175} can accept substitute amino acids yielding a protein which is still at least partially active. The trpA9813 mutation was induced by ultraviolet light. It does not recombine with trpA-446 (See Table 2). The region of altered amino acid sequences in three bradytrophs obtained from this mutant either spontaneously (R8 and R11) or after ICR-191A mutagenesis (R13) [46] are shown beneath the wild type sequence. Amino acid substitutions are underlined.

And finally, two different mutations at one site, trpA23 and trpA46^{Asp}, containing arg_{211} and asp_{211} respectively, were found to give slow-growing revertants with a very labile α subunit when treated with the frameshift mutagen. Further treatment of these with the same mutagen induced fully active proteins with the sequences shown in Figure 1. The trpA46^{Asp} and trpA23 mutations are missense changes at position α_{211} (Table 2). The two revertants shown beneath the sequence of the former and the single revertant shown beneath the sequence of the latter were obtained by two-step spontaneous reversions [43]. The wild type sequence in this region is given in the last line for comparison. The inescapable conclusion is that not only is considerable latitude permitted in residues 209-213, but the labile, weakly active protein in the first stage revertant resulted from a base addition between codons 208 and 209, which provoked out-of-frame reading until a chain-terminating codon appeared somewhere in the subsequent 60 codons [43]. This abbreviated α-chain, too labile to be isolated, possessed enough activity to relieve the tryptophan requirement partially even though several missense and nonsense sites for complete auxotrophy exist between residues 211 and 243. Reference [43] contains a description of the wild type and mutant codon assignments deduced from these studies.

The presence of only a single entry for missense mutants in the β-chain in Table 2 merely indicates the paucity of effort expended along this line. Many trpB missense mutants exist [47, 48], but for most of them the amino acid substitutions have not yet been deduced. It is worth noting, however, that the ability to select mutants whose β_2 subunits are inactive by themselves but regain activity when α-chains are present, the so-called repairable class of mutants [1, 47], may permit the investigation of many sites that would be overlooked if only complete auxotrophs in the wild type were selectable.

It has been reported that for a hybrid molecule containing one β-chain from each of two different trpB mutants to be active, one of the mutant proteins must be repairable [49]. Till now such complementing mutant pairs have been studied only in vitro.

Strain	Sequence Found								
	173	174	175	176	177	178	179	180	181
wild type	- Tyr -	Thr -	Tyr -	Leu -	Leu -	Ser -	Arg -	Ala -	Gly -
trpA9813R8	- Tyr -	Thr -	Phe -	Cys -	Cys -	His -	Gly -	Ala -	Gly -
" R11	- Tyr -	Asn -	Leu -	Leu -	Leu -	Ser -	Arg -	Ala -	Gly -
" R13	- Tyr -	Thr -	Phe -	Cys -	Cys -	His -	Glu -	Gln -	Gly -

Strain	Sequence Found								
	206	207	208	209	210	211	212	213	214
trpA46^{Asp}	- Ala -	Pro -	Pro -	Leu -	Gln -	Asp -	Phe -	Gly -	Ile -
" R3-20	- Ala -	Pro -	Pro -	Ile -	Ala -	Gly -	Phe -	Gly -	Ile-
" R3-7	- Ala -	Pro -	Pro -	Ile -	Ala -	Gly -	Phe -	Cys -	Ile-
trpA23	- Ala -	Pro -	Pro -	Leu -	Gln -	Arg -	Phe -	Gly -	Ile -
" R1-6	- Ala -	Pro -	Pro -	Ile -	Ala -	Gly -	Phe -	Gly -	Ile -
wild type	- Ala -	Pro -	Pro -	Leu -	Gln -	Gly -	Phe -	Gly -	Ile -

FIG. 1. Sequences of the E. coli α-chain altered by frameshift events.

B. Chemical Modification

Both the α and β_2 subunits of the E. coli enzyme can be inactivated by sulfhydryl-reactive reagents. This work has been reviewed [1]. The tentative conclusions with respect to the α chain are: (1) that two of the three sulfhydryl groups (cys_{81} and cys_{118}) are located near each other and in proximity to the binding site for InGP, and (2) that the third sulfhydryl, cys_{154}, is more inaccessible in the native subunit than the other two to the alkylating reagent N-ethylmaleimide, but becomes more exposed in the presence of the substrate indole. Hardman and Hardman [50] strengthened and extended the first conclusion in a study utilizing the bifunctional reagent 1,5-difluoro-2,4-dinitrobenzene. This compound inactivates the α subunit while crosslinking cys_{81} with lys_{109}. From this and earlier work the authors conclude that the sulfhydryl group of cys_{81} lies 5-7 Å from the ε-amino group of lys_{109} and 10-13 Å from the sulfhydryl group of cys_{118}.

Myers and Hardman [51] also investigated covalent cross-linkages formed within the α-chain by formaldehyde. This agent does not inactivate the α-chain but introduces methylene bridges between asn_{157} and ser_{215} and between gln_{219} and ser_{233}.

To summarize the conclusions adduced so far from mutational and chemical modifications of the α-chain of the enzyme, the residue at α_{174} influences α_{211} and that at α_{176} influences α_{213}. Position α_{215} can be formaldehyde cross-linked to α_{157}. Position α_{22} is strongly influenced by α_{211}. As most of these relationships were determined by analysis of in-activating missense substitutions, it seems likely that the region $\alpha_{211-215}$ is intimately involved with the active site. Two of the cysteines, α_{81} and α_{118}, are about 10 Å apart and concerned with InGP binding. The amino group of lys_{109} is only about 5 Å from cys_{81}. Cys_{154} lies buried within the molecule unless indole is present, and the configuration of the α chain between gln_{219} and ser_{233} must form a loop so that these two residues can be cross-linked by formaldehyde. All these relationships should become subject to confirmation and refinement by x-ray crystallographic analysis in the future.

The β_2 subunit is photoinactivated in the presence of pyridoxal 5'-phosphate and L-serine as a result of the destruction of one histidyl residue per chain [52]. This essential, photosensitive histidyl residue appears to be one of the two histidyl residues which occur in the pyridoxyl peptide [53]. Diethylpyrocarbonate modification of two histidyl residues of the β-chain also entirely inactivates the enzyme [52].

IV. REACTION MECHANISM

Most of the information concerning reaction intermediates in the tryptophan synthetase reaction has been recently reviewed [1, 4]. Using stopped-flow kinetic methods and taking advantage of some of the side reactions of the β_2 subunit with L-serine, York [54] defined the order of appearance of three spectrally distinguishable intermediates involving PLP. In order, these are the aqua species (absorption at 4200 Å, strong fluorescence emission at 5000 Å), the pale species (absorption at 3300 Å, nonfluorescent), and the amber species (absorption at 4700 Å, nonfluorescent). The conversion of the aqua to the pale species appears to be rate-limiting during serine deamination by the β_2 subunit [reaction (11), Table 1], but upon addition of the α subunit the deamination reaction stops with the accumulation of the pale species, as though its conversion to the amber species is blocked. If a substrate for addition across the 2-3 double bond is added to the $\alpha_2\beta_2$-L-serine complex blocked at the pale species (either indole or a thiol will serve, see Sec. 1), the reaction proceeds but the predominant intermediate observed is the amber species, as though its conversion to free enzyme and products has become rate-limiting.

Ammonium ions interact with the β_2 subunit to influence both reaction rates and specificity [14, 24]. This ion influences the appearance of spectral intermediates in a fashion similar to the α subunit, although York [54] points out some important differences.

Miles and McPhie [55] have found that substitution of deuterium for the α-hydrogen of serine decreases the rate of pyruvate formation from serine fourfold. This large kinetic isotope effect indicates that the dissociation of the α-C-H bond of L-serine is the rate-determining step in the formation of pyruvate in the absence of added NH_4^+ ion. Miles and McPhie conclude that the aqua species which accumulates before the rate-limiting proton abstraction under these conditions has the structure of the enzyme-bound Schiff base formed between L-serine and PLP.

The results described above were obtained using conventional steady state enzymology and stopped flow spectrophotometric and fluorometric methods. Faeder and Hammes [56, 57] have employed stopped-flow temperature-jump absorption methods and reached conclusions compatible with those above. Till now, little advantage has been taken of the rich supply of mutant forms of the enzyme available for extending these studies (see Sec. III).

Recently Matchett has investigated the mode of inhibition of N. crassa tryptophan synthetase by indoleacrylic acid [58]. He determined that it interferes with reactions (3) and (5) involving L-serine, and its effects can be reversed by L-serine. In view of the similarity of the reaction mechanisms found for the fungal and bacterial enzymes and the well-known ability of indoleacrylic acid to cause derepression of the trp genes of E. coli [59], it might be valuable to investigate the effect of this compound on the appearance of reaction intermediates with the E. coli enzyme.

V.   REGULATION OF ENZYME SYNTHESIS

In the enteric bacteria (but not in all bacteria; see Sec. VI, A), the structural genes for the enzymes devoted exclusively to the synthesis of tryptophan are clustered on the chromosomes in one operon and regulated coordinately. The order of these genes and the reactions their products catalyze are shown in Figure 2. Abbreviations used in Figure 2 are: Pyr, pyruvate; PRPP, 5-phosphoribosyl-pyrophosphate; PP, pyrophosphate; CDRP, 1-(o-carboxyphenylamino)-1-deoxyribulose phosphate; P1 and P2, approximate sites of the normal and low-efficiency promoters, respectively. The numbered reactions are: 1, anthranilate synthetase; 2, phosphoribosyl transferase; 3, phosphoribosylanthranilate (PRA) isomerase; 4, indoleglycerolphosphate (InGP) synthetase; 5 and 6, the A and B reactions of tryptophan synthetase. The isolation of specific transducing

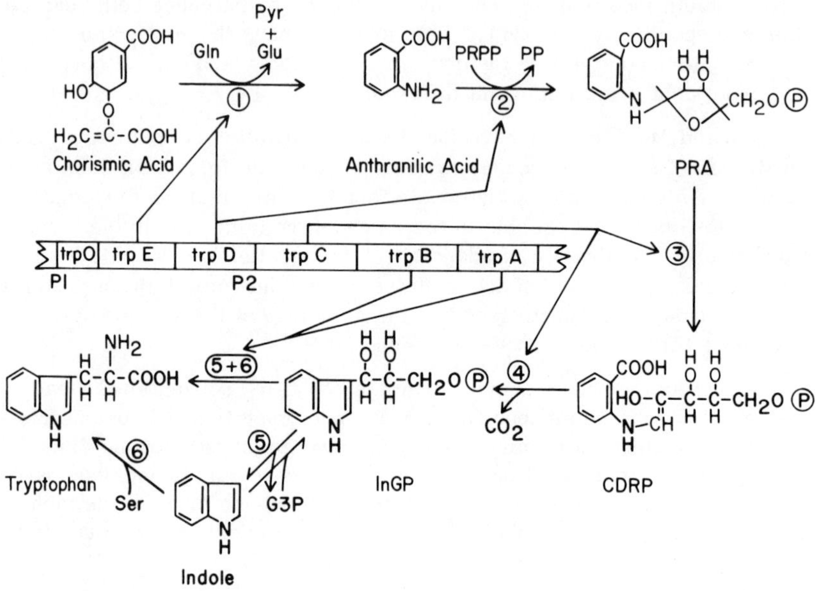

FIG. 2. The sequence of reactions in the pathway from chorismic acid to tryptophan and the trp genes of E. coli (and S. typhimurium) concerned with each reaction.

phages carrying all or parts of this bacterial operon inserted in their chromosome has permitted detailed study of the messenger RNA (mRNA) transcribed in vivo from the trp gene cluster [3]. Regulation of enzyme level is achieved by varying the number of mRNA transcripts of the operon.

Some biosynthetic enzymes are clearly regulated in response to the cell's level of transfer RNA for that amino acid, but the enzymes of the trp operon are not among these [59, 60]. In the trp case the regulation of transcription has now been studied in vitro using purified RNA polymerase and partially purified trp repressor preparations [61-63] as well as with partially purified extracts containing the endogenous forms of these constituents [64]. In both cases it is apparent that charged tryptophanyl tRNA cannot be the effector molecule, and it is highly probable that tryptophan itself, rather than something derived from it, is the effector [61, 62, 64]. The participation of the tryptophan-activating enzyme in the repressor complex, suggested earlier on genetic grounds [65], has not been confirmed in these in vitro experiments [61, 63, 64].

The structural gene for the trp repressor protein, designated trpR, is located at 1 min on the standard E. coli chromosome map, far from the genes of the trp operon (27 min) or the locus for the tryptophanyl tRNA activating enzyme (65 min). The product of the trpR gene seems to be a simple protein [66] with a molecular weight near 58,000 [61, 63], although it is not certain that it does not contain subunits or combine with other proteins to become fully functional. On combination with the effector, L-tryptophan, this protein recognizes and binds to the operator region (trpO, Fig. 2), a segment of the trp operon lying between the site of RNA polymerase attachment (promoter) and the first structural gene, trpE. This inhibits transcription of the structural genes by RNA polymerase [3, 62-64, 66]. The trp operon thus seems to exemplify the operon model for genetic organization and regulation proposed by Jacob and Monod over a decade ago [67]. Recently rather detailed studies of the promoter and operator regions have been accomplished (reviewed in Ref. 3), and it is likely that the sequence of nucleotides in this region will become known shortly [68].

One unusual aspect of the transcription of the trp operon is the occurrence, both in S. typhimurium and E. coli, of a second, weak promoter in the distal portion of the second gene. This second site of initiation for RNA polymerase is not noticeable when the operon is derepressed, for all five gene products are then produced equimolarly. When the operon is repressed, however, and transcription initiations at the normal promoter are aborted, the last three gene products are produced at five times the basal rate of the first two gene products [69]. This internal promoter has been located in the distal portion of trpD [70], not between trpD and trpC, and it does not respond to the repressor. Its maximum level of function is only about 2% that of the normal promoter at the beginning of the operon.

It is not clear at present whether the second, internal promoter just described has some selective value for the organism or whether it is the chance occurrence of an initiating nucleotide sequence in the distal portion of trpD. It is known that other promoter sites can be created by mutation within the first or second genes of the operon [71-73]. Mutants of this type are easily obtained when the conditions for their selection have been established. Perhaps a similar event occurring at the site of the second, internal promoter proved to have some selective value because the increase in basal activity of the genes for tryptophan synthetase, trpB and trpA, permitted utilization of indole as a source of tryptophan without the necessity of derepressing the synthesis of the earlier genes. If this were the case, the higher basal level of the trpC product (which catalyzes two reactions in the conversion of anthranilate to InGP) would be an unnecessary accompaniment, determined perhaps by the fact that the nucleotide sequence most favorable to mutation to a suitable internal promoter lay proximal to rather than within the third gene. Such a hypothesis seems more plausible

in an organism such as E. coli, having an active and inducible tryptophanase, than in S. typhimurium, for when E. coli is presented with a large
amount of tryptophan, this enzyme converts it to indole whence it can be
recovered for use later in protein synthesis only through the action of
tryptophan synthetase. S. typhimurium, lacking tryptophanase, might
still contain the internal promoter as a useless legacy from an ancestor
having it.

In a detailed study of transcription after rifampicin treatment, Rose and
Yanofsky [74] concluded that there are no genes of unknown function at the
extremities of the trp operon, i.e., that the five known structural genes
account for all the mRNA synthesized except for a very short region (less
than 200 nucleotides long) prior to the first gene and thought to comprise
the operator and any associated controlling elements [74-76].

A nonsense mutation in the proximal portion of a gene within an operon
can have deleterious effects on the amounts of protein produced from those
genes distal to it. This effect, termed polarity, was first noticed in the
histidine operon of S. typhimurium [77]. The polarity phenomenon has
been reviewed [78] and thoroughly discussed with respect to the trp
operon in S. typhimurium and E. coli [3]. There is a recent evaluation
of the polar effects of a large number of nonsense mutations scattered
throughout the trp operon [79]. The degree of diminished function of
genes distal to that occupied by the polar mutation appears to be related to
the absence of mRNA for these genes [80]. The precise molecular mechanism responsible for polarity remains a topic for investigation [3]. The
two strongest hypotheses are that the failure of ribosomes to traverse the
segment of mRNA distal to a nonsense mutation results in susceptibility to
cleavage by an endonuclease and subsequent rapid exonucleolytic degradation of the distal fragment(s) [81-83], or that there is premature termination of transcription when a translation termination codon is not immediately
followed by an initiation sequence [84, 85]. There is as yet no agreement
between workers in the area concerning which, if either, of these two
hypotheses is correct.

Early studies of polarity in the trp operon revealed an apparent decrease
in the level of function of the gene immediately proximal to that occupied by
the nonsense mutant, as well as distal genes. Unlike the distal polarity
effects, this was usually a moderate inhibition and never extended back
proximally more than one, or at most two, genes [86-88]. It was termed
a short range antipolar effect and was noticed with nonsense point mutants
and deletions but never with missense mutants. The amount of antipolarity
shown by a nonsense mutant in the trpA gene is related to the distance
separating it from the distal end of the gene [87], just as is the case with
polar mutants in the remainder of the operon. More recent work has
failed to demonstrate significant antipolarity from nonsense mutations in

any but trpA, the last gene in the operon [79]. When sought in the second and third genes of the nine-gene histidine operon, no antipolarity was found [89]. Thus the status of antipolarity as a regulatory phenomenon is anything but clear. The varied results obtained in different studies may be a reflection of variations in strains, growth conditions, and techniques used for derepression. Factors such as these have been suggested as reasons for discrepancies in studies of polarity and mRNA decay [3, 90, 91]. It is important that it be established whether antipolar effects are confined to the trp operon or are more general, and whether they are restricted to the last gene of that operon or not. If they are restricted to the trpA gene, it seems possible that in the absence of an α-chain of normal length the β-chain is either synthesized more slowly or degraded more rapidly than normal. Such an interpretation does not easily explain the gradient of antipolarity seen by Yanofsky and Ito [87], however. If the phenomenon of antipolarity is more general but less obvious under some growth conditions than others, it might reflect an exonucleolytic attack from the 3' end on the abbreviated message seen with polar nonsense mutants. In this case there are no immediately obvious reasons, other than technical ones, why the phenomenon should not be equally prominent in all the genes of the operon. And finally, if instability of the gene product or the mRNA is not responsible for antipolarity, it is possible that reluctance of ribosomes to detach at the nonsense codon could cause a "log jam" effect that would delay translation of the immediately proximal gene. Although the magnitude of this effect might be quite dependent on conditions, it is hard to see why it also would not be generalized to all genes and operons. Clearly, more work needs to be done before a choice can be made among these or other hypotheses.

In addition to polar and antipolar effects, one other condition is known to cause a disproportion in the amounts of various trp operon gene products in E. coli. During starvation for tryptophan in an auxotroph, the products of trpE and trpD as well as the α-chain of tryptophan synthetase continue to be slowly synthesized, while amounts of trpC gene product and tryptophan synthetase β-chain remain constant [92]. The reason for this incoordination remains unknown, but it may result from the fact that those few molecules of tryptophan that do become available (from proteolysis, for example) are used preferentially by ribosomes translating the proximal genes. Because the α-chain lacks tryptophan, ribosomes initiating near the trpB-trpA intercistronic region could nevertheless synthesize the entire trpA gene product. In support of this hypothesis, Brammar [193], who studied a double frameshift trpA mutant containing tryptophan at residue α_{213}, found that it did not synthesize the mutant α subunit under conditions of tryptophan starvation. E. Murgola [93] has obtained the same result for a missense mutant containing tryptophan at position α_{211}.

Stetson and Somerville [94] studied the regulation of an E. coli trp
operon carried on an episome, the F'colV colB particle of Fredericq [95].
Under conditions of derepression, the episome-borne version appears to
produce about twice as much tryptophan synthetase as the same segment on
the chromosome. Although direct measurements of episomal DNA by
cesium chloride density gradients give an episome-chromosome ratio of
1:1 [96], an analysis of mutation rates for a trpA marker on the episome
and chromosome suggests a 3:1 ratio [94]. Measurements of tryptophan
operon mRNA in haploids and merodiploids suggest a 2:1 preponderance
of mRNA from the episome [97], which is concordant with the enzymatic
assays. Whether the episome is in a 1:1 ratio with the chromosome and
is somewhat preferentially transcribed, or whether there is in fact a 2:1
ratio of episome to chromosome and transcription is equivalent, is not yet
clear, but in general the functioning of the operon is quite similar regardless of its chromosomal or extrachromosomal location. It is clear that
merodiploids, constructed so that the chromosome produces the β-chain
and the episome the α-chain of tryptophan synthetase, produce normal
amounts of catalytically efficient enzyme [94].

Finally, it may be noted that although it may prove common that genes
for the two subunits of a multimeric protein are located next to each other
and regulated coordinately, it is not an invariable rule. In B. subtilis
the genes for the large and small subunits of anthranilate synthetase are
unlinked [98], although all the trp genes except that for the small subunit
are in a coordinately controlled operon [99] with many similarities to the
E. coli one (see Sec. VI, A). Recent work has demonstrated that the
genes for the large and small subunits of anthranilate synthetase are also
unlinked in Acinetobacter calcoaceticus, a gram negative bacterium with
a different chromosomal organization of its trp genes [100, 101]. In
eukaryotic organisms such as the fungi, where in general the clustering
of genes of related function is less prominent than in the enteric bacteria,
loci controlling an enzyme aggregate with five activities in the common
aromatic acid biosynthetic pathway are linked in an operon-like arrangement [102-104]. Another enzyme aggregate, however, the one catalyzing
the first, third, and fourth reactions in the tryptophan pathway in N. crassa,
is determined by structural genes located on two different chromosomes
[105]. Thus the simplification in regulation afforded by an operon-like
arrangement of the structural genes for a multimeric protein is not always
utilized, either by prokaryotic or eukaryotic cells.

In at least one bacterial group, the pseudomonads, where the trp genes
are dispersed to several chromosomal locations, the genes for the α-and
β-chains of tryptophan synthetase are separate from any other structural
trp genes and are regulated by induction (with InGP as the effector)
rather than by tryptophan repression [106, 107]. Recent studies suggest
that in this system of regulation the α subunit of the enzyme forms part of
the repressor [108].

VI. EVOLUTIONARY RELATIONSHIPS

A. Chromosomal Organization

Although most of the foregoing discussion of tryptophan synthetase has been based on work with E. coli or its close relatives, other studies of the enzyme have been done in a wide variety of organisms. Before considering these I will summarize current information on the chromosomal organization of the structural genes for the tryptophan pathway and the aggregates of these enzymes found in various bacteria, fungi, algae, and plants. Then, after comparing primary and quaternary structural variations in the tryptophan synthetases from these different sources, as far as is presently known, the section will conclude with a reappraisal of speculations made about a decade ago concerning the direction taken in the evolution of this multimeric enzyme.

All members of the enteric bacterial group studied so far appear to have the structural genes for the tryptophan pathway clustered in one operon on the chromosome [109]. The genetic and biochemical evidence for this in E. coli and S. typhimurium was described in Section V. Less direct, but suggestive, evidence is available for Serratia marcescens [110], Aeromonas formicans [111], and several members of each of the genera Citrobacter, Enterobacter (Aerobacter), Erwinia, and Proteus [109, 112]. Most of the indirect evidence for an operon structure rests on coordinate regulation of the tryptophan enzymes [109-111], evidence of polarity and antipolarity [109, 111], and evidence for the same internal, low-level promoter described for E. coli and S. typhimurium in Section V [109, 111, 112]. Although the number of enteric bacteria examined is still small, the sample seems representative, for it includes several species from each of the three major taxonomic subgroups, Escherichia-Shigella-Salmonella-Citrobacter (GC content 50-52%), Enterobacter-Serratia-Klebsiella-Erwinia (GC content 54-59%), and Proteus-Providencia (GC content 39-50%), as well as the taxonomically controversial Aeromonas. This indirect evidence for the existence of an intact operon should be testable by genetic studies soon, as techniques for genetic exchange seem to be developing rapidly in Proteus [113], Klebsiella [114], Citrobacter [115], and Aeromonas [116], among others.

The enzyme aggregates found in various enteric bacteria are indicated in Figure 3. The nomenclature used for trp genes is based on the reaction(s) catalyzed by the gene product, as follows: A, tryptophan synthetase reaction A; B, tryptophan synthetase reaction B; C, InGP synthetase; D, phosphoribosyl transferase; E, anthranilate synthetase amination reaction; F, PRA isomerase; G, glutamine amidotransferase function for anthranilate synthetase. Those genes believed to be contiguous and transcribed coordinately are underlined. A dot between two adjacent gene symbols indicates a translation stop and restart signal. Genes producing bifunctional polypeptides are given two letters, one for each function (e. g., trpGD and trpCF in line 1). Genes in parentheses are assigned without certainty at present. Enzymes are designated by the numbers used in Fig. 2, except that when distinct the anthranilate synthetase amidotransferase activity is designated 1a and the glutamine amidotransferase activity coupled to anthranilate synthetase is 1b. Those enzyme activities found in an aggregate are surrounded by a box. Within the box the activities present on a single polypeptide are separated by a dot; activities on separate polypeptides are separated by a plus; unknown situations are designated by a question mark.

In all enteric bacteria tryptophan synthetase has an $\alpha_2\beta_2$ structure, and the α and β_2 subunits of different species are quite interchangeable [117], although some heterologous combinations bind with less affinity than the homologous ones [28, 29]. The two middle reactions of the pathway, phosphoribosylanthranilate (PRA) isomerase and InGP synthetase, are catalyzed by a single bifunctional polypeptide of about 45,000 mol wt [109, 111, 118].

	Chromosomal Organization	Enzyme Aggregation
Bacteria		
Enteric-type I	trpE·trpGD·trpCF·trpB·trpA	[1a+1b·2] [3·4] [5+6]
Enteric-type II	unknown	[1a+1b] [2] [3·4] [5+6]
Pseudomonas	trpE·(trpG?)·trpD·trpC trpF trpB·trpA	[1a+1b] [2] [3] [4] [5+6]
Acinetobacter	trpE trpG·trpD·trpC trpF·trpB·trpA	[1a+1b] [2] [3] [4] [5+6]
Bacillus	trpE·trpD·trpC·trpF·trpB·trpA trpG	[1a+1b] [2] [3] [4] [5+6]
Staphylococcus	trpE·trpD·trpC·trpF·trpB·trpA	[1] [2] [3] [4] [5+6]
Micrococcus	trpE·trpC·trpB·trpA trpD(trpF?)	unknown
Fungi		
Neurospora	trpEG trpCF trpD trpBA	[1+4·3] [2] [5·6]
Saccharomyces	trpEG trpC trpF trpD trpBA	[1+4] [2] [3] [5·6]
Saprolegnia	unknown	[1] [2] [3?4] [5?6]
Algae		
Anabaena	unknown	[1] [2] [3?4] [5+6]
Chlorella	unknown	unknown [5+6]
Euglena	unknown	[1] [2?3?4?5?6]
Plants	unknown	[1] [2] [3] [4] [5+6]

FIG. 3. Genes and enzyme aggregates in the tryptophan pathway of diverse organisms.

Enteric bacteria fall into two groups with respect to aggregation of the enzymes for the first two reactions of the pathway, anthranilate synthetase and phosphoribosyl transferase [109, 110]. In Escherichia sp., Salmonella sp., Citrobacter sp., Enterobacter aerogenes and cloacae, and Erwinia dissolvens, both reactions are catalyzed by an enzyme complex consisting of at least two α-and two β-chains, each chain approximately 62,000 in mol wt [119-121]. It is now apparent that one of these chains, the one encoded by trpD, is actually bifunctional, bearing the glutamine amidotransferase function for anthranilate synthetase in its operator-proximal one-third and the phosphoribosyl transferase function in its distal two-thirds [79, 122]. In the remaining enteric bacteria studied, Proteus morganii and vulgaris, Erwinia carotovora and hafniae, Serratia marcescens and marinorubra, Enterobacter liquifaciens, and A. formicans, this bifunctional trpD polypeptide is not found. Instead, one finds an independent phosphoribosyl transferase, ranging in molecular weight from 45,000 to 67,000 in various species, and an anthranilate synthetase that is a multimer of a large 60,000 mol wt chain and a small, 20,000 mol wt chain, the latter contributing the glutamine amidotransferase activity [109, 123, 124]. The obvious conclusion, that in the second group of bacteria the small anthranilate synthetase chain is homologous to the first one-third of the bifunctional chain in the first group, was suggested by the observation that the small chain of S. marcescens binds to and confers glutamine amidotransferase activity upon the trpE gene product of E. coli [125]. (A similar synergism involving the small subunit of E. hafniae anthranilate synthetase and the large E. coli anthranilate synthetase subunit has also been reported [109]). Recently the amino-terminal sequence of the small subunit of S. marcescens anthranilate synthetase was determined and found to be closely similar for 61 residues to the sequence at the amino-terminus of the trpD gene product from E. coli [126]. This result confirms the common derivation of the small, glutamine amidotransferase subunit in Serratia and its relatives and the first one-third of the trpD polypeptide in E. coli and its relatives.

In the enteric group, then, studies of the regulation of the trp enzymes have failed to uncover any remarkable differences suggesting the occurrence of more than a single operon homologous to the one in E. coli and S. typhimurium. Studies of enzyme aggregates show that there are two major subgroups, however, one in which the glutamine amidotransferase and phosphoribosyl transferase functions are fused into a single polypeptide and one in which they are not. The obvious interpretation is that an additional translational stop and restart signal occurs in the second group which is absent in the first group. Another possibility, however, is that the first one-third of the bifunctional second cistron has been duplicated and has achieved independent existence, either within or outside the operon. The bifunctional trpD polypeptide then may have evolved into an independent phosphoribosyl transferase by loss of the glutamine

amidotransferase activity and loss of the ability to bind to the product of
the trpE cistron. It should be noted parenthetically that in these and all
other bacterial anthranilate synthetases studied the large subunit, when
separated from its glutamine amidotransferase portion, can form anthranilate from chorismate and ammonia at high pH values and is just as
sensitive to feedback inhibition by L-tryptophan as the complex.

Before proceeding to other bacterial groups, it should perhaps be mentioned that studies of the tryptophan genes in different wild types of E. coli
have not revealed any striking differences within the operon, although one
difference between the frequency and extent of tonB deletions extending
into the trp region, noted early in comparisons of E. coli strains B and
K12, has been attributed to a specific, still undefined nucleotide sequence
lying between cysB and trpE, on the opposite side of the operon from tonB
[127]. One oddity known for some time is that the trp operons of E. coli
and S. typhimurium have opposite orientations, the former being transcribed counterclockwise and the latter clockwise as the chromosome is
conventionally represented [128]. The present estimate is that the inverted segment containing the trp operon actually consists of about 10% of
the genome [129], making it the most prominent topological difference
between the chromosomes of these two close relatives.

The first bacterial exception to the single operon configuration of the
trp genes was discovered in Pseudomonas aeruginosa [130]. In this
organism, and in P. putida as well [131, 132], the trp genes are dispersed
to three locations. (One should say "at least three" locations, for the
precise location of the gene for the small component of anthranilate synthetase remains unknown, though indirect evidence suggests it may be linked
to that for the large component). In this and subsequent paragraphs
these genes will be referred to by the proposed "universal" nomenclature [100] defined earlier, despite the fact that the original gene
designations for Salmonella, Pseudomonas, and the fungi were quite
different. The genes for anthranilate synthetase, phosphoribosyl transferase, and InGP synthetase, trpE, trpD and trpC, are in one cluster in that
order; the genes for tryptophan synthetase, trpB and trpA, are in another
cluster; and the gene for PRA isomerase, trpF, is separate from either of
the foregoing clusters. Interestingly, the regulation of each group is different, for the trpEDC group is repressible by tryptophan or a near
relative, just as the enteric operon is; trpBA is inducible by InGP; and
trpC is unregulated [106, 107, 133]. Pseudomonas also differs from the
enteric bacteria in having two separable, small proteins to catalyze the
PRA isomerase and InGP synthetase reactions. The molecular weight of
the isomerase in P. putida is about 39,000, and that of InGP synthetase is
32,000 [134]. Anthranilate synthetase contains a large and a small subunit [135], the latter serving the glutamine amidotransferase function as
was later found to be the case for some of the enteric bacteria, as well as

A. calcoaceticus and B. subtilis. In P. putida and aeruginosa anthranilate
synthetase appears to have an $\alpha\beta$ structure; in Pseudomonas acidovorans
and testosteroni it is probably an $\alpha_2\beta_2$ molecule [135]. Tryptophan
synthetase has the usual bacterial $\alpha_2\beta_2$ subunit structure [134].

A. calcoaceticus belongs to another large group of gram-negative bacteria, the Acinetobacter-Moraxella-Neisseria group. The genes for the
tryptophan pathway are distributed in three chromosomal locations in this
organism also, but the grouping differs from that in Pseudomonas [100].
In A. calcoaceticus the gene for the large subunit of anthranilate synthetase, trpE, is separate from the other trp genes. The gene for the small
anthranilate synthetase subunit (trpG) is linked to those for phosphoribosyl
transferase (trpD) and InGP synthetase (trpC), while that for PRA isomerase (trpF) is linked to those for tryptophan synthetase (trpB and trpA)
(Fig. 3). One interesting feature of A. calcoaceticus is that trpG mutants
require p-aminobenzoate or folate as well as tryptophan for growth, showing that the same small subunit is used for anthranilate synthetase and
p-aminobenzoate synthetase [100, 101], a situation foreshadowed in
B. subtilis [98] but not proved to be the case in the enteric bacteria or
elsewhere. Although Twarog and Liggins [136] studied the regulation and
molecular size of the tryptophan enzymes in A. calcoaceticus, the subsequent demonstration that anthranilate synthetase consists of two polypeptides controlled by unlinked genes [100, 101] makes it of great interest to
know whether both trpE and trpG respond coordinately to the same regulatory stimuli or whether trpG, in view of its dual usage, is normally either
produced in excess or regulated in response to another effector such as
p-aminobenzoate. The molecular weights of the anthranilate synthetase
multimer and all the other enzymes of the pathway except tryptophan
synthetase are very close to those reported for P. putida [136]. The
molecular weight reported for tryptophan synthetase, detected by reaction
(5) activity in zonal centrifugation experiments with wild type extracts, is
58,000 [136], but the zone of activity was quite broad. In view of the
difficulties experienced in determining the size and subunit structure of
B. subtilis tryptophan synthetase (see below), it seems possible that this
low molecular weight may have resulted from dissociation [Eq. (1) and
(2)] and that tryptophan synthetase in A. calcoaceticus, as in all other
bacteria, is a conventional $\alpha_2\beta_2$ molecule.

Among the gram-positive bacteria, B. subtilis has the longest history
as a subject for tryptophan pathway investigations. The first report of
genetic transformation in this species [137] employed an InGP synthetase
mutant. Subsequent studies with extensive collections of tryptophan auxotrophs showed that, with one exception, the tryptophan genes in this
organism are in a single, operon-like cluster [99, 138-140]. The exception is the gene for the small subunit of anthranilate synthetase, called
trpX by Kane et al. [98] but designated trpG in the universal nomenclature

used here. The gene for this 20,000 mol wt peptide, serving the glutamine amidotransferase function for p-aminobenzoate synthetase as well as anthranilate synthetase, is unlinked to the remaining six trp genes (Fig. 3). Although earlier studies were hampered by instability of some of the enzymes, particularly phosphoribosyl transferase and InGP synthetase, as soon as methods for their stabilization were found [99] it became clear that the molecular sizes are similar to those of P. putida and totally different from those of enteric bacteria. The regulation of the enzymes of the pathway is coordinate for the most part [99], although at the peak of derepression anthranilate synthetase activity lags somewhat. Kane et al. [98] suggest that in this case activity becomes limited by the amount of small component available. The synthesis of the small component is also under tryptophan control [98], however, and its regulation and that of the trp operon [141] are both modulated by a locus (termed mtr for 5-methyl-tryptophan resistance) analogous to the trpR locus in E. coli and the fir locus (for 5-fluoroindole resistance) affecting the trpEDC group in P. putida [133]. An interesting phenomenon noted by Roth and Nester [142] is that several genes adjacent to the trp operon, one on the operator side encoding an enzyme of the common aromatic pathway and two on the opposite side encoding enzymes in the histidine and tyrosine pathways, are also derepressed somewhat when the trp operon is derepressed. This effect is noted whether the derepression is brought about by tryptophan deprivation or mutation at the mtr locus. A precise molecular explanation of this effect probably must await successful in vitro studies of transcription and translation of this region.

Studies of B. subtilis tryptophan synthetase were hindered at first by instability and poor derepression conditions [143, 144]. Hoch recently solved these problems and has obtained both the α and β_2 subunits in homogeneous form [12, 13]. Under certain conditions the dissociation of the β_2 subunit [reaction (2)] can lead to spuriously low molecular weight estimates, the dissociated β-chains presumably reaggregating to active molecules under the assay conditions. Careful estimates of the molecular weights of the α-and β-chains of the B. subtilis enzyme do show them to be slightly shorter than their E. coli counterparts, however, at 26,000 and 41,800 mol wt, respectively [12, 13].

Proctor and Kloos [145, 146] described the genetics and some enzymology of the trp pathway in Staphylococcus aureus. Six trp genes are located in a cluster in the same order as those of B. subtilis. It is not yet known whether anthranilate synthetase is a multimer; if it contains a small glutamine amidotransferase subunit, the position of the gene for this polypeptide is unknown. It is certain, however, that anthranilate synthetase is not in a molecular aggregate with phosphoribosyl transferase or any other enzyme in the pathway. At a molecular weight of 30,000, the phosphoribosyl transferase seems remarkably small. Otherwise, the size of the trp enzymes and their behavior in general resembles B. subtilis. All activities of the pathway derepress similarly in tryptophan-deprived strains [146]. The instability of tryptophan synthetase in extracts has prevented any significant studies of that enzyme to date.

The trp genes in a morphologically similar but taxonomically distinct gram-positive coccus, Micrococcus luteus, are not in a single cluster [147]. Unfortunately, enzymological analysis in this species has not progressed to the point where one can tell whether the separate gene is trpD, trpF, or both of these. Also, it is not known whether there is a small, glutamine-binding subunit of anthranilate synthesis and, if so, where its gene is located.

Enzymological studies have been performed in an unusual gram-positive, rod-shaped organism, Chromobacterium violaceum [148]. The insoluble purple pigment produced by this organism is derived from tryptophan, making the selection of tryptophan auxotrophs relatively simple. The molecular weights and behavior of the tryptophan enzymes were found to be like those of B. subtilis. No evidence was found for repression or derepression of the rate of enzyme synthesis under condition of tryptophan excess or deprivation, however. Unfortunately, no system for genetic analysis exists in this organism.

Baskerville and Twarog [149] studied the regulation of trp enzymes in the anaerobic spore-forming bacterium, Clostridium butyricum. All the activities tested (phosphoribosyl transferase was not studied) responded coordinately to tryptophan repression and were constitutively derepressed in certain 5-methyltryptophan resistant mutants. Molecular weight estimates for anthranilate synthetase, InGP synthetase, and tryptophan synthetase are very similar to those of B. subtilis [98, 99]. Although no genetic information is available on this species, there is a general resemblance in tryptophan pathway enzymology to the three aerobic spore formers studied so far [150].

The survey by Hütter and DeMoss [151] of the enzymology of the tryptophan pathway in 22 conventional fungal species plus the myxomycete Physarum polycephalum represents a landmark in studies of this group, where Tatum and Bonner in 1944 [152] were the first to use tryptophan synthetase as a gene-enzyme system. There are three major groups of fungi with respect to trp gene organization and trp enzyme aggregates. Basidomycetes, ascomycetes except for the yeasts, myxomycetes, and some of the phycomycetes conform to the pattern seen in N. crassa. There are four structural genes, all of them unlinked. One determines a phosphoribosyl transferase which sediments independently of the other trp enzymes at approximately 6 S in sucrose gradients. Another encodes a tryptophan synthetase whose molecular nature will be discussed in Sec. VI, B. The polypeptides produced by the remaining two genes aggregate to form a multimer catalyzing the anthranilate synthetase, PRA isomerase, and InGP synthetase reactions. Hütter and DeMoss found two subclasses in this group, in one of which L-glutamine and ethylenediamine tetraacetate were required to maintain the integrity of the aggregate. The precise subunit structure of the aggregate has been the subject of several studies [105, 153, 154]. The most recent work [155] suggests that the aggregate has an $\alpha_2 \beta_2$ structure, a result consistent with genetic studies in both N. crassa [105] and Aspergillus nidulans [156]. Mutations affecting PRA isomerase and InGP synthetase are found in only one of the two genes, but the product of this bifunctional locus aggregates with and is required for expression of the entirely separate anthranilate synthetase locus.

A similar genetic and enzymological situation exists in the yeast Saccharomyces cerevisae and in two other yeast genera [151], but there are five unlinked trp loci. The PRA isomerase activity is distinct from the anthranilate synthetase-InGP synthetase aggregate [157]. These results suggest that what is the product of one gene in the rest of the fungi, a bifunctional polypeptide catalyzing PRA isomerase and InGP synthetase and associating with an anthranilate synthetase polypeptide, has become two polypeptides by the separation of the segment catalyzing PRA isomerization, leaving the remaining functions intact.

One group of phycomycetes, the Oomycetes group, including Saprolegnea sp. and Pythium sp., differs quite substantially from other fungi in having an anthranilate synthetase unassociated with other trp enzymes. In this group, the phosphoribosyl transferase is also independent, but PRA isomerase and InGP synthetase cosediment at 3.5 S. Whether these two activities are associated on a single polypeptide chain or not remains unclear. Previous biochemical taxonomic investigations of the phycomycetes had already established the uniqueness of the Oomycetes. Vogel [158], for example, showed that this group shares the diaminopimelic acid pathway of lysine formation with bacteria, blue-green algae, green algae, and plants, whereas all other fungi and the euglenoids make lysine via α-aminoadipate. Unfortunately, genetic analysis of the trp genes in this group of nonchitinous fungi has not yet been achieved.

Tryptophan pathway studies are just beginning in the algae. No genetic analyses have been performed, but in one of the blue-green, prokaryotic species, Anabaena variabilis, the aggregation of the enzymes has been studied [151, 159] and in another, Agmenellum quadruplicatum, a tryptophan auxotroph has been obtained and used in studies of the regulation of enzyme levels [160]. A. variabilis has an enzyme pattern very like that of the oomycetous fungi, with independent anthranilate synthetase and phosphoribosyl transferase molecules and cosedimentation of PRA isomerase and InGP synthetase at about 3.5 S [151]. All enzymes of the pathway in A. quadruplicatum increase in specific activity in response to tryptophan

deprivation [160], but additional studies will be required to determine the
precise mechanisms and effectors involved.

Studies of the tryptophan enzymes in conventional, eukaryotic, green
algae such as Chlorella ellipsoidea have so far been restricted to tryptophan synthetase [159], but it is clear that this organism has a two-compon-
ent, dissociating enzyme like the prokaryotic bacteria and blue-green algae
rather than a nondissociating one like the fungi (see Sec. VI, B). The
euglenoid eukaryotic microbe, Euglena gracilis, is unique among all the
organisms studied so far in having all the enzymes of the tryptophan pathway beyond anthranilate synthetase combined in a single aggregate of about
234,000 mol wt [161]. The anthranilate synthetase of this organism has a
molecular weight of 80,000 [194]. The subunit structure of anthranilate
synthetase, as well as that of the large trp enzyme complex in E. gracilis,
remains a subject for future study. Even at the present stage, however, it
is clear that although euglenoids share the α-aminoadipic acid pathway of
lysine formation with higher fungi, there are major differences in the
tryptophan enzymes of the two groups.

And finally, the tryptophan synthetic pathway of higher plants has been
the subject of some enzymological investigations [162-168]. It is clear
that the pathway is chemically the same as in microorganisms [163], that
feedback inhibition of anthranilate synthetase is primarily responsible for
control of tryptophan production [165], that feedback resistant mutants of
anthranilate synthetase can be selected by their resistance to growth
inhibition by tryptophan analogs [166], and that tryptophan synthetase is
dissociable into two unlike components [164, 167, 168]. The early enzymes
of the pathway all appear to reside on separate molecules. The molecular
weights of these enzymes, observed in corn and peas, are: anthranilate
synthetase, 110,000; phosphoribosyl transferase, 70,000-75,000; PRA
isomerase, 25,000; and InGP synthetase, 55,000 [162]. The latter is
probably a dimer, as active material at half that molecular weight is also
found [162]. Tryptophan synthetase, at 140,000 mol wt, apparently dissociates into two unlike subunits [167, 168]. Nothing is known at present
about the genes controlling these enzymes in plants.

B. Amino Acid Sequence and Structure

Primary sequence analysis of the α-chain of tryptophan synthetase has
been accomplished for several bacteria. The results are summarized in
Figures 4 and 5. Symbols used in Figure 4 are: Es, Escherichia coli;
Sh, Shigella dysenteriae; Sa, Salmonella typhimurium; En, Enterobacter
(Aerobacter) aerogenes; Se, Serratia marcescens; Ps, Pseudomonas
putida; Ba, Bacillus subtilis; Δ, no homologous residue was found; ↓,
identical residues at that position; X, residue not identified. Complete
sequences are available for three enteric bacteria, as well as partial

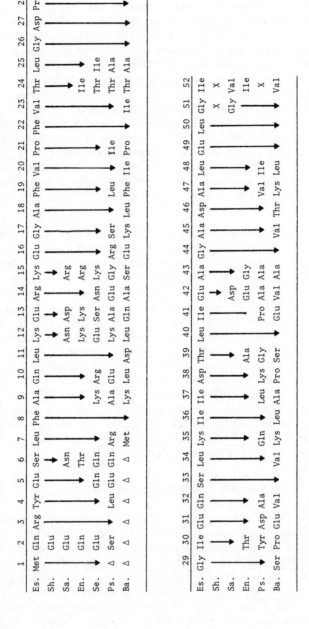

FIG. 4. Sequences known at the N-terminus of five enteric bacterial species, a pseudomonad and a bacillus.

sequences for two additional enteric species, P. putida and B. subtilis. Inspection of the data concerning the N-terminal 52-residue segment (Fig. 4) shows several things of interest. E. coli and S. dysenteriae are nearly identical, differing only in the second amino acid of the chain out of the first 50 [6]. S. typhimurium and E. aerogenes α-chains are known in their entirety and will be considered again later. In comparing their sequences with that of E. coli over the first 52 residues, they can be seen to be approximately equal in divergence, each with seven substitutions [169, 170]. Of the 28 residues known for S. marcescens, nine differ from E. coli [171], making it by far the most divergent of the enteric species so far described. This is consistent, for it is the only member of type II of the enteric bacteria in the comparison (see part A of this section). Incidentally, it is not obvious which of the four type I species is most closely related to S. marcescens.

2	6	12	13	15	24	30	39	42	43	52	56	66	68	91
Gln	Ser	Lys	Glu	Lys	Thr	Ile	Thr	Glu	Ala	Ile	Asp	Asn	Thr	Lys
Glu	Asn	Asn	Asp	Arg	Thr	Ile	Thr	Asp	Ala	Val	Asn	Asn	Asn	Asn
Gln	Thr	Lys	Lys	Arg	Ile	Thr	Ala	Glu	Gly	Ile	Asn	Gly	Ala	Asn

97	104	108	109	113	114	117	119	120	134	144	148	166	180	189
Ile	Asn	Asn	Lys	Glu	Phe	Gln	Glu	Lys	Gln	Leu	Val	Ile	Ala	Ala
Val	Lys	Ser	Pro	Glu	Leu	Arg	Glu	Gln	Gln	Leu	Ile	Val	Ser	Gly
Ile	Lys	Ser	Pro	Ala	Phe	Gln	Ala	Arg	Glu	Met	Ile	Ile	Ala	Ala

194	197	198	201	204	208	216	218	221	225	243	244	245	246	247
Asn	Val	Ala	Lys	Asn	Pro	Ala	Asp	Lys	Asp	Gln	His	Asn	Ile	Glu
His	Ile	Glu	Lys	His	Ala	Ser	Glu	Val	Arg	Lys	Asn	Leu	Ala	Ser
His	Val	Glu	Ala	His	Pro	Ala	Glu	Ser	Asp	Arg	His	Leu	Asp	Glu

249	250	253	254	256	257	260	261	262	266	267	268	269
Glu	Lys	Ala	Ala	Lys	Val	Gln	Pro	Met	Thr	Arg	Ser	X
Lys	Gln	Ala	Glu	Arg	Ser	Ser	Ala	Met	Ser	Arg	Ala	X
Gln	Thr	Asp	Glu	Lys	Ala	Gln	Ser	Leu	Thr	Lys	Thr	Ala

FIG. 5. Positions at which amino acid differences are found in the complete α-chain sequences of Escherichia coli (top line), Salmonella typhimurium (middle line) and Enterobacter (Aerobacter) aerogenes (bottom line). (From Ref. 7, courtesy of the publisher.)

All of the amino acid substitutions in the type I group and five of the nine in S. marcescens can be obtained from E. coli by a single-base codon change. This, along with the similarities in chromosomal organization presented in part A of this section, may be taken as evidence that the enteric bacteria represent a natural taxonomic group, not an artificial one.

P. putida lacks the N-terminal methionine residue present in all the enteric bacterial α-chains, but this is probably due to an intracellular peptidase capable of cleaving the methionyl-serine bond rather than a fundamental genetic difference [172]. P. putida differs from E. coli in 24 of 50 residues, but the basic homology in sequence is readily apparent. No internal deletions or insertions are required to maintain register throughout the length of the known sequence. The sequence seems best conserved in residues 22-29 and 43-51 (E. coli numbering). Thirteen of the 24 amino acid substitutions require more than a single-base change in the codon. The data seem to show that the Pseudomonas trpA gene is derived from the same ancestor as the enteric bacterial one, despite the fact that the chromosomal organization of the trp genes in the two groups of organisms is quite different.

And finally, the lowest line in Figure 4 presents the sequence recently obtained for the α-chain of B. subtilis tryptophan synthetase [173]. Here only 20 of 47 residues are identical, and it would appear that the first six residues of the chain are missing. Fifteen of the 27 amino acid substitutions require a two-base change in the codon. Again, two regions of the sequence seem better conserved than the rest, residues 19-28 and 48-52 (E. coli numbering). These two conserved areas overlap areas in the P. putida sequence. The only two positions in the first 52 residues of the E. coli α-chain known to give rise to inactive molecules by missense mutations are phe_{22} and glu_{49} (see Sect. III, A). It is noteworthy that these residues are unchanged in all organisms studied, and they lie within the two conserved areas. A nonsense mutation at lys_{15} can be suppressed by several ochre suppressors, indicating that considerable latitude is permitted in this unconserved region of the sequence. The conclusion that the two conserved areas lie either near the active site or deep within the three-dimensional structure, so that extreme conformational deformation attends their alteration, seems inescapable.

As discussed earlier, the molecular weight of the B. subtilis α-chain is about 10% less than that of E. coli [12, 173]. Thus, B. subtilis probably will have other deleted segments somewhere in the α-chain sequence.

Figure 5 shows those positions in the total α-chain sequence where differences are seen when the entire E. coli α-chain is compared with that of S. typhimurium and E. aerogenes [169, 170]. In toto, throughout 268 residues there are 44 differences between E. coli and S. typhimurium, eight requiring a two-base change in the codon; 38 differences between E. coli and E. aerogenes, six requiring a two-base change; and 44 between S. typhimurium and E. aerogenes, nine requiring a two-base change. In addition, E. aerogenes has an extra alanine added at the C-terminus. There is a noticeable clustering of substitutions in the C-terminal region of the protein, between residue 243 and the end. Incidentally, the extra alanine at the end cannot be obtained by a single-base change from any of the three termination codons.

Although amino acid substitutions appear to be randomly distributed throughout the first 90% of the sequence, it is again noteworthy that all the positions where inactivating missense mutations occur (Table 2) are completely conserved, as are the residues immediately adjacent to them. One certainly may anticipate much additional delineation of important functional areas as the complete sequences of more distantly related organisms become available.

In considering the taxonomic and evolutionary implications of these results, it must be admitted that what is known represents only a beginning. But the beginning seems highly promising, with enough sequence homology, encompassing such dissimilar organisms as the gram-negative enteric and pseudomonad groups and the gram-positive, spore-forming bacilli, to allow quantitative comparisons and to indicate that the contemporaneous trpA genes in these diverse organisms must in fact have arisen from the same ancestral gene.

From the amino acid sequences of their respective α-chains, the minimum number of DNA base changes in the trpA genes of E. coli, S. typhimurium and E. aerogenes can be calculated. When this is compared with the calculated number of base differences in intergeneric DNA-RNA hybrids by a measurement of thermal stability, a proportional but much larger number is found [7, 174]. From this result the authors have concluded that approximately as many "silent" base changes to synonymous codons have occurred and been fixed during evolution as changes leading to amino acid differences. Whether these are random events or whether they have conferred some selective advantage upon their bearers is an open question. Their occurrence does shed some light on the paradox that rather close bacterial relatives having similar chromosomal organizations and related gene products (Proteus sp. and Serratia sp. in the enteric bacteria, for example) may have quite different DNA base ratios.

At present only a single comparison of the amino acid sequence of a region of the β-chain in two different bacteria as available. The segment 23 residues long surrounding the lysine that forms a pyridoxylidene bond with the cofactor was found to be highly conserved between P. putida and E. coli [175]. Eight of the 23 residues are different; five of these require a two-base change in the codon, but none of the ionized groups in the side chains has been varied.

There are several useful methods for comparing the relatedness of proteins without obligatory purification and primary structure determination. Immunological cross-reactivity may be evaluated with any suitable antiserum, and in addition a multimer such as tryptophan synthetase allows inquiry into the ability of subunits from one organism to combine with and stimulate subunits from another. Immunological comparisons of the α and β_2 subunits from selected enteric bacteria [28, 29] have given results wholly consistent with the amino acid sequence data. One surprising result, however, is that β_2 subunits of any pair show greater cross-reactivity by microcomplement fixation than α subunits from the same pair [29]. This might be interpreted to indicate that α-chains will show more amino acid differences than β-chains of the same organisms. Chemical confirmation or refutation of this inference is highly desirable.

Immunological analyses of organisms more distantly related than E. coli and S. marcescens have been accomplished in several cases by neutralization of enzymic activity. Thus, antisera prepared against the β_2 subunits of E. coli and P. putida and against the N. crassa enzyme will all neutralize the β_2 subunit of B. subtilis, although each to less than 5% of its homologous titer [12]. The E. coli antiserum also weakly neutralizes P. putida β_2 subunits [176]. Antiserum prepared against the S. typhimurium β_2 subunit weakly neutralizes Anabaena tryptophan synthetase, while antiserum against the N. crassa enzyme does not [159]. Enzyme from a higher plant is weakly inhibited by antiserum to either the E. coli β_2 subunit or the Neurospora enzyme, but not by antiserum to the E. coli α subunit [164]. In fact, α subunit cross-reactions are rarely observed, which may be a corollary to the microcomplement fixation results just described for the Enterobacteriaceae, or may indicate that that portion of the active site bearing PLP, and thus obligatorily similar in all tryptophan synthetases, is a particularly good antigenic determinant.

For those enzymes that dissociate freely, interstrain subunit complementation would seem to be a highly specific indication of relatedness. This is an experiment easily performed even with impure preparations. One caveat is that when mutants are used to provide one subunit in the absence of the other, it must first be established that an inactive missense protein is not produced as the result of the mutation, as this could have greater affinity for the active member than its heterologous, normal counterpart and thus block a positive reaction. Excellent interspecies complementation is accomplished with the α and β_2 subunits of enteric bacteria as diverse as E. coli and S. marcescens [107], although the affinity is detectably impaired [29]. No stimulation of the activity of a P. putida subunit by an E. coli partner is found [11], and there is no evidence of α-β_2 binding between heterologous components from these sources [176]. P. putida and E. coli β_2 subunits have also been dissociated and reaggregated in the same solution [Eq. (2)] without the

appearance of an interspecific β - β hybrid [176]. B. subtilis β₂ component, on the other hand, can be stimulated to 30% maximal activity by α
subunits from either E. coli or P. putida, and the affinity of the heterologous subunits seems quite good [12]. Even the β₂ subunit from N. tabaccum
can be stimulated, although weakly, by E. coli α subunit, but the converse
combination, E. coli β₂ subunit and plant α subunit, is inactive [164].

It is reported that the β₂ subunit of plants [110, 167, 168] and algae
[159], in contrast to that of bacteria [1, 4], is as active in converting
indole to tryptophan [reaction (5)] in the absence of α subunits as in its
presence. In such a system, α-β₂ complementation can best be measured
by the appearance of activity in the overall reaction [reaction (3)]. It will
be desirable to confirm these reports with more purified preparation, for
in early experiments with partially purified B. subtilis enzyme [143, 144]
it was concluded that α subunits did not stimulate reaction (5) activity; by
hindsight it appears that α subunits had remained bound to the β₂ subunits
throughout the purification procedure, since pure β₂ subunit preparations
can be stimulated twentyfold in reaction (5) by the addition of α subunits
[12]. Nevertheless, tryptophan synthetase from the blue-green algae and
from nonfungal eukaryotic cells may have evolved toward greater inde-
pendence of the α and β₂ subunits in reactions (4) and (5), though reaction
(3) still requires the α₂β₂ complex [164, 167]. Responses of tryptophan
synthetase from Anabaena [159], plants [164, 167], Euglena [161], and
the fungi [161, 177] to monovalent cations are quite different from those
of the bacterial enzymes and from each other, but the molecular significance of these differences is not yet clear.

The subunit structure of the enzyme from the fungi and yeast has been
a source of some controversy over the years. Genetic analysis in both
N. crassa and S. cerevisae has been accomplished to the fine structure
level with large numbers of mutants, and analysis of heterokaryons in the
first case and diploids in the second case allows large scale complementation experiments. In Neurospora, mutations affecting either or both tryptophan synthetase subreactions are confined to a single area of the chromosome having all the characteristics of a single gene [178-180]. Complementation, when it occurs, is of the interallelic type between mutants
having full-sized, missense proteins [180]. Those missense proteins
lacking only reaction (4) activity are confined to one segment of the recombinational map of this locus, and those lacking reaction (5) are confined to another portion [179, 180], as is also true for the bifunctional
trpD and trpC loci in E. coli. These genetic results are paralleled in the
S. cerevisae system [181]. Here, however, Manney [182] found three
unusual nonsense mutations mapping between the two major functional
regions and producing a fragment of 35,000 mol wt possessing reaction (4)
activity. These unusual mutants do not complement mutants lacking reaction (4) activity, and in other ways are unlike trpB nonsense mutants in

E. coli. Manney's conclusion that they produce just that portion of a single polypeptide chain containing the reaction (4) active site, unattached to the remainder of the chain, seems logical and is in agreement with all the genetic evidence available.

These detailed genetic experiments suggest that in yeast and most fungi the tryptophan synthetase molecule, at 140,000 mol wt, should have the structure of a homodimer, with each 70,000 mol wt polypeptide chain participating in the active sites for both reactions (4) and (5); for many years this interpretation was at variance with the only detailed chemical analysis of a fungal enzyme, that of Carsiotis et al. which appeared in 1965 [183]. These authors reported that guanidine hydrochloride dissociates the N. crassa enzyme into subunits of 35,000 mol wt, and that, although the N-terminal amino acid is blocked, C-terminal amino acid analysis by hydrazinolysis or carboxypeptidase A digestion showed 2 mol of phenylalanine and 2 mol of leucine per mole of enzyme. The number of tryptic peptides they observed by fingerprinting was compatible with either an α_2 or an $\alpha_2\beta_2$ structure, of course.

Since this work, there have been several reports of proteolysis of tryptophan synthetase and other PLP enzymes in yeast [178, 184, 185] and Neurospora [186]. The protease involved is itself modulated by an inhibitory protein [186, 187], as was accurately predicted by Manney [188]. The degree of proteolysis is thus a very complex variable in the purification methods used earlier [176, 177, 183], and may have resulted at times in preparations that were quite active and homogeneous in molecular weight, but contained many molecules with one or a small number of proteolytic cleavages. It would not be surprising if these cleavages took place primarily near the center of the polypeptide chain.

Very recently Matchett [189] has purified tryptophan synthetase from N. crassa under conditions avoiding proteolytic cleavage. When sized in the presence and absence of dissociating agents, this molecule reduces in molecular weight no more than one-half, i.e., it appears to consist of only two subunits of 70,000 mol wt. This result is compatible with the genetic analysis just described and suggests that during evolution of the fungi the two polypeptides seen in all other organisms have been fused into one. This is reminiscent of the fusion of the trpG and trpD polypeptides in type I enteric bacteria and the fusion of the trpC and trpF polypeptides shown by all the enteric bacteria.

C. Evolutionary Trends

Probably the first person to give a coherent framework to the rapidly developing information about the tryptophan synthetase molecules in various organisms in the early 1960s was David M. Bonner. In a characteristically daring and wide-ranging paper published posthumously [190],

Bonner along with DeMoss and Mills perceived that the tryptophan synthetase of N. crassa and yeast might be the result of a fusion of elements resembling the α- and β-chains of the bacterial enzyme. Taking this surmise and the little that was known at the time about the tryptophan synthetases of B. subtilis, Anabaena, and Chlorella, these authors imagined a course of evolution beginning with a hypothetical primogenitor having two independent enzymes, one for reaction (4) and the other for reaction (5). They supposed that in successive stages the two enzymes would become increasingly more dependent on each other until, going beyond the stage seen in the enteric bacteria, they would fuse into a single component as seen in the fungi. The driving force for this evolutionary progression was thought to be an increased stability of the interdependent system and improved economy and coordination achieved by having one genetic locus in place of two. They were not unaware of the fact that increases in efficiency resulting from conservation of intermediates, in this case indole, may also be of selective advantage in such a progression.

How has their scheme fared in light of the results of the succeeding decade? Surprisingly well, it seems to me. The principle they perceived, that dissimilar polypeptides with different active sites can come together in a cooperative complex and eventually fuse into a single polypeptide, seems firmly established, not by the discovery of the hypothetical primogenitor with independent enzymes for reactions (4) and (5), but by the elucidation of similar events involving other elements of the tryptophan pathway. The best authenticated examples are the fusion of the glutamine amidotransferase portion of anthranilate synthetase with phosphoribosyl transferase in type I enteric bacteria [120-122], the fusion of PRA isomerase and InGP synthetase in enteric bacteria [79, 109, 111, 118], N. crassa [105, 153, 155], and perhaps elsewhere, and the contraction of the small and large subunits of anthranilate synthetase into a single functional locus in the fungi, dependent on association with the bifunctional PRA isomerase-InGP synthetase molecule in N. crassa or the InGP synthetase enzyme in yeast (see Fig. 3). In at least one case in the histidine pathway in E. coli, a fusion of adjacent genes to produce a bifunctional polypeptide has occurred in the laboratory [191].

It is not so obvious now as it seemed 10 years ago that the direction of evolution is toward fusion of genetic elements having related functions. Plants, presumably the most highly evolved organisms having to make their own tryptophan, have a tryptophan synthetase which is not only not fused but may show less dependence on interaction than the known bacterial enzymes. Are enteric bacteria farther along the evolutionary progression than the spore-forming bacilli? There are many such contradictions. In short, the past decade has confirmed some of the options available during evolution, but it has not wholly clarified the nature and direction of the forces influencing their employment.

One aspect of evolutionary change made more obvious by recent work is the extent to which structural genes have been translocated and attached to different regulatory elements during evolution. So far, no two major groups of bacteria have their trp genes in the same chromosomal disposition, yet the basic plan of organization of the pathway is visible in all combinations, and amino acid sequences prove the identity of these units. Recently, Jackson and Yanofsky [192] described a selective system in which duplications and translocations of multigenic segments of the trp operon to another section of the chromosome can be sought and quantified. Surprisingly, this type of event is found to be nearly as frequent as point mutation. The mechanism of this process can be studied in laboratory situations now, limited only by the investigator's ingenuity.

This observer sees several immediate and long-term goals to be attained by additional comparative studies extending the inroads begun by the work described in this section. Obviously, the natural relationships of microorganisms as well as certain higher organisms can be clarified by analyses like the simple genetic and biochemical ones described. Second, insights into the intimate mechanism of catalysis in coordinated, multifunctional enzyme complexes may be revealed more easily by comparative studies than by narrowly focusing on a single system. Finally, an understanding of the effect of environmental and internal variables on the genotype of an organism can be approached only through comparative evaluations; these should give rich insights into the nature and strength of factors shaping the appearance and behavior of ever more complex and highly developed organisms.

VII.   RECENT DEVELOPMENTS

The publication by O. Adachi, L. D. Kohn, and E. W. Miles [194] of their studies of both native and reconstituted crystalline $\alpha_2\beta_2$ complexes of E. coli tryptophan synthetase completes the triad of crystalline forms mentioned in Sec. II. It is perhaps noteworthy that each member of this triad exhibits a different crystalline shape — rectangular plates for the α subunit [31], hexagonal plates for the β_2 subunit [30], and rods for the $\alpha_2\beta_2$ complex.

A recent finding concerning regulation of synthesis of the enzymes of the tryptophan operon deserves mention. The length of trp messenger RNA preceeding the initiation codon for the first polypeptide chain is more than 150 nucleotides [195], much longer than the 26 nucleotides reported for gal messenger RNA [196] or the 38 found for lac messenger RNA [197] in the same organism. This extensive "leader" sequence is believed to have a regulatory function, based on studies of deletion mutants lacking all or part of it [198].

Recently a new technique for artificially inserting the normal E. coli trp operon into the DNA of the Col E1 plasmid was reported [199]. The number of copies of the plasmid present per cell is under experimental control and can be quite large. In this way it has been ascertained that the chromosomal trpR gene normally produces sufficient repressor molecules to inactivate only about 30 trp operators. Using cells that contain either large numbers of artificial plasmids or deletions of the leader sequence, extracts can be prepared in which tryptophan synthetase comprises more than 20% of the soluble protein. Such strains are obviously the preferred starting material for most kinds of experiments with the E. coli enzyme.

REFERENCES

1. C. Yanofsky and I. P. Crawford, The Enzymes, 3rd ed., Vol. 7 (P. D. Boyer, ed.), Academic Press, New York, 1972, p. 1.

2. P. Margolin, Metabolic Pathways, 3rd ed., Vol. 5 (H. J. Vogel, ed.), Academic Press, New York, 1971, p. 389.

3. F. Imamoto, Progress in Nucleic Acid Research and Molecular Biology, Vol. 13 (J. N. Davidson and W. E. Cohn, eds.), Academic Press, New York, 1973, p. 339.

4. K. Kirschner and R. Wiskocil, Protein-Protein Interactions, (23rd Colloquium der Geselschaft fur Biologische Chemie, R. Jaenekke, ed.) Springer-Verlag, Berlin, 1972, p. 245.

5. J. R. Guest, G. R. Drapeau, B. C. Carlton, and C. Yanofsky, J. Biol. Chem., 242, 5442 (1967).

6. S. L. Li and C. Yanofsky, J. Biol. Chem., 247, 1031 (1972).

7. S. L. Li, R. M. Denney, and C. Yanofsky, Proc. Natl. Acad. Sci., U. S., 70, 1112 (1973).

8. D. A. Wilson and I. P. Crawford, J. Biol. Chem., 240, 4801 (1965).

9. R. Fluri, L. E. Jackson, W. E. Lee, and I. P. Crawford, J. Biol. Chem., 246, 6620 (1971).

10. R. G. H. Cotton and I. P. Crawford, J. Biol. Chem., 247, 1883 (1972).

11. T. Enatsu and I. P. Crawford, J. Bacteriol., 108, 431 (1971).

12. S. O. Hoch, J. Biol. Chem., 248, 2992 (1973).

13. S. O. Hoch, J. Biol. Chem., 248, 2999 (1973).

14. M. Hatanaka, E. A. White, K. Horibata, and I. P. Crawford, Arch. Biochem. Biophys., 97, 596 (1962).

15. T. E. Creighton, Eur. J. Biochem., 13, 1 (1970).

16. G. M. Hathaway, S. Kida, and I. P. Crawford, Biochemistry, 8, 989 (1969).

17. J. A. DeMoss, Biochim. Biophys. Acta, 62, 279 (1962).

18. C. Yanofsky and M. Rachmeler, Biochim. Biophys. Acta, 28, 640 (1958).

19. W. A. Held and O. H. Smith, J. Bacteriol., 101, 209 (1970).

20. O. H. Smith and C. Yanofsky, Methods in Enzymology, Vol. 5 (S. P. Colowick and N. O. Kaplan, eds.), Academic Press, New York, 1962, p. 794.

21. A. N. Hall, D. J. Lea, and H. M. Ryden, Biochem. J., 84, 12 (1962).

22. I. P. Crawford and J. Ito, Proc. Natl. Acad. Sci., U. S., 51, 390 (1964).

23. H. Kumagai and E. W. Miles, Biochem. Biophys. Res. Comm., 44, 1271 (1971).

24. E. W. Miles, M. Hatanaka, and I. P. Crawford, Biochemistry, 7, 2742 (1968).

25. M. E. Goldberg and R. L. Baldwin, Biochemistry, 6, 2113 (1967).

26. T. E. Creighton and C. Yanofsky, J. Biol. Chem., 241, 980 (1966).

27. R. F. Dicamelli, E. Balbinder, and J. Lebowitz, Arch. Biochem. Biophys., 155, 315 (1973).

28. T. M. Murphy and S. E. Mills, J. Bacteriol., 97, 1310 (1969).

29. V. Rocha, I. P. Crawford, and S. E. Mills, J. Bacteriol., 111, 163 (1972).

30. O. Adachi and E. W. Miles, J. Biol. Chem., 249, 5430 (1974).

31. G. E. Schultz and T. E. Creighton, Eur. J. Biochem., 10, 195 (1969).

32. T. E. Creighton, personal communication.

33. O. Adachi and E. W. Miles, personal communication.

34. S. E. Mills, J. Baron-Murphy, V. Rocha, and I. P. Crawford, Arch. Biochem. Biophys., 156, 365 (1973).

35. M. J. Schlesinger, J. Biol. Chem., 240, 4293 (1965).

36. W. L. Albritton and A. P. Levin, J. Bacteriol., 111, 597 (1972).

37. D. T. Browne, G. L. Kenyon, E. L. Packer, D. M. Wilson, and H. Sternlicht, Biochem. Biophys. Res. Comm., 50, 42 (1973).

38. C. Yanofsky, H. Berger and W. J. Brammar, Proc. 12th Intern. Congr. Genet., Vol. 3, Tokyo (1969), p. 155.

39. C. Yanofsky and V. Horn, J. Biol. Chem., 247, 4494 (1972).

40. C. Yanofsky, J. Ito, and V. Horn, Cold Spring Harb. Symp. Quant. Biol., 31, 151 (1966).

41. C. Yanofsky, V. Horn and D. Thorpe, Science, 146, 1593 (1964).

42. G. R. Drapeau, W. J. Brammar, and C. Yanofsky, J. Mol. Biol., 35, 357 (1968).

43. H. Berger, W. J. Brammar, and C. Yanofsky, J. Mol. Biol., 34, 219 (1968).

44. C. Yanofsky, B. C. Carlton, J. R. Guest, D. R. Helinski, and U. Henning, Proc. Natl. Acad. Sci., U. S., 51, 266 (1964).

45. B. N. Ames and H. J. Whitfield, Jr., Cold Spring Harb. Symp. Quant. Biol., 31, 221 (1966).

46. W. J. Brammar, H. Berger, and C. Yanofsky, Proc. Natl. Acad. Sci., U. S., 58, 1499 (1967).

47. I. P. Crawford, S. Sikes, N. O. Belser, and L. Martinez, Genetics, 65, 201 (1970).
48. I. P. Crawford and S. Sikes, Genetics, 66, 607 (1971).
49. S. Kida and I. P. Crawford, J. Bacteriol., 118, 551 (1974).
50. J. K. Hardman and D. F. Hardman, J. Biol. Chem., 246, 689 (1971).
51. J. S. Myers and J. K. Hardman, J. Biol. Chem., 246, 3863 (1971).
52. E. W. Miles and H. Kumagai, J. Biol. Chem., 249, 2843 (1974).
53. E. W. Miles, Biochem. Biophys. Res. Commun., 57, 849 (1974).
54. S. S. York, Biochemistry, 11, 2733 (1972).
55. E. W. Miles and P. McPhie, J. Biol. Chem., 249, 2852 (1974).
56. E. J. Faeder and G. G. Hammes, Biochemistry, 9, 4043 (1970).
57. E. J. Faeder and G. G. Hammes, Biochemistry, 10, 1041 (1971).
58. W. H. Matchett, J. Bacteriol., 110, 146 (1972).
59. W. F. Doolittle and C. Yanofsky, J. Bacteriol., 95, 1283 (1968).
60. R. D. Mosteller and C. Yanofsky, J. Bacteriol., 105, 268 (1971).
61. C. L. Squires, J. K. Rose, C. Yanofsky, H. L. Yang, and G. Zubay, Nature New Biol., 245, 131 (1973).
62. J. K. Rose, C. L. Squires, C. Yanofsky, H. L. Yang, and G. Zubay, Nature New Biol., 245, 133 (1973).
63. Y. Shimizu, N. Shimizu, and M. Hayashi, Proc. Natl. Acad. Sci., U. S., 70, 1990 (1973).
64. D. McGeoch, J. McGeoch, and D. Morse, Nature New Biol., 245, 137 (1973).
65. K. Ito, Mol. Gen. Genet., 115, 349 (1972).
66. D. E. Morse and C. Yanofsky, J. Mol. Biol., 44, 185 (1969).
67. F. Jacob and J. Monod, J. Mol. Biol., 3, 318 (1961).
68. P. T. Cohen, M. Yaniv and C. Yanofsky, J. Mol. Biol., 74, 163 (1973).
69. D. E. Morse and C. Yanofsky, J. Mol. Biol., 38, 447 (1968).
70. E. N. Jackson and C. Yanofsky, J. Mol. Biol., 69, 307 (1972).
71. D. E. Morse and C. Yanofsky, J. Mol. Biol., 41, 317 (1969).
72. G. W. Westhoff and R. H. Baurle, J. Mol. Biol., 49, 171 (1970).
73. R. Callahan and E. Balbinder, Science, 168, 1586 (1970).
74. J. K. Rose and C. Yanofsky, J. Bacteriol., 108, 615 (1971).
75. S. Hiraga and C. Yanofsky, J. Mol. Biol., 72, 103 (1972).
76. E. N. Jackson and C. Yanofsky, J. Mol. Biol., 76, 89 (1973).
77. B. N. Ames and P. E. Hartman, Cold Spring Harb. Symp. Quant. Biol., 28, 349 (1963).
78. P. Margolin, Amer. Naturalist, 101, 301 (1967).
79. C. Yanofsky, V. Horn, M. Bonner, and S. Stasiowski, Genetics, 69, 409 (1971).
80. F. Imamoto and C. Yanofsky, J. Mol. Biol., 28, 1 (1967).
81. D. E. Morse and C. Yanofsky, Nature, 224, 329 (1969).
82. M. Kuwano, D. Schlessinger, and D. E. Morse, Nature New Biology, 231, 214 (1971).

83. D. Apirion, Mol. Gen. Genet., 122, 313 (1973).
84. F. Imamoto, Y. Kano, and S. Tani, Cold Spring Harb. Symp. Quant. Biol., 35, 471 (1970).
85. F. Imamoto and S. Tani, Nature New Biology, 240, 172 (1972).
86. J. Ito and I. P. Crawford, Genetics, 52, 1303 (1965).
87. C. Yanofsky and J. Ito, J. Mol. Biol., 24, 143 (1967).
88. E. Balbinder, A. J. Blume, A. Weber, and H. Tamaki, J. Bacteriol., 95, 2217 (1968).
89. G. R. Fink and R. G. Martin, J. Mol. Biol., 30, 97 (1967).
90. N. Morikawa and F. Imamoto, Nature, 223, 37 (1969).
91. D. E. Morse, R. Mosteller, R. F. Baker, and C. Yanofsky, Nature, 223, 40 (1969).
92. C. Yanofsky and J. Ito, J. Mol. Biol., 21, 313 (1966).
93. E. J. Murgola and C. Yanofsky, J. Mol. Biol., 86, 775 (1974).
94. H. Stetson and R. L. Somerville, Mol. Gen. Genet., 111, 342 (1971).
95. P. Fredericq, Zentralbl. Bakteriol. Parasitenk. Infektionskr. Abt. I Orig., 196, 142 (1965).
96. R. F. Hickson, T. F. Roth, and D. R. Helinski, Proc. Natl. Acad. Sci., U. S., 58, 1721 (1967).
97. T. Yura, M. Imai, T. Okamoto, and S. Hiraga, Biochem. Biophys. Acta, 169, 494 (1968).
98. J. F. Kane, W. M. Holmes, and R. A. Jensen, J. Biol. Chem., 247, 1587 (1972).
99. S. O. Hoch, C. Anagnostopoulos, and I. P. Crawford, Biochem. Biophys. Res. Commun., 35, 838 (1969).
100. R. Sawula and I. P. Crawford, J. Bacteriol., 112, 797 (1972).
101. R. Sawula and I. P. Crawford, J. Biol. Chem., 248, 3573 (1973).
102. N. H. Giles, M. E. Case, C. W. H. Partridge, and S. I. Ahmed, Proc. Natl. Acad., U. S., 58, 1453 (1967).
103. S. I. Ahmed and N. H. Giles, J. Bacteriol., 99, 231 (1969).
104. M. B. Berlyn and N. H. Giles, Genet. Res., 19, 261 (1972).
105. J. A. DeMoss, R. W. Jackson, and J. H. Chalmers, Jr., Genetics, 56, 413 (1967).
106. I. P. Crawford and I. C. Gunsalus, Proc. Natl. Acad. Sci., U. S., 56, 717 (1966).
107. D. H. Calhoun, D. L. Pierson, and R. A. Jensen, Mol. Gen. Genet., 121, 117 (1973)
108. A. R. Proctor and I. P. Crawford, Proc. Natl. Acad. Sci., U. S., 72, in press (1975).
109. M. T. Largen, Thesis, Univ. of California, Riverside, 1972.
110. M. A. Hutchinson and W. L. Belser, J. Bacteriol., 98, 109 (1969).
111. I. P. Crawford, S. Sikes, and D. K. Melhorn, Arch. für Mikrobiol., 59, 72 (1967).
112. M. T. Largen and W. L. Belser, Genetics, 75, 19 (1973).
113. P. Gemski, Jr., J. A. Wohlheiter, and L. S. Baron, Proc. Natl. Acad. Sci., U. S., 58, 1461 (1967).

114. H. Matsumoto, Genetical Research, 21, 47 (1973).
115. J. deGraff, W. Barendsen, and A. H. Stouthamer, Mol. Gen. Genet., 121, 259 (1973).
116. T. Aoki, S. Egusa, Y. Ogata, and T. Watanabe, J. Gen. Microbiol., 65, 343 (1971).
117. E. Balbinder, Biochem. Biophys. Res. Commun., 17, 770 (1964).
118. J. F. McQuade and T. E. Creighton, Eur. J. Biochem., 16, 199 (1970).
119. J. Ito and C. Yanofsky, J. Bacteriol., 97, 734 (1969).
120. E. J. Henderson and H. Zalkin, J. Biol. Chem., 246, 6891 (1971).
121. A. F. Egan and F. Gibson, Biochem. J., 130, 847 (1972).
122. M. Grieshaber and R. Bauerle, Nature New Biol., 236, 232 (1972).
123. H. Zalkin and L. H. Hwang, J. Biol. Chem., 246, 6899 (1971).
124. F. Robb, M. A. Hutchinson, and W. L. Belser, J. Biol. Chem., 246, 6908 (1971).
125. F. Robb and W. L. Belser, Biochem. Biophys. Acta, 285, 243 (1972).
126. S. L. Li, J. Hanlon, and C. Yanofsky, Nature, 248, 48 (1974).
127. J. A. Spudich, V. Horn, and C. Yanofsky, J. Mol. Biol., 53, 49 (1970).
128. K. E. Sanderson and C. A. Hall, Genetics, 64, 215 (1970).
129. F. Casse, M.-C. Pascal, and M. Chippaux, Mol. Gen. Genet., 124, 253 (1973).
130. B. Fargie and B. W. Holloway, Genet. Res., 6, 284 (1965).
131. I. C. Gunsalus, C. F. Gunsalus, A. M. Chakrabarty, S. Sikes, and I. P. Crawford, Genetics, 60, 419 (1968).
132. B. W. Holloway, V. Krishnapillai, and V. Stanisich, Ann. Rev. Genet., 5, 425 (1971).
133. R. Maurer and I. P. Crawford, J. Bacteriol., 106, 331 (1971).
134. T. Enatsu and I. P. Crawford, J. Bacteriol., 95, 107 (1968).
135. S. F. Queener and I. C. Gunsalus, Proc. Natl. Acad. Sci., U. S., 67, 1225 (1970).
136. R. Twarog and G. L., Liggins, J. Bacteriol., 104, 254 (1970).
137. J. Spizizen, Proc. Natl. Acad. Sci., U. S., 44, 1072 (1958).
138. C. Anagnostopoulos and I. P. Crawford, Proc. Natl. Acad. Sci., U. S., 47, 378 (1961).
139. C. Anagnostopoulos and I. P. Crawford, Compt. Rend. Acad. Sci. Fr., 265, 93 (1967).
140. B. C. Carlton and D. D. Whitt, Genetics, 62, 445 (1969).
141. S. O. Hoch, C. W. Roth, I. P. Crawford, and E. W. Nester, J. Bacteriol., 105, 38 (1971).
142. C. W. Roth and E. W. Nester, J. Mol. Biol., 62, 577 (1971).
143. A. K. Schwarz and D. M. Bonner, Biochem. Biophys. Acta, 89, 337 (1964).
144. J. W. Meduski and S. Zamenhof, Biochem. J., 112, 285 (1969).

145. A. R. Proctor and W. E. Kloos, J. Gen. Microbiol., 64, 319 (1970).

146. A. R. Proctor and W. E. Kloos, J. Bacteriol., 114, 169 (1973).

147. W. E. Kloos and N. E. Rose, Genetics, 66, 595 (1970).

148. J. Wegman and I. P. Crawford, J. Bacteriol., 95, 2325 (1968).

149. E. N. Baskerville and R. Twarog, J. Bacteriol., 112, 304 (1972).

150. S. O. Hoch and I. P. Crawford, J. Bacteriol., 116, 685 (1973).

151. R. Hütter and J. A. DeMoss, J. Bacteriol., 94, 1896 (1967).

152. E. L. Tatum and D. M. Bonner, Proc. Natl. Acad. Sci., U. S.,
     30, 30 (1944).

153. F. H. Gaertner and J. A. DeMoss, J. Biol. Chem., 244, 2716
     (1969).

154. F. H. Gaertner and J. L. Leef, Biochem. Biophys. Res. Commun.,
     41, 1192 (1970).

155. F. M. Hulett and J. A. DeMoss, Abs. Annu. Meet. Amer. Soc.
     Microbiol., 73rd (1973), p. 148.

156. C. F. Roberts, Genetics, 55, 233 (1967).

157. J. A. DeMoss, Biochem. Biophys. Res. Commun., 18, 850 (1965).

158. H. J. Vogel, Evolving Genes and Enzymes (V. Bryson and H. J.
     Vogel, eds.), Academic Press, New York, 1965, p. 251.

159. K. Sakaguchi, Biochem. Biophys. Acta, 220, 580 (1970).

160. L. O. Ingram, D. Pierson, J. F. Kane, C. van Baalen, and
     R. A. Jensen, J. Bacteriol., 111, 112 (1972).

161. J. C. Lara and S. E. Mills, J. Bacteriol., 110, 1100 (1972).

162. C. Hankins and S. E. Mills, in preparation.

163. D. P. Delmer and S. E. Mills, Plant. Physiol., 43, 81 (1968).

164. D. P. Delmer and S. E. Mills, Biochem. Biophys. Acta, 167, 431
     (1968).

165. W. L. Belser, J. Baron-Murphy, D. P. Delmer, and S. E. Mills,
     Biochem. Biophys. Acta, 237, 1 (1971).

166. J. M. Widholm, Biochem. Biophys. Acta, 261, 52 (1972).

167. R. T. Nagao and T. C. Moore, Arch. Biochem. Biophys, 149, 402
     (1972).

168. J. Chen and W. G. Boll, Can. J. Bot., 50, 587 (1972).

169. S. L. Li and C. Yanofsky, J. Biol. Chem., 248, 1830 (1973).

170. S. L. Li and C. Yanofsky, J. Biol. Chem., 248, 1837 (1973).

171. S. L. Li, G. R. Drapeau, and C. Yanofsky, J. Bacteriol., 113,
     1507 (1973).

172. I. P. Crawford and C. Yanofsky, J. Bacteriol., 108, 248 (1971).

173. S. L. Li and S. O. Hoch, J. Bacteriol., 118, 187 (1974).

174. R. M. Denney and C. Yanofsky, J. Mol. Biol., 64, 319 (1972).

175. R. Maurer and I. P. Crawford, J. Biol. Chem., 246, 6625 (1971).

176. R. Maurer and I. P. Crawford, Arch. Biochem. Biophys., 144,
     193 (1971).

177. R. G. Meyer, J. Germershausen, and S. R. Suskind, Methods in
     Enzymology, Vol. 17A (H. Taber and C. W. Tabor, eds.),
     Academic Press, New York, 1970, p. 406.

178. M. Ahmed and D. Catcheside, Heredity, 15, 55 (1960).
179. S. Kaplan, Y. Suyama, and D. M. Bonner, Genetics, 49, 145 (1964).
180. A. M. Lacy, Biochem. Biophys. Res. Commun., 18, 812 (1965).
181. T. R. Manney, W. Duntze, N. Janosko, and J. Salazar, J. Bacteriol., 99, 590 (1969).
182. T. R. Manney, Genetics, 60, 719 (1968).
183. M. Carsiotis, E. Apella, P. Provost, J. Germershausen, and S. R. Suskind, Biochem. Biophys. Res. Commun., 18, 877 (1965).
184. E. G. Afting, T. Katsunuma, H. Holzer, N. Katunuma, and E. Kominami, Biochem. Biophys. Res. Commun., 47, 103 (1972).
185. T. Katsunuma, E. Schött, S. Elsässer, and H. Holzer, Eur. J. Biochem., 27, 520 (1972).
186. P. H. Yu, M. R. Kula, and H. Tsai, Eur. J. Biochem., 32, 129 (1973).
187. A. R. Gerguson, T. Katsunuma, H. Betz, and H. Holzer, Eur. J. Biochem., 32, 444 (1973).
188. T. R. Manney, J. Bacteriol., 96, 403 (1968).
189. W. H. Matchett and J. A. DeMoss, J. Biol. Chem., 250, in press.
190. D. M. Bonner, J. A. DeMoss, and S. E. Mills, Evolving Genes and Proteins (V. Bryson and H. J. Vogel, eds.), Academic Press, New York, 1965, p. 305.
191. J. Yourno, T. Kohno, and J. R. Roth, Nature, 228, 820 (1970).
192. E. N. Jackson and C. Yanofsky, J. Bacteriol., 116, 33 (1973).
193. W. J. Brammar, J. Gen. Microbiol., 76, 395 (1973).
194. O. Adachi, L. D. Kohn, and E. W. Miles, J. Biol. Chem., 249, 7756 (1974).
195. M. J. Bronson, C. Squires, and C. Yanofsky, Proc. Natl. Acad. Sci., U. S., 70, 2335 (1973).
196. R. E. Musso, B. deCrombrugge, I. Pastan, J. Sklar, P. Yot, and S. Weissman, Proc. Natl. Acad. Sci., U. S., 71, 4940 (1974).
197. N. M. Maizels, Proc. Natl. Acad. Sci., U. S., 70, 3585 (1973).
198. E. N. Jackson and C. Yanofsky, J. Mol. Biol., 76, 89 (1973).
199. V. Hershfield, H. W. Boyer, C. Yanofsky, M. A. Lovett, and D. R. Helinski, Proc. Natl. Acad. Sci., U. S., 71, 3455 (1974).

Chapter 7

SQUALENE AND STEROL CARRIER PROTEINS

Mary E. Dempsey

Department of Biochemistry
University of Minnesota
Medical School
Minneapolis, Minnesota

I.   INTRODUCTION   268

II. DISCOVERY OF SQUALENE AND STEROL CARRIER
     PROTEIN   270

III. STRUCTURAL CHARACTERISTICS OF SQUALENE AND
     STEROL CARRIER PROTEIN   274
     A. Purification   274
     B. Molecular Characteristics; Protomer Form   277
     C. Lipid Binding; Oligomer Form   279

IV. FUNCTIONS OF SQUALENE AND STEROL CARRIER
     PROTEIN   282
     A. Role in Cholesterol Synthesis; Proposed Mechanism
         of Action   286
     B. Relationship to Plasma Lipoproteins   287
     C. Role in Steroid Hormone Synthesis   290
     D. Role in Bile Acid Synthesis   291
     E. Role in Membrane Formation by Protozoa   292
     F. Relationship to Other Lipid-carrying Proteins and
         Soluble Protein Cofactors   294

267

V.   UBIQUITOUS OCCURRENCE OF SQUALENE AND STEROL
      CARRIER PROTEIN   296

VI.   REGULATORY ROLE OF SQUALENE AND STEROL CARRIER
      PROTEIN   297

VII. CONCLUDING REMARKS   301

I. INTRODUCTION

Progress in elucidation of mechanisms of sterol biosynthesis, metabolism, and transport, as well as regulation of these processes, has been hampered primarily due to water-insolubility of the substrate molecules and the particulate nature of the enzymes involved. The recent discovery of an ubiquitous, versatile, soluble protein which binds sterols and other lipids offers hope for accelerated clarification of important unsolved problems in this area of research. The soluble protein was named <u>squalene and sterol carrier protein</u> (SCP) to indicate its function in cholesterol synthesis, metabolism, and transport [1, 2].

It is appropriate to include current knowledge of SCP in this volume on subunit proteins; all evidence indicates that the functional form of SCP is a high molecular weight complex of SCP monomers and lipid molecules (Fig. 1). Pro-SCP has a molecular weight of 16,000 daltons; the cholesterol precursor-SCP complex is a high molecular weight oligomer (> 150,000 daltons). Current studies indicate that SCP is not only required for conversion of water-insoluble precursors to cholesterol by microsomal enzymes (Fig. 2), but is also required for cholesterol

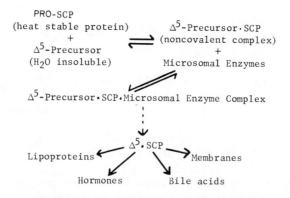

FIG. 1. Proposed biological roles of SCP in cholesterol (Δ^5) biosynthesis and metabolism. (From Ref. 70, courtesy of Annual Reviews, Inc.)

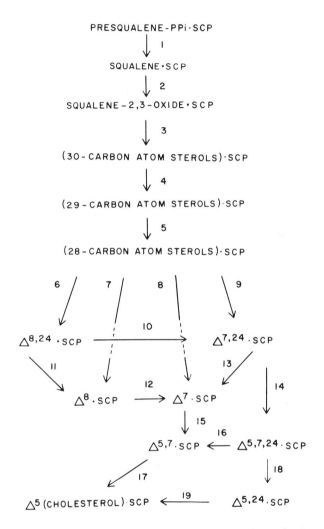

FIG. 2. Outline of pathways and intermediate compounds in the later stages of cholesterol biosynthesis; role of SCP. (From Ref. 70, courtesy of Annual Reviews, Inc.)

metabolism to bile acids and steroid hormones by microsomal and mito-chondrial enzymes (Fig. 1). In Figure 2, intermediates are shown as SCP complexes, and the positions of unsaturation are indicated by the delta (Δ) symbol. In addition, SCP has functional and possibly structural similarities to one of the major human plasma, high density lipoprotein polypeptides and is involved in membrane biosynthesis (as indicated by studies with protozoa) (Fig. 1). Preliminary evidence further suggests that SCP may play a regulatory role in sterol synthesis and metabolism. It thus appears that SCP is an answer to a lipid biochemist's prayer. However, a note of caution should be sounded, i.e., all studies to date with SCP were performed in vitro. It is essential, although difficult to establish in future work that the apparent functional roles of SCP are operative in vivo.

The purposes of this chapter are to summarize observations which led to the detection of SCP; to describe the known structural and functional properties of SCP, as well as its general biological occurrence; and to speculate on possible regulatory roles of SCP in lipid metabolism. A number of findings from the author's laboratory, not yet published, are also included.

II. DISCOVERY OF SQUALENE AND STEROL CARRIER PROTEIN

Several comprehensive reviews are available concerning pathways and intermediates in sterol biosynthesis, e.g., Clayton and Frantz and Schroepfer [3, 4], as well as properties of microsomal enzymes catalyzing specific steps, e.g., Gaylor [5]; See also the volume edited by Clayton [6].

During the 1950s and early 1960s many investigators studying cholesterol biosynthesis in vitro noted the requirement for the soluble fraction of liver homogenates, in addition to the microsomal fraction (e.g., Bucher and her colleagues [7-9]; Frantz et al. [10]; Kandutsch [11]; Goodman et al. [12]). Our first entrance into this field was through attempts to purify and characterize microsomal enzymes catalyzing some of the final steps in cholesterol synthesis, e.g., Δ^5-dehydrogenase and Δ^7-reductase (Fig. 3) [13, 14]. We found that the lower half of the high-speed supernatant fraction of liver homogenates contained both these enzymic activities. In addition, a combination of the upper half of the high-speed supernatant fraction and the washed microsomal fraction was also catalytically active, whereas the same fractions alone were only slightly active (Table 1) [14]. The lower half of the high-speed supernatant fraction apparently contains a mixture of the soluble fraction and slower sedimenting microsomes [15, 16]. This mixture was useful in demonstrating the intermediary role of $\Delta^{5,7}$-sterols in cholesterol biosynthesis, the requirement for NADPH by Δ^7-reductase, (step 2, Fig. 3), and the sites of action of various inhibitors of cholesterol synthesis [17-19].

The discovery of SCP resulted from attempts to identify the material present in the soluble liver fraction which was required for activity by

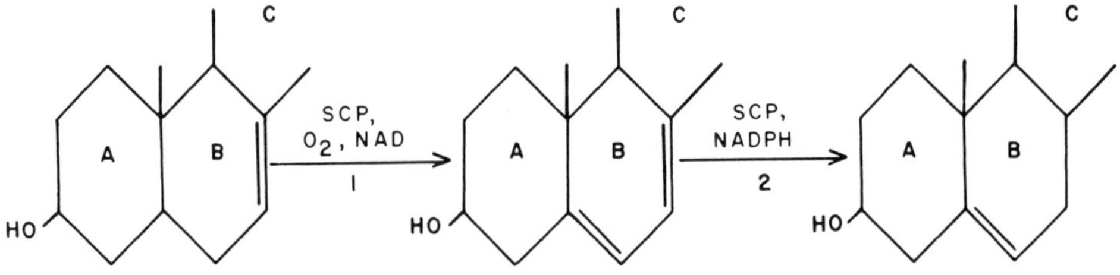

Δ^7-CHOLESTENOL $\Delta^{5,7}$-CHOLESTADIENOL   CHOLESTEROL

1. Δ^7-STEROL-Δ^5-DEHYDROGENASE (Δ^5-DEHYDROGENASE)

2. $\Delta^{5,7}$-STEROL-Δ^7-REDUCTASE (Δ^7-REDUCTASE)

FIG. 3. Reactions catalyzed by microsomal Δ^5 - dehydrogenase and Δ^7 - reductase; cofactor requirements.

TABLE 1

Liver Cell Fractions Required for Cholesterol Synthesis [a]

Cell Fractions	Conversion to cholesterol μmol/mg microsomal protein/hour	
	$\Delta^{5,7}$ - Cholestadienol	Δ^7 - Cholestenol
Microsomal [b]	0.4	0.1
Microsomal [b] +soluble [c]	1.7	1.4
Soluble [c]	0	0

[a]Adapted from Ref. 14.

[b]The 105,000 x g sediment obtained after removal of the nuclear and mitochondrial fractions.

[c]The slower sedimenting (upper-half) of the 105,000 x g supernatant fraction.

microsomal fraction enzymes, in addition to low molecular weight cofactors, e.g., NADPH. A careful study of the effects of increasing levels of soluble fraction protein to microsomal Δ^7 - reductase (in the presence of excess NADPH) (Step 2, Fig. 3) revealed a marked activation of enzymatic activity until a maximum was reached (Fig. 4) [1]. The pattern of activation suggested that subunit interactions were occurring. The absolute requirement for the soluble fraction by microsomal enzymes catalyzing cholesterol synthesis was demonstrated by using partially purified microsomal preparations [16], washed extensively by centrifugation to remove endogenous soluble fractions (e.g., Table 2) [20].

More than 10 years ago we noted that the material required, in addition to NADPH, for microsomal Δ^7- reductase activity (step 2, Fig. 3) was heat-stable and nondialyzable [14]. The material also remained with the

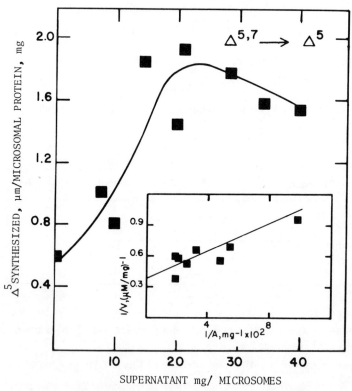

FIG. 4. Requirement of the high-speed supernatant fraction of liver homogenates for maximum activity of Δ^7-reductase. (From Ref. 1, courtesy of Academic Press.)

TABLE 2

Absolute Requirement for the Soluble Fraction by
Microsomal Δ^5 - Dehydrogenase[a]

Enzyme Preparation	Washes[b]	Activation of Enzyme by Soluble Fraction (-fold)
Microsomal	1	4.0
	2	5.4
	3	9.0
Purified microsomal[c]	1	15.3
	3	∞

[a] Adapted from Ref. 20.

[b] By centrifugation in buffer at 105,000 x g.

[c] By method described in Ref. 16.

protein fraction during gel filtration. It was precipitated by ammonium sulfate and perchloric acid and inactivated by trypsin [1, 16]. At first this heat-stable protein did not appear to be a general requirement for enzymes catalyzing cholesterol synthesis, e.g., it could not be shown to function with microsomal Δ^5 - dehydrogenase (Step 1, Fig. 3) [15, 16]. A key observation was that heat treatment of the soluble fraction caused loss of a low level of pyridine nucleotide which survives gel filtration of the soluble fraction and is required for dehydrogenase activity [21, 22]. Thus, a combination of the heat-stable protein and pyridine nucleotide (NAD) was necessary to observe dehydrogenase activity. The next key finding was that the soluble, heat-stable protein bound insoluble precursors of cholesterol and that bound precursors were readily converted to cholesterol by microsomal enzymes [1, 2, 23]. The name squalene and sterol carrier protein (SCP) was coined to describe the function of the heat-stable protein (Fig. 1) [2]. Most aspects of our initial results were quickly confirmed by other laboratories [24-27], and the SCP story began. It is hoped it will not end as a tale of an interesting artifact.

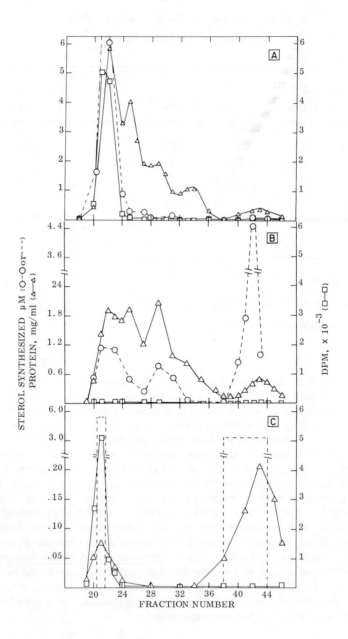

FRACTION NUMBER

III. STRUCTURAL CHARACTERISTICS OF SQUALENE AND STEROL CARRIER PROTEIN

In this section, current methods of purifying SCP from the soluble fraction of liver homogenates are described, as well as some of the known structural properties of SCP.

A. Purification

The first method developed for purification of SCP capitalized on the known heat-stability of this soluble protein (Fig. 5A, B; Fig. 6, Method 1) [1, 2]. In Figure 5, protein content is designated by $\triangle$———$\triangle$; labeled sterol binding by, $\square$———$\square$; and functional (SCP) activity of protein fractions with microsomal enzymes by $\bigcirc$---$\bigcirc$. In Figure 6, protein content is designated by $\bullet$———$\bullet$; SCP functional activity determined using the Δ^5-dehydrogenase assay (Fig. 7) by $\bigcirc$---$\bigcirc$. Method 1 involves heat treatment of the liver soluble fraction at high ionic strength followed by sizing on Sephadex G-75 [(1-2) and Fig. 5-B]; Method 2 involves sizing of the soluble fraction without prior heating. At least two molecular forms of SCP are produced. One is a low molecular weight protomer form (sterol free); the other is a high molecular weight oligomer form containing bound sterols and lipids (cf. also part C of this section). Both of these forms are functionally active with microsomal enzymes. Often molecular forms intermediate in size between these two major forms are observed, no doubt due to the marked tendency of SCP to aggregate (cf. Fig. 5B and Fig. 6, method 1). The second method for purification of SCP avoids the heat step at high ionic strength, i.e., the soluble fraction is simply concentrated and sized by gel filtration (Fig. 6, Method 2) [28, 29]. The yield of SCP is markedly increased (e.g., five- to sixfold) by the latter technique. Homogeneous pro-SCP is obtained by treatment of the low molecular weight form of SCP (isolated by either the heat or nonheat method) with an additional ion exchange chromatographic step [29]. Our evidence indicates that the level of homogeneous pro-SCP in liver may be as high as 1% of the soluble protein. The structural characteristics of pro-SCP preparations obtained by these methods are outlined in part B of this section. Scallen et al. [25] have also described a method of purifying SCP from liver using alumina-cellulose columns, followed by gel filtration.

FIG. 5. Gel filtration (Sephadex G-75) patterns illustrating the purification and molecular properties of SCP isolated from the liver soluble fraction. A, sterol binding to the oligomer form of SCP (fractions 21-24); SCP activity of the oligomer; B, formation of protomer-SCP (fractions 40-44) by heat treatment at high ionic strength; occurrence of higher molecular weight aggregates of the protomer; lack of sterol binding to the protomer; C, conversion of protomer-SCP to the oligomer; sterol binding to the oligomer. (From Ref. 2, courtesy of the Journal of Biological Chemistry.)

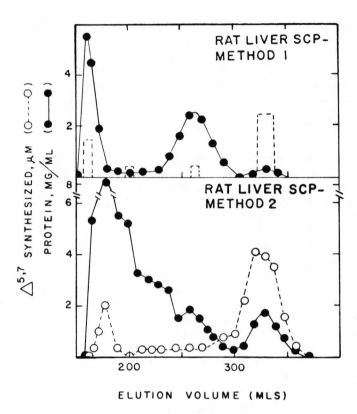

FIG. 6. Gel filtration patterns illustrating the SCP protein and activity yields obtained by two methods. (From Refs. 28 and 29.)

Their yield was approximately 0.4% of the soluble protein; the degree of purification or homogeneity of their preparation has not been reported.

It should be noted that the estimated degree of purification of SCP depends on the method used for assay of SCP functional activity. For example, using the Δ^5 - dehydrogenase assay [16, 22] (Step 1, Fig. 3) we reported a 720-fold purification, based on the activity of a Δ^7-cholestenol-SCP complex with the dehydrogenase [20] (cf. also Sect. IV). It is hoped that workers in this area will soon agree on a standard method for determining SCP functional activity. The Δ^5 - dehydrogenase assay may be appropriate for this purpose; it is rapid and does not require labeled substrate, i.e., the characteristic ultraviolet absorption spectrum of a $\Delta^{5,7}$ -sterol [17] reflects the overall activity of the sterol substrate–SCP–enzyme complex (Fig. 7) [22].

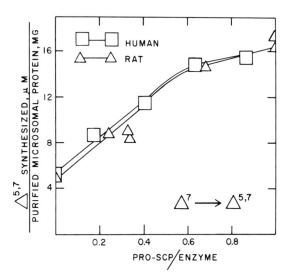

FIG. 7. Assay of SCP activity by the Δ^5-dehydrogenase technique [22]; functional similarity of human-and rat-SCP.

B. Molecular Characteristics; Protomer Form

The protomer form of SCP (pro-SCP or apo-SCP) prepared by either Method 1 or Method 2 (Fig. 6, and see part A of this section) exhibits a molecular weight of 16,000 daltons, as determined by gel filtration and SDS polyacrylamide gel electrophoresis (Figs. 8 and 9). Prior to ion exchange chromatography, pro-SCP also migrates as a single band during polyacrylamide gel electrophoresis in urea at pH 9 (Fig. 10). However, at pH 4.3 in this same electrophoretic system, several minor impurities appear; these are removed by the ion exchange chromatographic step (Fig. 11). The amino acid compositions of rat, human, and protozoan SCP (Sect. IV, D) are compared in Table 3 [30]. Numerous similarities in composition are apparent. It will be important in future studies to assess the possible functional role of the sulfhydryl or disulfide groups in these molecules (cf. also section IV). The availability of homogeneous SCP offers opportunities for definitive studies on its primary structure (cf. Sect. IV, B) and for the development of immunochemical assays for its occurrence in vivo under various physiological conditions (cf. Sect. VI).

At present there is a controversy in the literature regarding the molecular weight of sterol-free SCP. Scallen et al. [24, 25] reported molecular weights varying from 5×10^3 to 6×10^4 daltons, depending on the method of measurement used (e.g., ultracentrifugation, gel filtration). Although

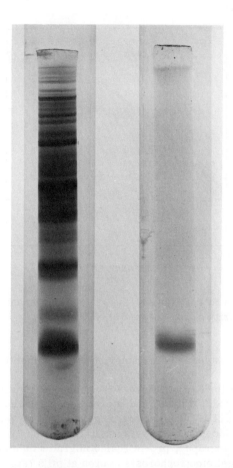

FIG. 8. Polyacrylamide gel electrophoresis in sodium dodecyl sulfate of the high-speed supernatant (soluble) fraction of a liver homogenate (left) and purified pro-SCP (right). (From Ref. 20, courtesy of the National Academy of Sciences.)

we have not observed an SCP species of less than 16,000 daltons, higher molecular weight aggregates are common in the unpurified soluble fraction of liver homogenates (cf. Fig. 5B and Fig. 6) [1, 2, 20, 29]. Furthermore, as will be discussed in part C of this section, multiple forms of SCP containing bound lipid have also been identified by density gradient ultracentrifugation techniques [31].

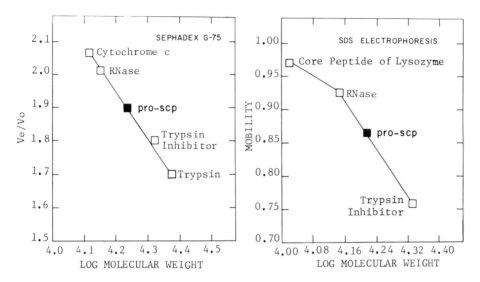

FIG. 9. Migration of the protomer form of SCP (M. W. 16,000 daltons) relative to proteins of known molecular weight during gel filtration (left) and sodium dodecyl sulfate (SDS) polyacrylamide gel electrophoresis (right). (From Ref. 20, courtesy of the National Academy of Sciences.)

C. Lipid Binding; Oligomer Form

As indicated previously, the form of SCP containing bound lipid is a high molecular weight oligomer detected in the void volume of a gel filtration column (e.g., Sephadex G-75 or G-100) (cf. Fig. 5). The oligomer is considered to be the form of SCP which functions with microsomal enzymes (cf. Fig. 1 and Sect. IV). We originally expected that one stoichiometric complex of sterol and SCP would be present in the oligomer form. However, recent studies by Carlson et al. [31], employing density gradient centrifugation techniques, showed that a spectrum of sterol-SCP complexes $(2.3 \times 10^5$ to 2×10^6 daltons) are formed when pro-SCP is converted to its oligomer (Fig. 12). Formation of the oligomer from pro-SCP is enhanced by the presence of phospholipid, in addition to sterols and phospholipid is also bound to SCP [20]. Presence of phospholipid in the oligomer does not appear to interfere with the function of the oligomer with microsomal enzymes [20, 22].

To facilitate comparison of the degree of binding of various compounds to SCP (Table 4), we developed a rapid technique involving separation of protein-bound and unbound species by gel filtration [2, 20, 22].

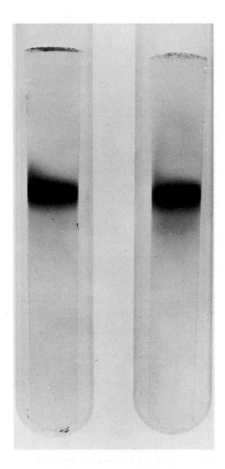

FIG. 10. Migration of pro-SCP prepared by Method 1 (Fig. 6) (left) and by Method 2 (Fig. 6) (right) during polyacrylamide gel electrophoresis in urea at pH 9.5. (Adapted from Refs. 28 and 29.)

SCP not only binds water insoluble precursors of cholesterol and phospholipids but steroid hormone and bile acid precursors, fatty acids, protozoan sterols and plant sterols as well (Table 4) [1, 2, 20]. In addition, the affinity of SCP for pyridine nucleotides (Table 4) suggests that an SCP-bound cofactor may participate in the oxidation or reduction of sterol substrates occurring during cholesterol synthesis (cf. Fig. 13, Sect. IV). Water soluble precursors of cholesterol, steroid hormones, bile acids, cholesterol esters, cholestane, and keto derivatives of cholestane (not cholesterol precursors) are not bound or are poorly bound to SCP (Table 4).

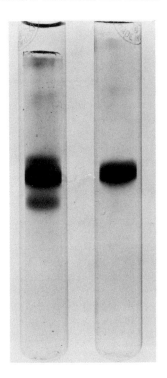

FIG. 11. Migration of pro-SCP during polyacrylamide gel electrophoresis in urea at pH 4.3; prior (left) and following (right) ion-exchange chromatography. (Adapted from Refs. 28 and 29.)

Hydroxy derivatives of cholestane (especially 3 β , 5 α , 6 β -cholestanetriol) are bound to SCP and also inhibit the later stages of cholesterol synthesis [33]. Phenethybiguanide, a weak inhibitor of Δ^7 - reductase activity (Step 2, Fig. 3) [34] is not bound to SCP, whereas the powerful Δ^7 - reductast inhibitor, AY-9944 [35, 36], is bound to SCP at levels sufficient to completely inhibit the reductase (Table 4) [36].

The data in Table 4 indicate that both nonpolar and polar groups are important in lipid binding to SCP, e.g., esterification of the sterol nucleus decreases the degree of binding to SCP. Apparently SCP must contain hydrophobic regions to accommodate the sterol nucleus and its side chain or the long hydrocarbon chains of fatty acids; SCP also must contain hydrophilic regions for binding hydroxyl groups and other polar areas in lipid molecules [20]. We have suggested that a more appropriate name for SCP is lipid carrier protein (LCP) with SCP describing the function of

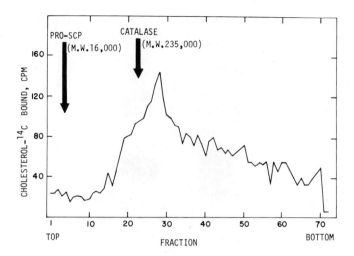

FIG. 12. Density gradient centrifugation pattern of the cholesterol-SCP complexes formed during conversion of pro-SCP to the oligomer form. The molecular weights of the complexes range from 2.3 x 10^5 to 2 x 10^6 daltons. (Adapted from Ref. 31.)

LCP in cholesterol biosynthesis, metabolism, and transport. Furthermore, as will be discussed in later sections of this chapter, many of the compounds bound to SCP (e.g., bile acid precursors) appear to be metabolized as SCP complexes similarly to cholesterol precursors.

It is apparent that extensive further studies are needed to obtain binding constants and a full understanding of the nature of the interaction of SCP with lipid molecules. One approach we are exploring is the use of fluorescent probes which show an enhanced quantum yield in a lipophilic environment [37] and often reflect conformational changes occurring at the active site of an enzyme. (e.g., luciferase [38]). In preliminary studies, Carlson [39] found that the dye anilino-naphthalene-sulfonate (ANS) binds (mole per mole) to pro-SCP with an enhanced quantum yield and blue wavelength shift in its fluorescent spectrum. Whether the binding of ANS occurs at or near the lipid binding site of SCP is currently under investigation.

IV.   FUNCTIONS OF SQUALENE AND STEROL CARRIER PROTEIN

In this section the various established and proposed biological roles of SCP in sterol synthesis and metabolism are discussed (cf. Fig. 1), as well as probable mechanisms of action of SCP.

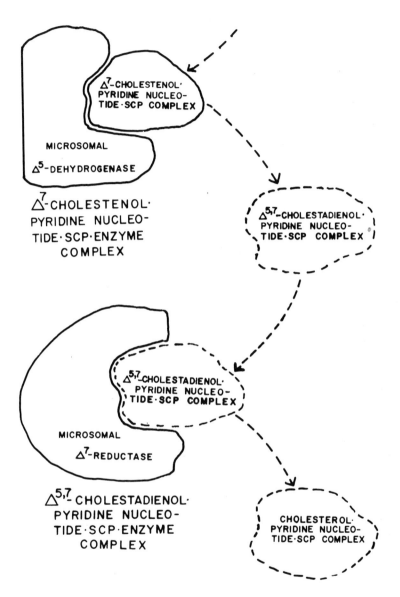

FIG. 13. Scheme outlining the proposed molecular events occurring during final steps in cholesterol synthesis, conversion of Δ^7- cholestenol to cholesterol. For further details, see text (Sec. III, A). (Adapted from Ref. 32.)

TABLE 3

Amino Acid Compositions of Rat-, Human-, and Protozoan SCP[a]

Amino acid	Rat SCP[b]	Human SCP[b]	Protozoan SCP[b]
Aspartic acid	10.4	8.1	8.4
Threonine	7.5	7.3	3.9
Serine	4.5	4.4	3.4
Glutamic acid	12.2	12.4	11.5
Proline	3.7	1.6	2.5
Glycine	10.7	7.7	11.4
Alanine	3.5	3.5	8.4
Half-cystine	3.3	1.9	1.7
Valine	8.5	6.1	4.3
Methionine	4.7	2.2	1.6
Isoleucine	7.0	5.2	4.1
Leucine	6.5	5.6	4.6
Tyrosine	3.8	1.9	1.2
Phenylalanine	5.9	3.7	2.9
Lysine	12.0	11.0	6.4
Histidine	1.6	1.2	1.6
Arginine	2.2	1.7	1.5

[a] Adapted from Ref. 30.

[b] μ moles amino acid/μ mole pro-SCP (16,000 daltons, mol wt).

TABLE 4
Relative Binding of Lipids and Other Compounds to SCP [a]

Compound	Relative binding
Cholesterol	1.0
Cholesterol precursors	
Mevalonate	< 0.1
Farnesyl pyrophosphate [b]	< 0.1
Presqualene pyrophosphate [b]	1.0
Squalene	1.0
Lanosterol (30-carbon atom)	1.0
29-Carbon atom precursors	1.0
28-Carbon atom precursors	1.0
27-Carbon atom precursors	1.0
Steroid hormones and precursors	
Pregnenolone	0.3
Progesterone	0.2
Testosterone	< 0.1
Estradiol	< 0.1
Bile acids and precursors	
7α-Hydroxy-cholesterol	1.0
Other 27-carbon atom precursors (cf. Fig. 16)	1.0
Cholic acid	< 0.1
Taurocholic acid	< 0.1
Chenodeoxycholic acid	< 0.1
Cholesterol esters	
Cholesteroyl-acetate	0.3
Cholesteroyl-palmitate	0.1
Cholesteroyl-stearate	< 0.1
Cholesteroyl-oleate	0.4
Fatty acids	
Acetic acid	< 0.1
Caproic acid (6-carbon atom)	0.8
14- to 18-Carbon atom, saturated and unsaturated	1.0
Phospholipids	
Lecithin	1.0
Phosphatidyl serine	1.0

TABLE 4 (Cont'd)

Relative Binding of Lipids and Other Compounds to SCP[a]

Compound	Relative binding
Protozoan sterols [c]	
$\Delta^{5,7,22}$ - Cholestratrienol	1.0
$\Delta^{5,7,22}$ - Cholestratrienyl-acetate	0.6
Tetrahymenol	< 0.1
Plant sterols [d]	
β-Sitosterol	1.0
Cycloartenol	0.9
24-Methylene-cholesterol	1.0
Cholestane and its derivatives (not cholesterol precursors)	
Cholestane	0.2
Cholestanol	0.3
3β, 5α-Cholestanediol	0.5
3β, 5α-Cholestanetriol	1.0
3-Ketone derivatives of cholestane	< 0.1
Δ^7-Reductase inhibitors	
Phenethylbiguanide	< 0.1
AY-9944	0.1
Pyridine nucleotides	
NADPH	1.0
NAD	1.0

[a] The majority of these findings are taken from Refs. 2 and 20.

[b] See Ref. 26.

[c] In collaboration with Drs. R. Conner, M. J. Karoly, and J. Landrey.

[d] In collaboration with Dr. W. R. Nes.

A.   Role in Cholesterol Synthesis; Proposed Mechanism of Action

It is now generally accepted that all known water-insoluble precursors of cholesterol are capable of complexing with SCP (cf. Table 4) and are readily converted to products by microsomal enzymes (Fig. 2) [1, 2]. Our data showed that conversion to products by microsomal enzymes of a

sterol present in a sterol-SCP complex (oligomer form) is markedly
faster than conversion of initially unbound sterol (e. g. , Table 5) [20].
Related findings indicate that microsomal enzymes have a high affinity for
a preformed sterol-SCP complex (oligomer) [20]. We suggest that during
cholesterol synthesis an early water-insoluble (e. g. , presqualene pyrophosphate or squalene) forms a noncovalent complex with SCP yielding the
oligomer. This complex combines with the first microsomal enzyme, and
the precursor-SCP is converted to its product, the following precursor-
SCP. The latter complex combines with the microsomal enzyme next in
the sequence of cholesterol synthesis. Finally, cholesterol-SCP results
(Fig. 1). This proposal is also shown schematically in Figure 13 for final
steps in cholesterol synthesis, conversion of Δ^7-cholestenol to cholesterol
(Fig. 3). The SCP complex is portrayed in Figure 13 as containing bound
pyridine nucleotide required for the oxidation of one substrate and the reduction of the next. Further evidence is needed to substantiate that these
molecular events occur in vivo. If this proposal is valid, it is plausible
that SCP levels or lipid-SCP complexes could regulate sterol synthesis
and metabolism, as discussed in Sec. VI.

It should be mentioned that there are suggestions in the literature for
the occurrence of more than one liver SCP protein, e. g. , one heat stable
and another heat labile [25, 27]. Our work does not support this proposal;
it is likely that these differences will be resolved when assay conditions
for SCP are standardized (cf. Sec. III). For example, loss of an essential
but unrecognized cofactor on heating a soluble fraction from liver could
lead to the proposal that certain enzymatic steps require a heat-labile
SCP. see Sec. IV, B, C, and E and Sec. V.

B. Relationship to Plasma Lipoproteins

The characteristics of SCP outlined in previous sections of this chapter
are in accord with the suggestion that SCP could be a precursor of certain
plasma lipoproteins or, indeed, could be present in circulating lipoproteins [1, 2, 20]. To test this hypothesis, we first studied the effects of
human plasma lipoprotein fractions (very low density [VLDL], low density
[LDL], and high density [HDL]) on microsomal enzymes requiring liver
SCP for activity, e. g. , Δ^5-dehydrogenase (Step 1, Fig. 3). The most
striking finding was that SCP activity is correlated with HDL levels [20,
40, 41]; VLDL and LDL fractions are either inhibitory or without effect.
The nomenclature and molecular characteristics of the two major HDL
polypeptides are summarized in Tables 6 and 7. Both polypeptides have
been sequenced: apoLP-Gln-I (apo-A-I) by Baker et al. [45, 46], and
apoLP-Gln-II (apo-A-II) by Brewer et al. [47]. Our studies with these
purified polypeptides indicate that SCP activity resides primarily with

TABLE 5

Increased Conversion Rate to Products of a Cholesterol
Precursor-SCP Complex [a]

Cholesterol precursor	Total sterol synthesized, μ moles
Dihydrolanosterol (C-30)	
Dihydro-SCP	0.8
Dihydro + SCP	0.1
Dihydro; no SCP	0.0
Cholesta-7, 24-dienol (C-27)	
$\triangle^{7,24}$ - SCP	1.6
$\triangle^{7,24}$ + SCP	0.7
$\triangle^{7,24}$; no SCP	0.2

[a] Adapted from Ref. 20.

TABLE 6

Nomenclature for the Two Major Apopolypeptides of Human Plasma
High Density Lipoproteins (HDL)

C-Terminal amino acid [a]	Gel filtration in urea [b]	Other [c]
ApoLP-Gln-I	Fraction III	Apo-A-I
ApoLP-Gln-II	Fraction IV	Apo-A-II

[a] Ref. 42.
[b] Ref. 43.
[c] Ref. 44.

apoLP-Gln-II (apo-A-II) (e.g., Fig. 14 [48]. The Gln-II molecule has a
molecular weight of 16,000 daltons (as does pro-SCP; see Sec. III) and
two identical polypeptide chains joined by a single disulfide bond (Table 7).
Although our preliminary findings indicate that functionally Gln-II and
SCP may not be identical (Fig. 14), the possible structural similarity or
identity of these molecules is worthy of further detailed investigation. Now

TABLE 7

Molecular Characteristics of the Major HDL Apopolypeptides [a]

Apopeptide	Missing amino acids	N-Terminal amino acid	Molecular weight
ApoLP-Gln-I (Apo-A-I)	CySH, CyS-SCy, Ile	Asp	28,000
ApoLP-Gln-II (Apo-A-II)	CySH, His, Arg, Try	Blocked (PCA)[b]	16,000 (8,000)[c]

[a] Summarized from Refs. 45-47.

[b] Pyrrolidone carboxylic acid.

[c] By conversion of R-Cys-SCy-R to 2 R-CySH, yielding two identical polypeptide chains.

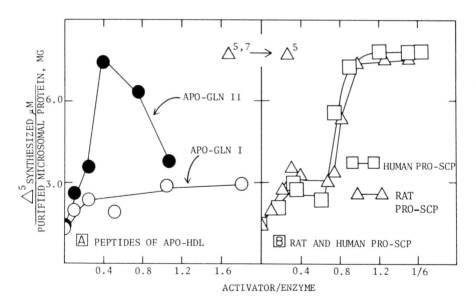

FIG. 14. A-functions of human plasma high density lipoprotein apo-poly-peptides in cholesterol synthesis; B-functions of rat and human liver pro-SCP in the same reaction. (Adapted from Ref. 48.)

that pro-SCP is homogeneous (see Sec. III), it should be feasible to obtain
definitive answers to this portant question and also insights into regulation
of cholesterol synthesis and metabolism (Sec. VI).

C. Role in Steroid Hormone Synthesis

In accord with the roles proposed for SCP (Fig. 1), we recently demonstrated that liver SCP will function with an adrenal mitochondrial enzyme
complex catalyzing the conversion of cholesterol to pregnenolone (Fig. 15)
[49]. The soluble fraction of adrenal tissue also contains an SCP which
functions with liver microsomal enzymes catalyzing cholesterol synthesis
[49] (see also Sec. V). In further work, Kan and Ungar [50] isolated
from adrenal tissue a heat-stable protein structurally similar (e.g., molecular weight) and perhaps identical to liver SCP. This adrenal SCP

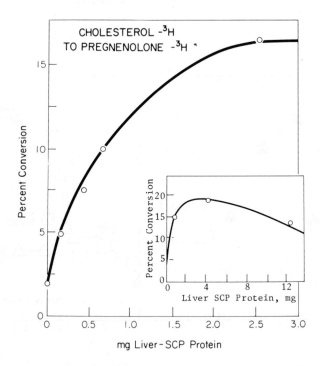

FIG. 15. Function of liver SCP in the side-chain cleavage of cholesterol
catalyzed by adrenal mitochondrial enzymes. (From Ref. 49 by courtesy
of Academic Press.)

functions in the side chain cleavage of cholesterol by adrenal enzymes. Again, additional studies are necessary to determine the relationships (metabolic, structural, regulatory) between liver SCP and adrenal SCP (cf. Sec. VI).

D. Role in Bile Acid Synthesis

With regard to the role of SCP in bile acid synthesis (Fig. 1), our initial efforts have centered on the reaction catalyzed by liver microsomal 12α–hydroxylase (see Fig. 16 for an outline of the enzymatic steps in bile acid synthesis). The 12α-hydroxylase was shown to require the soluble fraction of liver for maximum activity and, specifically, purified SCP (Fig. 17) [51, 52]. It was also demonstrated that during purification of liver SCP the functional activity of various protein fractions for enzymatic steps in

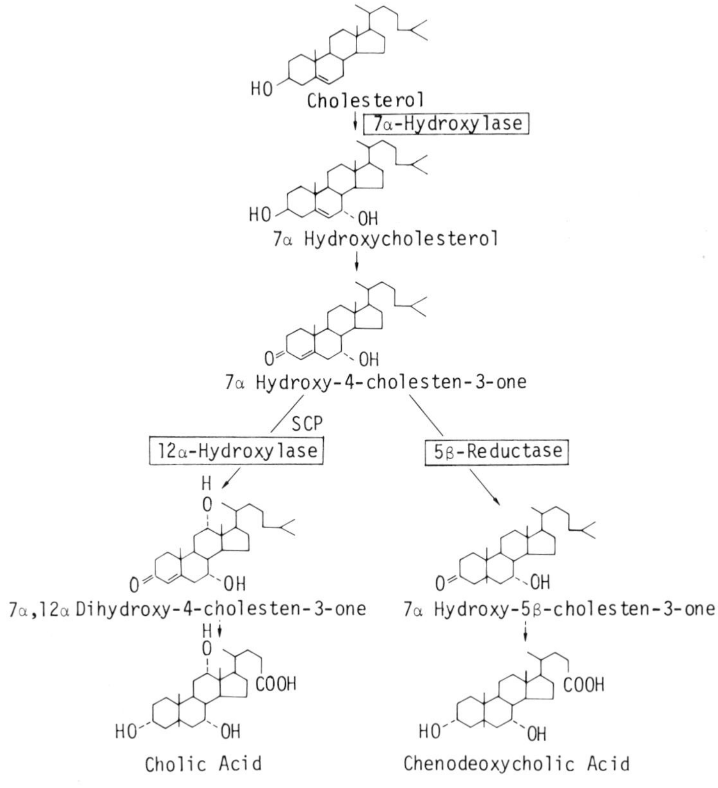

FIG. 16. Pathways of bile acid synthesis from cholesterol; site of SCP function. (Adapted from Refs. 51 and 52.)

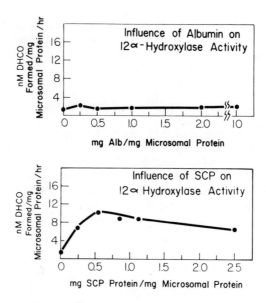

FIG. 17. Influence of albumin (top) and pro-SCP (bottom) on liver 12α-
hydroxylast activity (conversion of 7α-hydroxy-4-cholesten-3-one to 7α,
12α-dihydroxy-4-cholesten-3-one (DHCO); cf. Fig. 16). (Adapted from
Refs. 51 and 52.)

cholesterol and bile acid synthesis were identical (e.g., Fig. 18) [52].
As mentioned previously (Sec. II, C), all the 27-carbon atom bile acid precursors (cf. Fig. 16) bind to SCP. It appears plausible that reactions, in
addition to 12α-hydroxylase, will require SCP. Of special interest is the
7α-hydroxylase, considered to be a major regulatory enzyme in bile acid
synthesis (cf. Sec. VI).

E. Role in Membrane Formation by Protozoa

One approach we are developing to elucidate the proposed role of SCP
in membrane formation (Fig. 1) is the study of sterol interconversions by
a protozoan, Tetrahymena pyriformis. Conner, Mallory, and their colleagues [53, 54] showed that when cholesterol is supplied to the growth
media of a Tetrahymena culture, synthesis of tetrahymenol (the normal
pentacyclic triterpenoid alcohol produced by this strain from mevalonate)
is inhibited; cholesterol is converted to $\Delta^{5,7,22}$-cholestatrienol; and this
sterol is incorporated into cellular membranes of the animal. Calimbas
[30, 55, 56] studied the time course of induction of microsomal Δ^7- and
Δ^{22}-dehydrogenases (Fig. 19) and also demonstrated the requirement for

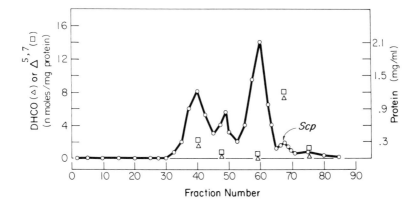

FIG. 18. Purification by gel filtration (Fig. 5b or Fig. 6, Method 1) and function of SCP in sterol (Δ^5-dehydrogenase reaction; cf. Step 1, Fig. 3 and Fig. 7) and bile acid (12α-hydroxylase reaction; cf. Figs. 16 and 17) synthesis. (Adapted from Refs. 51 and 52.)

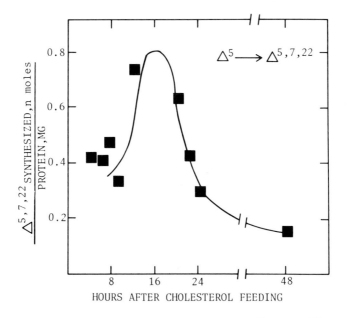

FIG. 19. Time course of induction of microsomal Δ^7- and Δ^{22}-dehydrogenase (conversion of cholesterol (Δ^5) to $\Delta^{5,7,22}$-cholestatrienol [$\Delta^{5,7,22}$]) by a <u>Tetrahymena</u> culture after cholesterol feeding. (Adapted from Refs. 30, 55 and 56.)

the soluble fraction by these enzymes (Table 8). In further studies, Calimbas [30, 55, 56] obtained evidence that protozoan SCP is present in the soluble fraction of Tetrahymena homogenates (Fig. 20); protozoan SCP will also substitute for liver SCP in reactions catalyzed by liver enzymes, and liver SCP will substitute in the same manner for protozoan SCP (e.g., Fig. 21). Protozoan SCP may be purified by the same techniques used for liver SCP; the amino acid composition of protozoan SCP is strikingly similar to that of mammalian liver SCP (Table 3). Protozoan SCP also undergoes protomer to oligomer polymerization during lipid binding (cf. Sec. III, C). As will be indicated in the following section, the requirement for an SCP-like molecule appears to be widespread in nature.

F. Relationship to Other Lipid-carrying Proteins and Soluble Protein Cofactors

Recently, several other lipid-carrying proteins have been isolated from mammalian tissues. For example, Ehnholm and Zilversmit [57] purified a protein (molecular weight 21,000 daltons) from the soluble fraction of beef heart. This protein participates in the exchange of phospholipid between mitochondria and microsomes. A second protein (molecular weight

TABLE 8

Cellular Fractions Required for Protozoan Δ^7 - and Δ^{22}-Dehydrogenase Activity [a]

Fraction(s)	Conversion of cholesterol to $\Delta^{5,7,22}$ - Cholestatrienol
Microsomal (Δ^5) [b] + soluble (Δ^5) [b]	19.2%
Microsomal (Δ^5) [b] + soluble	20.1%
Microsomal + soluble (Δ^5) [b]	< 5%
Microsomal + soluble	< 5%
Microsomal	0%
Soluble (Δ^5) [b]	0%
Soluble	0%

[a] Adapted from Refs. 30, 55 and 56.

[b] Δ^5 = cholesterol present during growth of Tetrahymena pyriformis.

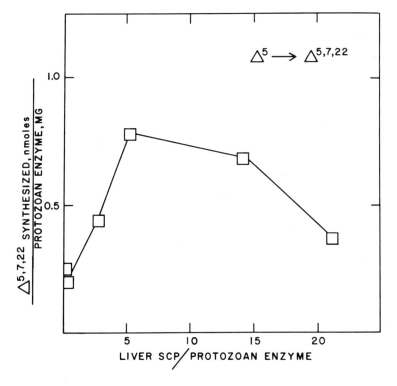

CHOLESTEROL (Δ^5) $\Delta^{5,7,22}$-CHOLESTATRIEN-3β-OL

FIG. 20. Role of protozoan–SCP in sterol conversions by protozoan microsomal enzymes. (Adapted from Refs. 30, 55, and 56.)

FIG. 21. Function of liver SCP with protozoan microsomal Δ^7- and Δ^{22}-dehydrogenases. (Adapted from Refs. 30, 55, and 56.)

12,000 daltons), which binds long-chain fatty acids and other anions, has been obtained from the soluble fraction of intestine and other tissue homogenates [58, 59]. Whether or not these proteins, differing somewhat in molecular weight from pro-SCP, are related to SCP remains to be elucidated. In this regard there are various reports of soluble proteins required for steps in drug metabolism (e.g., Ref. 60). Of more pertinence to the subject of this chapter is the report by Bloch's group [61, 62] that a soluble protein is required for microsomal squalene epoxidase. The latter protein is apparently heat labile and of higher molecular weight than pro-SCP. As we are able to show conversion of squalene to cholesterol and other sterols by microsomal enzymes in the presence of purified SCP and appropriate cofactors [29], a relationship of SCP to the epoxidase protein is plausible.

V. UBIQUITOUS OCCURRENCE OF SQUALENE AND STEROL CARRIER PROTEIN

The occurrence of SCP in human and rat liver is well established (Sec. II-III). Recently SCP or an SCP-like protein was also detected in various other mammalian tissues, e.g., heart, kidney, intestine, lung, spleen, brain, muscle, adrenal (Figs. 22 and 23) [28, 49, 50], and human term placenta [63]. Several authors (e.g., see Refs. 64-66) apparently were unable to detect SCP in nervous and other extrahepatic tissues. Again, these discrepancies will be resolved when assay methods for SCP functional activity are standardized, i.e., functional activity is best assayed using one enzymatic reaction rather than a coupled series of many steps (cf. Sec. IV, A). As it is widely accepted that cholesterol synthesis occurs in all mammalian tissues (cf. Ref. 67), SCP, if it is to fulfill the role we have assigned to it (Figs. 1 and 2), must also occur in these tissues. SCP is present in mouse fibroblasts (L-cells) [68] which actively synthesize sterols, and this synthesis is regulated by cholesterol in the growth medium [69]. These and other cultured cells offer valuable tools for defining the role of SCP in regulation of lipid synthesis and metabolism (Sec. VI).

The purification and characterization of protozoan SCP was summarized in Sec. IV, E. SCP was also partially purified from yeast by Rilling [26]. To date, the isolation of SCP from other lower forms of life has not been reported. However, it is likely that a plant-type SCP exists, e.g., plant sterols readily bind mammalian SCP (Table 4).

In view of the apparent ubiquitous occurrence of SCP or SCP-like molecules, we propose that a requirement for a protein with SCP properties is a general biological requirement for synthesis, metabolism, and transport of sterols (and perhaps other water-insoluble molecules) — plus regulation of these processes [70].

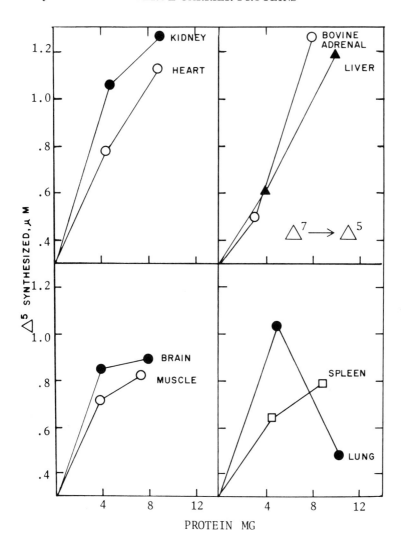

FIG. 22. Occurrence of SCP activity in mammalian tissues; assayed by function in conversion of Δ^7-cholestenol (Δ^7) to cholesterol (Δ^5) by liver enzymes (cf. Fig. 3). (Adapted from Refs. 28 and 29.)

VI. REGULATORY ROLE OF SQUALENE AND STEROL
CARRIER PROTEIN

The characteristics of SCP described in previous sections suggest that this versatile protein molecule could regulate the processes in which it

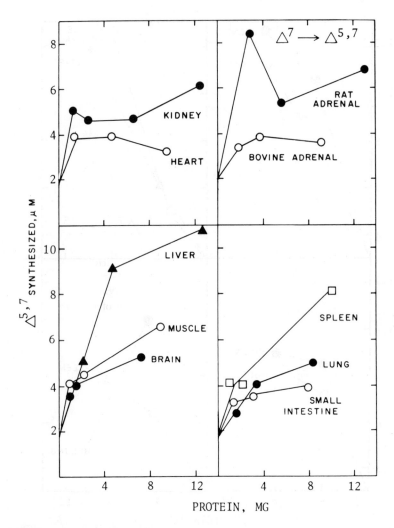

FIG. 23. Occurrence of SCP activity in mammalian tissues; assayed by function in the Δ^5-dehydrogenase reaction (Step 1, Fig. 3 and Fig. 7). (Adapted from Refs. 28 and 29.)

participates. Preliminary evidence supporting this proposal, as well as speculations regarding yet undiscovered regulatory roles for SCP, are presented here.

Current information on the regulation of cholesterol synthesis and metabolism is summarized in a recent review [70]. Briefly, there is

considerable evidence that sterol synthesis is regulated at several stages
(cf. Fig. 24). Prior to squalene, synthesis may be regulated by cytoplasmic enzymes (acetoacetyl-CoA thiolase and ß-hydroxy-ß-methyl-glutaryl-
CoA [HMG-CoA] synthase) catalyzing HMG-CoA synthesis from acetyl-CoA
(Stage 1, Fig. 24) [71, 72]. The next step in cholesterol synthesis, HMG-
CoA reductase (Stage 2, Fig. 24), is widely accepted as a major regulatory
site [73]. HMG-CoA reductase undergoes a diurnal rhythm in activity
[73]; it is probably regulated in response to physiological conditions by
several mechanisms, some involving turnover of enzyme protein [74, 75]
and others being direct effects on enzyme activity [76-78]. Additional
evidence indicates that sites of regulation exist following mevalonate and
prior to squalene (e.g., Stages 3 and 4, Fig. 24) [79, 80]. Indeed,
Edmond and Popják [81] demonstrated the existence of a pathway (the
transmethylglutaconate shunt) leading intermediates arising from
mevalonate back to HMG-CoA, ketone bodies, and acetyl-CoA, rather
than to squalene and cholesterol.

As suggested by the scheme in Fig. 24, SCP or a cholesterol-SCP complex could regulate sterol synthesis at some or all the stages of sterol
synthesis. In support of this proposal, we recently reported that SCP, as
well as the HDL fraction of human plasma, is capable of modifying microsomal HMG-CoA reductase in vitro (cf. Fig. 25 and Table 9) [31, 82].
These direct effects on enzymatic activity could result from dissociation of
the subunut structure of HMG-CoA reductase or from effects on the membrane structure of the microsomal enzyme. Now that HMG-CoA is available

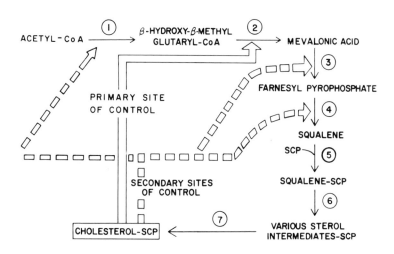

FIG. 24. Proposed sites of regulation by SCP in cholesterol synthesis.
(From Ref. 70, courtesy of Annual Reviews, Inc.)

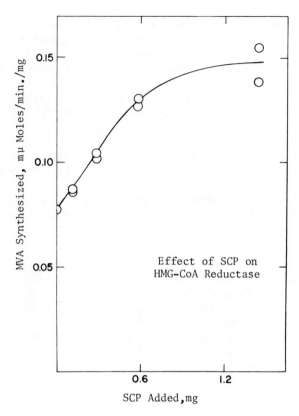

FIG. 25. Activation of microsomal HMG-CoA reductase by pro-SCP.
(Adapted from Refs. 31 and 82.)

in solubilized form (e.g., see Refs 83-85), it should be possible to define
the nature of the SCP effect (Fig. 25). With regard to other possible
regulatory effects of SCP, Ritter et al. [86] discovered that the level of
cellular SCP could control the pathways by which cholesterol is synthesized
from sterol precursors (cf. Fig. 2). For example, at high levels of SCP
the pathway involving saturated side-chain sterol intermediates would
predominate (Fig. 26). Related evidence indicates that SCP levels are
elevated in fetal and newborn rat livers, conditions where rapid sterol
synthesis is occurring [87]. Furthermore, the possible relationship of
SCP to plasma high density lipoproteins (Sec. IV, B) suggests another area
where SCP could regulate cholesterol synthesis and metabolism. In this
regard, Frnka and Reiser [88] reported that the synthesis rate of HDL
polypeptides is reduced by cholesterol feeding, indicating that this process
could function in regulation of cholesterolgenesis. Still another area for a

TABLE 9

Effects of Human Lipoprotein Fractions on HMG-CoA Reductase [a]

Fraction added	Protein, μg	MVA synthesized, $m\mu$moles/30 min.
None	—	1.5
HDL	50	1.8
	210	2.5
	980	2.7
LDL	28	1.5
	140	1.9
	560	1.9
VLDL	15	3.2
	77	2.0
	310	1.8

[a] Adapted from Refs. 31 and 82.

possible regulatory function of SCP is in metabolism of cholesterol to steroid hormones and bile acids. Several authors have invoked the need for a cholesterol-binding protein to carry free cholesterol from lipid stores to adrenal mitochondrial enzymes catalyzing steroid synthesis in response to ACTH stimulation [89-91].

Clearly, there is need for extensive further efforts to delineate effects of SCP and lipid-SCP complexes on known regulatory enzymes in vitro. Of potentially greater importance are studies designed to obtain evidence for a regulatory role of SCP in vivo. For example, data are needed on possible changes in SCP levels in response to physiological conditions known to affect rates of cholesterol synthesis (e.g., sterol feeding, fasting, biliary diversion). The development of a specific immunochemical assay for SCP would facilitate this effort.

VII. CONCLUDING REMARKS

Although considerable progress has been made in understanding sterol biosynthesis and metabolism, it is certain that future studies should center on attempts to elucidate molecular events involved in regulation of these processes. These challenging investigations are essential to further

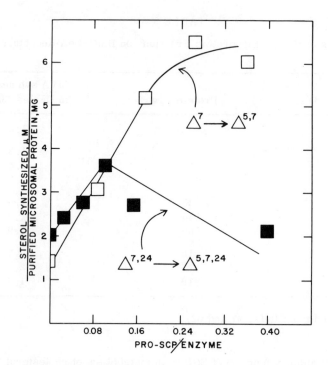

FIG. 26. Influence of pro-SCP levels on the conversion of Δ^7-cholestenol (Δ^7) to $\Delta^{5,7}$-cholestadienol ($\Delta^{5,7}$) and $\Delta^{7,24}$-cholestadienol ($\Delta^{7,24}$) to $\Delta^{5,7,24}$-cholestatrienol ($\Delta^{5,7,24}$) by Δ^5-dehydrogenase (Step 1, Fig. 3). (Adapted from Ref. 86.)

basic knowledge and also to prevent pathological conditions (e.g., atherosclerosis, coronary disease) probably resulting from defective regulation. The future appears bright for assignment of a major regulatory role to squalene and sterol carrier protein.

ACKNOWLEDGEMENTS

Studies from the author's laboratory mentioned here were supported by NHLI grants HL-8634 and 72-2915 (The Lipid Research Clinics Program).

REFERENCES

1. M. C. Ritter and M. E. Dempsey, Biochem. Biophys. Res. Commun., 38, 921 (1970).
2. M. C. Ritter and M. E. Dempsey, J. Biol. Chem., 246, 1536 (1971).
3. R. B. Clayton, Quart. Rev., 19, 168 (1965).
4. I. D. Frantz, Jr. and G. J. Schroepfer, Jr., Ann. Rev. Biochem., 36, 691 (1967).
5. J. L. Gaylor, Adv. Lipid Res., 10, 89 (1972).
6. R. B. Clayton (ed.), Methods in Enzymology, Vol. 15, Academic Press, New York, 1969.
7. N. L. Bucher, J. Am. Chem. Soc., 75, 498 (1953).
8. I. D. Frantz, Jr. and N. L. Bucher, J. Biol. Chem., 206, 471 (1954).
9. N. L. Bucher and K. McGarrahan, J. Biol. Chem., 222, 1 (1956).
10. I. D. Frantz, Jr., A. G. Davison, E. Dulit, and M. L. Mobberley, J. Biol. Chem., 234, 2290 (1959).
11. A. A. Kandutsch, J. Biol. Chem., 237, 358 (1962).
12. D. S. Goodman, J. Avigan and D. Steinberg, J. Biol. Chem., 238, 1287 (1963).
13. M. E. Dempsey, Fed. Proc., 21, 299 (1962).
14. M. E. Dempsey, J. D. Seaton and R. W. Trockman, Fed. Proc., 22, 529 (1963).
15. M. E. Dempsey, M. C. Ritter, K. J. Bisset, and S. E. Stone, Fed. Proc., 27, 524 (1968).
16. M. E. Dempsey, in Methods in Enzymology, Vol. 15 (R. B. Clayton, ed.), Academic Press, New York, 1969, p. 505.
17. M. E. Dempsey, J. D. Seaton, G. J. Schroepfer, Jr., and R. W. Trockman, J. Biol. Chem., 239, 1381 (1964).
18. M. E. Dempsey, J. Biol. Chem., 240, 4176 (1965).
19. M. E. Dempsey, Ann. N. Y. Acad. Sci., 148, 631 (1968).
20. M. C. Ritter and M. E. Dempsey, Proc. Nat. Acad. Sci., U. S., 70, 265 (1973).
21. M. E. Dempsey, K. J. Bisset, and M. C. Ritter, Circulation, 38, VI-5 (1968).
22. M. E. Dempsey, in Pharmacology of Hypolipidemic Agents (D. Kritchevsky, ed.), Springer-Verlag, Berlin, 1975, in press.
23. M. E. Dempsey, in Advances in Experimental Medicine and Biology, Vol. 13 (R. Paoletti and A. N. Davison, eds.), Plenum, New York, 1971, p. 31.
24. T. J. Scallen, M. W. Schuster, and A. K. Dhar, J. Biol. Chem., 246, 224 (1971).
25. T. J. Scallen, M. V. Srikantaiah, H. B. Skralant, and E. Hansbury, FEBS Letters, 25, 227 (1972).

26. H. C. Rilling, Biochem. Biophys. Res. Commun., 46, 470 (1972).

27. R. C. Johnson and S. N. Shah, Biochem. Biophys. Res. Commun., 53, 105 (1973).

28. K. E. McCoy, D. F. Koehler and J. P. Carlson, Fed. Proc., 32, 519 (1973).

29. M. E. Dempsey, K. E. McCoy, J. P. Carlson, and T. D. Calimbas, Fed. Proc., 33, 1429 (1974).

30. T. D. Calimbas, Ph.D. Thesis, Univ. of Minnesota, Minneapolis, 1974.

31. J. P. Carlson, K. E. McCoy, and M. E. Dempsey, Circulation, 48, IV-70 (1973).

32. M. C. Ritter and M. E. Dempsey, in preparation.

33. M. E. Dempsey, M. C. Ritter, D. T. Witiak, and R. A. Parker, in Atherosclerosis: Proc. 2nd Intl. Symp. (R. Jones, ed.), Springer-Verlag, Berlin, 1970, p. 290.

34. M. E. Dempsey, in Advances in Experimental Medicine and Biology, Vol. 4 (W. L. Holmes, L. A. Carlson, and R. Paoletti, eds.), Plenum, New York, 1969, p. 511.

35. D. Dvornik, M. Kraml, and J. F. Bagli, Biochemistry, 5, 1060 (1966).

36. M. E. Dempsey, in Progress in Biochemical Pharmacology, Vol. 2, (D. Kritchevsky, R. Paoletti and D. Steinberg, eds.), Karger, Basel, 1967, p. 21.

37. L. Brand and J. R. Gohlke, Ann. Rev. Biochem., 41, 843 (1972).

38. M. Deluca, Biochemistry, 8, 160 (1969).

39. J. P. Carlson, unpublished data.

40. D. F. Koehler and M. E. Dempsey, Circulation, 48, IV-246 (1973).

41. D. F. Koehler, M. S. Thesis, Univ. of Minnesota, Minneapolis, 1974.

42. S. E. Lux, R. I. Levy, A. M. Gotto, Jr., and D. S. Fredrickson, J. Clin. Invest., 51, 2505 (1972).

43. A. Scanu, J. Toth, C. Edelstein, S. Koga, and E. Stiller, Biochemistry, 8, 3309 (1969).

44. G. Kostner and P. Alaupovic, FEBS Letters, 15, 320 (1971).

45. H. N. Baker, R. L. Jackson, and A. M. Gotto, Jr., Biochemistry, 12, 3866 (1973).

46. H. N. Baker, A. M. Gotto, Jr., T. Delahunty, and R. L. Jackson, Circulation, 48, IV-80 (1973).

47. H. B. Brewer, S. E. Lux, R. Ronan, and K. M. John, Proc. Natl. Acad. Sci., U. S., 69, 1304 (1972).

48. M. E. Dempsey, M. C. Ritter, and S. E. Lux, Fed. Proc., 31, 430 (1972).

49. K. W. Kan, M. C. Ritter, F. Ungar, and M. E. Dempsey, Biochem. Biophys. Res. Commun., 48, 423 (1972).

50. K. W. Kan and F. Ungar, J. Biol. Chem., 248, 2868 (1973).

AUTHOR INDEX

Numbers in brackets are reference numbers and indicate that an author's work is referred to although his name is not cited in the text. Underlined numbers show the page on which the complete reference is cited.

A

Abraham, S., 182[35,37], 183[40], 216
Achter, E. K., 143[39], 176
Adachi, O., 229[33], 249[194], 258[30], 260, 265
Adair, G. S., 9, 40
Adair, W. L., Jr., 108[45,46], 109[45,46], 111[45,46], 115[63], 116[63], 119[46], 120[63], 122[84], 123[84], 127[46,63], 130[8'], 132, 133, 134
Adams, M. J., 20[27], 40
Adler, S. P., 59[66,71], 60[71], 61[71], 62[66], 80[66], 81[66], 84
Afting, E. G., 256[184], 265
Ahmed, M., 255[178], 256[178], 265
Ahmed, S. I., 240[102,103], 262
Alaupovic, P., 288[44], 304
Alberts, A. W., 183[48,49,60], 188[105,106], 190[105,106,114, 117], 191[105,106,114], 192[106], 193[105,106,114], 204[49], 217, 219
Alberty, R. A., 44[4], 81
Albritton, W. L., 229[36], 260
Allen, S. H. G., 186[82], 190[82], 218
Allmann, D. W., 299[78], 305

Ames, B. N., 230[45], 238[77], 260, 261
Anagnostopoulos, C., 240[99], 245[99,138,139], 246[99], 247[99], 262, 263
Anderson, L., 108[48], 111[48], 133
Anderson, R. C., 184[67,70], 217
Anderson, W. B., 44[13], 45[39], 58[59,60], 59[59,60,68], 60[60], 61[13], 79[13], 80[13], 81, 83, 84
Andrews, P., 149[64], 154, 155, 157, 158, 163[84], 164[64], 167, 177
Angeletti, R. H., 148[56], 176
Ankel, H., 109[44], 112, 118, 121[81], 132, 133, 134
Aoki, T., 241[116], 263
Apella, E., 256[183], 265
Apirion, D., 238[83], 262
Arias, I. M., 296[59], 305
Arnon, R., 147[52,53], 176
Aronson, N. N., 174[124], 179
Arth, G. E., 184[67,69,70], 217, 218
Aschaffenburg, R., 143[36,37], 176
Ashwell, G., 174[124], 179
Atassi, M. Z., 147[48], 176
Avigan, J., 270[12], 303

307

B

Babad, H., 138[9], 139[9], 141, 155, 156, 157, 175

Bagli, J. F., 281[35], 304

Bagshaw, W., 169[97], 178

Baker, H. N., 287, 289[45,46], 304

Baker, J. R., 167[93], 178

Baker, R. F., [91], 262

Balbinder, E., 229[27], 237[73], 238[88], 242[117], 260, 261, 262, 263

Baldwin, R. L., 228[25], 260

Barden, R. E., 189[112], 219

Barel, A. O., 147[44], 176

Barendsen, W., 241[115], 263

Barker, R., 149[63], 150[63], 151, 164[63], 177

Barman, T. E., 169[96,97,98], 173[111], 178

Barnes, C. C., 115[61], 133

Baron, L. S., 241[113], 262

Baron-Murphy, J., 229[34], 249[165], 260, 264

Baskerville, E. N., 247, 264

Battig, F. A., 58[61], 83

Baurle, R. H., 237[72], 243[122], 257[122], 261, 263

Beg, Z. H., 299[78], 305

Bell, J. D., 47[33], 82

Belser, N. O., 232[47], 261

Belser, W. L., 241[110,112], 243[110,124,125], 249[165], 255[110], 262, 263, 264

Berberich, M. A., 71[83], 84

Berger, H., 230[38], 231[38,43], 232[43], [46], 260

Berlyn, M. B., 240[104], 262

Bernhard, S. A., 8[15], 38[52,54], 40, 41

Bertland, A. U., II, 91[14], 94[14], 95[14], 97, 98[14], 99[24,25,26, 29], 102[29,33], 103[29], 104[29], 105, 111[26,29], 113[26], 127[29], 129[4'], 130[10'], 131, 132, 134

Bertland, L. H., 91[14], 94[14], 95[14], 97, 98[14], 131

Betz, H., 256[187], 265

Bevill, R. D., III, 109[51], 133

Beytia, E., 299[80], 305

Bhaduri, A., 90[5], 91[5,11], 92[5], 94[5], 99[5,11,28], 105[5], 113[11], 114[5], 131[11'], 131, 132, 135

Biederbick, K., 186[76], 190[76], 193[76], 218

Bisset, K. J., 270[15], 273[15,21], 303

Bittman, R., 19[57], 38[57], 41

Bloch, K., 296, 305

Blumberg, W. E., 13[22], 34[22], 40

Blume, A. J., 238[88], 262

Bohr, C., 8, 40

Boll, W. G., 249[168], 255[168], 264

Bonnemere, C., 184[72], 186[72], 218

Bonner, D. M., 246[143], 247, 255[143,179], 256[190], 263, 264, 265

Bonner, M., 238[79], 239[79], 243[79], 257[79], 261

Bortz, W. M., 186[84], 205[84], 210[84], 218

Bosmann, H. B., 173[115], 174, 178

Boyd, G. S., 301[90], 306

Boyer, H. W., 259[199], 265

Boyer, P. D., 139[19,20], 175

Bradbury, J. H., 143[40], 176

Bradshaw, R. A., 147[45], 148[56], 176

Brady, R. O., 182[16,25,30], 183[16,25], 216

Brammar, W. J., 230[38], 231[38,42,43], 232[43], 239[46'], 260, 265

Brand, L., 282[37], 304

Brew, K., 138, 139[16,17], 141, 142[26,27,28,29,30,31,32,33,

Brew, K. (cont'd)
34], 143[41,42], 146[28,32,33,
34], 147[27,30,31], 148[16,41],
156[16], 158[16], 163[26],
168[31], 170[16,32,33], 171,
173[116], 174, 175, 176, 178
Brewer, H. B., 287[47], 289[47],
304
Briedis, A., 299[80], 305
Brodbeck, U., 138[5,6], 139[11],
140[6,21], 141[5,6,21,22],
147[22], 149[21,60], 150[21],
158[60], 164[60], 170[6], 174,
175, 176
Bronson, M. J., 258[195], 265
Brown, C. M., 44[6], 81
Brown, M. S., 44[13,23], 58[23],
59[23], 60[23], 61[13], 62,
79[13], 80[13], 81, 82, 300[84],
306
Browne, D. T., 229[37], 260
Browne, W. J., 142, 147[31],
168[31], 175
Bruice, T. C., 186, 218
Brümmer, W., 186[76], 190[76],
193[76], 218
Bucher, N. L., 270[8], 303
Buehner, M., 20[27], 40
Bugge, B., 99[26,30,31], 100[31],
102[31], 104[30], 111[26],
113[26], 123[86], 130[10'], 132,
134
Burk, D., 23, 40
Burton, D., 214[145], 220
Buttin, G., 106[39], 132

C

Caban, C. E., 59[69], 60[69], 84
Calhoun, D. H., 240[107], 244[107],
254[107], 262
Calimbas, T., 292[55,56], 293[55,
56], 294, 295[55,56], 305
Calimbas, T. D., 274[29], 276[29],
277[30], 278[29], 280[29], 281[29],

284[30], 292[30,55,56], 293[30,
55,56], 294, 295[30,55,56], 296
[29], 297[29], 298[29], 304, 305
Callahan, R., 237[73], 261
Campbell, P. N., 142, 175
Carlson, D. M., 141, 156[23], 175
Carlson, J. P., 274[28,29],
276[28,29], 278[29,31], 279,
280[28,29], 281[28,29], 282[31],
296[28,29], 297[28,29], 298[28,
29], 299[31,82], 300[31,82],
301[31,82], 304, 306
Carlton, B. C., 224[5], 231[44],
245[140], 259, 260, 263
Carpenter, F. H., 202[137], 203,
220
Carsiotis, M., 256, 265
Case, M. E., 240[102], 262
Caspar, D. L. D., 6[13], 40
Casse, F., 244[129], 263
Castellino, F. J., 142[28],
146[28], 148[57], 169[57], 175,
176
Catcheside, D., 255[178], 256[178],
265
Chaberek, S., 55[52], 56[52], 83
Chaikoff, I. L., 182[35,37],
183[40], 216
Chakrabarty, A. M., 244[131],
263
Chalmers, J. H., Jr., 240[105],
248[105], 257[105], 262
Chang, H.-C., 183[51], 186[51],
188[51,93], 193[51], 194[51],
196[51], 197[51], 201[51],
202[51], 204[51], 207[51],
209[51,138], 210[51], 212[93],
213[93], 217, 219, 220
Changeux, J.-P., 3[4], 4[4,5,7],
5, 7[4], 13[4,5], 14, 19[5,25,
26], 34[5], 39, 40, 67[79,81],
84
Cheetham, R. D., 149[68],
171[68], 177
Chen, J., 249[168], 255[168], 264

Chippaux, M., 244[129], <u>263</u>
Cho, I. C., 33[45], <u>41</u>
Chock, P. B., 49[45], 52[49],
    54[49,50,51], 66[75,76],
    72[49,50], <u>83</u>, <u>84</u>
Christensen, A., 90[5], 91[5,11],
    92[5], 94[5], 99[5,11], 105[5],
    113[11], 114[5], <u>131</u>
Ciardi, J. E., 44[13], 47[27],
    52[48], 57[57], 58[57], 61[13],
    63[48], 72, 73[27,48,84], 74[48],
    75[48,85], 76[27], 77[27], 78,
    79[13], 80[13], <u>81</u>, <u>82</u>, <u>83</u>, <u>84</u>
Cimino, F., 45, 47[27], 73[27],
    74[27], 75[27,85], 76[27], 77[27],
    78[27], <u>82</u>, <u>83</u>, <u>84</u>
Clayton, R. B., 270, <u>303</u>
Cleland, W. W., 189[111], <u>219</u>
Clinkenbeard, K. D., 299[71,72],
    <u>305</u>
Coffey, R. G., 149[66,67], <u>177</u>
Cohen, P. T., 237[68], <u>261</u>
Colowick, S. P., 115[61], <u>133</u>
Colvin, B., 149[60], 158[60],
    164[60], <u>176</u>
Conner, R. L., 292, <u>305</u>
Conway, A., 37[49], 38[49], <u>41</u>
Cook, R. A., 5[8], 26[32], 38[60],
    <u>40</u>, <u>41</u>
Cooper, T. G., 183[62], 188[104],
    191[62], 192[104], 193[104], <u>219</u>
Cornish-Bowden, A., 34[46], <u>41</u>
Cotton, R. G. H., 225[10], 231[10],
    <u>259</u>
Cowburn, D. A., 143[41], 148[41],
    <u>176</u>
Cox, J. M., 13[21], <u>40</u>
Craig, M. C., 299[80], <u>305</u>
Crastes de Paulet, A., 296[63], <u>305</u>
Crawford, I. P., 224[1,8], 225[1,9,
    10,11,14], 226[16], 227[22,24],
    228[22,24], 229[8,29,34], 230[1],
    231[10], 232[1,47,48,49], 233[1],
    234[1], 235[14,24], 238[86],
    240[99,100,101,106,108], 241[111],
    242[29,111], 244[100,106,131,133,

134], 245[99,100,101,138,139],
    246[99,133,134], 247[99,148,
    150], 252[172], 253[175],
    254[11,29,176], 255[1,176],
    256[176], 257[111], [46], [141],
    <u>259</u>, <u>260</u>, <u>261</u>, <u>262</u>, <u>263</u>, <u>264</u>
Creighton, T. E., 226[15],
    227[26], 229[31,32], 242[118],
    257[118], 258[31], [141], <u>259</u>,
    <u>260</u>, <u>263</u>
Creveling, C. R., 90[5], 91[5,12],
    92[5], 94[5], 99[5], 105[5],
    114[5,60], <u>131</u>, <u>133</u>
Cynkin, M. A., 171[104], <u>178</u>

D

Dalziel, K., 115[62], <u>133</u>
Dana, S. E., 300[84], <u>306</u>
Darnall, D. W., 6[11], 8[11],
    31[11], <u>40</u>
Darrow, R. A., 90[6,7], 91[6,7,
    12,13], 92[6,13], 94[6,13],
    95[6,7,13], 96[6,7,21], 97,
    99[13,28], 105[6,7], 114[60],
    <u>131</u>, <u>132</u>, <u>133</u>
Dastoor, M. N., 47[33], <u>82</u>
Davies, G., 202[136], 203, <u>220</u>
Davis, J. E., 118[68], 129, <u>133</u>,
    <u>134</u>
Davis, L., 113, 114, 119, <u>133</u>
Davison, A. G., 270[10], <u>303</u>
DeCrombrugge, B., 258[196], <u>265</u>
DeGraff, J., 241[115], <u>263</u>
De Grombrugghe, B., 129[3'],
    <u>134</u>
Delahunty, T., 287[46], 289[4 6],
    <u>304</u>
Delmer, D. P., 249[163,164,165],
    254[164], 255[164], <u>264</u>
Deluca, M., 282[38], <u>304</u>
DeMoss, J. A., 226, 240[105],
    247, 248[105,151,153,155,157],
    256[189,190], 257[105,153,155],
    <u>259</u>, <u>262</u>, <u>264</u>, <u>265</u>

Dempsey, M. E. , 268[1,2,70],
269[70], 270[13,14,15,16,17,18,
19], 271[14], 272[1,14,16,20],
273[1,2,15,16,20,21,22,23],
274[1,2,29], 275[2], 276[16,17,
20,22,29], 277[22], 278[1,2,20,
29,31], 279[2,20,22,31], 280[1,2,
20,29], 281[20,29,33,34,36],
282[31], 283[32], 286[1,2,20],
287[1,2,20,40], 288[20,48],
289[48], 290[49], 291[51,52],
292[51,52], 293[51,52], 296[29,
49,68,70], 297[29], 298[29,70],
299[31,70,82], 300[31,82,86,87],
301[31,82], 302[86], 303, 304,
305, 306
Denny, R. M. , 224[7], 229[7],
251[7], 253[7,174], 259, 264
Denton, M. D. , 44[12], 48[40],
49[42], 52[12,42], 54[42], 55[42,
54], 60[12], 63[40,54], 64[40],
66[40,54], 67[40,54], 68[54],
70[42], 73[42], 74[40], 81, 83
Denton, W. L. , 138[5,6], 140[6],
141[5,6], 170[6,100], 175, 178
Dernanleau, D. A. , 147[45], 176
De Robichon Szulmajster, H. ,
90[3,4], 91[4], 95[4], 131
Derosier, D. J. , 6[14], 40
Descompe, B. , 296[63], 305
Deuel, T. F. , 58[58], 83
Dhar, A. K. , 273[24], 277[24], 303
Dicamelli, R. F. , 229[27], 260
Diedrich, D. F. , 108[48], 111[48],
133
Dietschy, J. M. , 296[67], 300[84],
305, 306
Dils, R. , 182[36,38], 216
Dimroth, P. , 183[61], 188[100],
190[61,100,116], 191[61,100],
192[61,100], 193[61,116], 217,
219
Dituri, F. , 182[33], 216
Dolmans, M. , 147[44], 176

Doolittle, W. F. , 235[59], 236[59],
261
Dorfman, A. , 167[94], 178
Drapeau, G. R. , 224[5], 231[42],
251[171], 259, 260, 264
Druzhinina, T. N. , 121[78], 134
Duée, E. , 38[56], 41
Dugan, R. E. , 299[80], 305
Dulit, E. , 270[10], 303
Dulley, J. R. , 152[76], 153[76],
177
Duntze, W. , 255[181], 265
du Vigneaud, V. , 184[64,65], 217
Dvornik, D. , 281[35], 304

E

Easton, N. R. , 184[67,68,70],
217, 218
Ebner, E. , 45[38], 59[38,67],
60[67], 82, 84
Ebner, K. E. , 120[76], 121[76],
134, 138[5,6], 139[11,12,13,14,
15], 140[6,21], 141[5,6,21,22],
147[22], 149[21,59,60], 150[21,
71,72], 151[71,73,74], 152[72,
74,75], 153[74,75], 156, 157[81],
158[60,81,85,86,87], 159[81,86,
87], 160[81,86,87], 161[81,86,
87], 162[81], 163[81,86,87],
164[60,81,86,87,89], 167[85],
169[99], 170[6,59,99,100,101],
171[105,108], 174, 174, 175,
176, 177, 178, 179
Edelstein, C. , 288[43], 304
Edmond, J. B. , 299, 305
Edwards, J. , 1852[5], 193[120],
201[5], 202[5], 206[5], 213[5,
120], 214[5,120], 210, 215,
220
Edwards, P. A. , 299[74,75], 305
Egan, A. F. , 243[121], 257[121],
263
Egusa, S. , 241[116], 263

Ehnholm, C., 294, 305
Eigen, M., 19[57], 38[57], 41
Eisenberg, D., 47[33], 82
Ellenrieder, G. V., 19[59],
    38[59], 41
Elsässer, S., 59[67], 60[67], 84,
    256[185], 265
Enatsu, T., 225[11], 244[134],
    245[134], 254[11], 259, 263
Estes, L. W., 149[70], 171[70],
    177

F

Faeder, E. J., 235, 261
Fall, R. R., 190[114,115], 191[114,
    115], 193[114], 219
Fan, D. F., 121[77], 134
Fargie, B., 244[130], 263
Faure, A., 147[51], 176
Feingold, D. S., 121[77], 134,
    149[70], 171[70], 177
Feldbruegge, D. H., 299[80], 305
Fenna, R. E., 143[36,37], 176
Fenselau, C., 187[91,92], 188[91,
    92], 193[91,92], 218
Filmer, D., 11[24], 13, 19, 34[24],
    40, 67[80], 84
Findlay, J. B. C., 142[32], 146[32],
    170[32], 176
Fink, G. R., 239[89], 262
Fitzgerald, D. K., 149[59,60],
    158[60], 164[60],170[59,101],
    176, 178
Fleischer, B., 149[69], 171[69],
    177
Fleischer, S., 149[69], 171[69], 177
Fluri, R., 225[9], 259
Folkers, K., 184[65,66,67,68,69,
    70], 217, 218
Ford, G. C., 20[27], 40
Formica, J. V., 182[16], 183[16],
    216
Frantz, I. D., Jr., 270, 300[86],
    302[86], 303, 306

Frazier, W. A., 148[56], 176
Fredericq, P., 240, 262
Fredrickson, D. S., 288[42], 304
Frey, P. A., 117, 118[68], 129[5'],
    130[7'], 133, 134
Fritz, I. B., 182[20], 183[20], 216
Frnka, J., 300, 306
Fung, C.-H., 189[112], 219
Fuyimoto, Y., 105[35,36], 132

G

Gabriel, O., 90[9], 91[16], 108[45,
    46], 109[45,46], 111[45,46],
    112[55], 113[55], 115[63], 116[63],
    117[65], 118[70,71], 119[46,55,
    72], 120[63], 122[84], 123[84,86],
    124[9,71,87,89], 126[89],
    127[9,46,63,71], 128[87], 130[8',
    9', 11'], 131, 132, 133, 134
Gaertner, F. H., 248[153,154],
    257[153], 264
Gallego, E., 19[58], 38[58], 41
Gancedo, C., 59[67], 60[67], 84
Gander, J. E., 139, 175
Ganguly, J., 182[26,27], 216
Garren, L. D., 301[89], 306
Gates, G. A., 197, 220
Gatmaitan, Z., 296[59], 305
Gaugler, R. W., 108[45], 109[45],
    111[45], 132
Gaunt, M. A., 121[81], 134
Gavey, K. L., 296[66], 305
Gaylor, J. L., 270, 303
Gemski, P., Jr., 241[113], 262
Geren, C. R., 174, 178
Geren, L. M., 174, 179
Gerguson, A. R., 256[187], 265
Gerhart, J. C., 4[6], 13, 19[25],
    40
Germershausen, J., 255[177],
    256[177,183], 264, 265
Gibson, D. M., 182[7,8,10,12,
    13,14,15,17,32], 183[12,14],
    215, 216, 299[78], 305

Gibson, F. , 243[121], 257[121], 263
Giles, N. H. , 240[102,103,104], 262
Gill, G. N. , 301[89], 306
Ginsburg, A. , 44[12,13,14,19,44],
    45[19,26,37], 47[19,26,28],
    48[40], 49[19,28,42,43], 50[43,44],
    51[14,28,44], 52[12,42,43,48],
    53[43], 54[19,42,43], 55[28,42,
    43,54,55,56], 56[43,55], 57[55,
    56], 58[37,58,59], 59[59,68,69],
    60[12,14,37,69], 61[13,14,37,
    72], 63[40,48,54], 64, 66[37,40,
    43,54], 67[40,54], 68[54], 70[28,
    42,55], 71[55,83], 73[19,42,48],
    74[40,48], 75[48], 78[19,26],
    79[13], 80[13,89], 81[89], 81, 82,
    83, 84
Ginsburg, V. , 165[92], 177
Glaser, L. , 113, 114, 117[66], 119,
    120, 121[80], 127[88], 131[12'],
    133, 134, 135
Goaman, L. C. G. , 13[21], 40
Gohlke, J. R. , 282[37], 304
Goldberg, M. E. , 228[25], 260
Goodall, D. , 19[58], 38[58], 41
Goodkin, P. , 47[33], 82
Goodman, D. S. , 270, 303
Goodwin, C. D. 299[77], 305
Gordon, E. , 183[58], 217
Gordon, S. G. , 188[105], 190[105],
    191[105], 193[105], 219
Gordon, W. G. , 138, 174
Goto, T. , 183[53,54], 212[53], 217
Gotto, A. M. , Jr. , 287[45,46],
    288[42], 289[45,46], 304
Gould, R. G. , 299[75,79], 300[85],
    305, 306
Gratzer, W. B. , 143[41], 148[41],
    176
Grawbowski, G. A. , 291[51,52],
    292[51,52], 293[51,52], 305
Green, N. M. , 193[118], 210[118],
    220
Gregolin, C. , 183[50,51,52], 186[51],
    188[51,52,93,94,95], 190[94],
    193[50,51], 194[50,51,52], 195[50],
    196[50,51,52], 197[50,52,94],
    199[50], 201[51,52], 202[51,52,
    94], 204[51], 205, 206, 207[50,
    51,52,94], 209[50,51,52],
    210[51,52], 212[52,93,94,95],
    213[93], 214[94], 217, 219
Grieshaber, M. , 243[122],
    257[122], 263
Grossman, A. , 186[85], 190[85],
    218
Guchhait, R. , 195[124], 199[124],
    220
Guchhait, R. B. , 182[6], 183[61,
    62], 187[91], 188[91,92,99,100,
    101,102,103,104], 190[61,99,
    100,116], 191[61,62,99,100],
    192[61,100,104], 193[61,91,92,
    93,101,104,116], 194, 195[124,
    125], 196[147], 197[134],
    199[124,125], 200[125], 201
    [125], 202[125,135], 204[125,
    135], 215, 217, 218, 219, 220,
    221
Guest, J. R. , 224[5], 231[44],
    259, 260
Gunsalus, C. F. , 244[131], 263
Gunsalus, I. C. , 240[106], 244[106,
    131,135], 245[135], 262, 263
Gurin, S. , 182[33], 216

                        H

Habeeb, A. F. S. A. , 147[48],
    176
Haber, J. E. , 30[34], 35[34], 41
Hall, A. N. , 227[21], 260
Hall, C. A. , 244[128], 263
Hamilton, J. A. , 184[72,73],
    186[72], 218
Hammes, G. G. , 55[53], 56[53],
    83, 235, 261
Handford, B. O. , 143[36], 176
Hankins, C. , 249[162], 264
Hanlon, J. , 243[126], 263
Hansbury, E. , 273[25], 274[25],
    277[25], 287[25], 303

Hanson, R. F., 291[51,52], 292[51, 52], 293[51,52], 305

Hardman, D. F., 233, 261

Hardman, J. K., 233, 234, 261

Harrington, K. T., 202[137], 203, 220

Harris, J. E., 37[50], 41

Harris, S. A., 184[65,66,67,68, 69,70], 217, 218

Hartman, P. E., 238[77], 261

Hashimoto, T., 188[107,109,110], 197[108], 219

Hasselbach, K., 8[16], 40

Hassid, W. Z., 138[9], 139[9], 141, 155, 156, 157, 175

Hatanaka, M., 225[14], 227[24], 228[24], 235[14,24], 259, 260

Hatch, M. D., 188[97], 214[97], 219

Hathaway, G. M., 226[16], 259

Hayashi, M., 236[63], 237[63], 261

Hegarty, A. F., 186, 218

Heidner, E. C., 47[33], 82

Heilmeyer, L., Jr., 44[10], 52[10], 81

Heinrich, C. P., 58[61], 83

Heinrikson, R. L., 45[34,36], 47[34,36], 82

Heinstein, P. F., 188[98], 214[98], 215[98], 219

Held, W. A., 226, 227[19], 260

Helinski, D. R., 107, 132, 231[44], 240[96], 259[199], 260, 262, 265

Heller, R. A., 300[85], 306

Helting, T., 165[91], 170[102], 177, 178

Henderson, E. J., 243[120], 257[120], 263

Henkens, R. W., 51[46], 83

Henley, K., 197[127], 220

Hennig, S. B., 45[37], 48[40], 52[48], 58[37,59], 59[59,68], 60[37], 61[37], 63[40,48], 64[40], 66[37,40], 67[40], 73[48], 74[40, 48], 75[48], 82, 83, 84

Henning, U., 231[44], 260

Henninger, G., 194[121], 197[121], 220

Herries, D. G., 173[116], 178

Hershfield, V., 259[199], 265

Heyl, D., 184[67,68], 217, 218

Hickson, R. F., 240[96], 262

Higgins, M., 299[76], 305

Hill, A. V., 8, 23, 40

Hill, E. A., 109[51], 133

Hill, R. L., 138[10], 139[16], 141[10], 142[26,27,28,29,30, 31,34], 146[28,34], 147[27,30, 31], 148[16,57], 149[61,63], 150[61,63], 151[61,63], 156[10], 158[26,61], 163[26], 164[61,63], 168[31], 169[57], 170[16], 175, 176, 177

Hindle, E. J., 173[112], 178

Hinze, H., 45[38], 59[38], 82

Hiraga, S., 238[75], 240[97], 261, 262

Ho, W. K., 296[58], 305

Hoagland, V. D., Jr., 90, 128[10], 131

Hoch, S. O., 225[12,13], 240[99], 245[99], 246[12,13,99], 247[99, 150], 252[12,173], 254[12], 255[12], [141], 259, 262, 263, 264

Hockert, M. I., 20[27], 40

Hofmann, K., 184[64], 217

Hogness, D. S., 90[8], 105[8], 106, 107[8], 108[41], 109[41], 119[41], 127[41], 129[2'], 131, 132, 134

Holcomb, G. N., [37], 132

Hollis, D., 187[92], 188[92], 193[92], 218

Holloway, B. W., 244[130,132], 263

Holmberg, N., 120[76], 121[76], 134

Holmes, L. G., 147[46], 176

Holmes, W. M., 240[98], 245[98], 246[98], 247[98], 262

Holzer, H. , 44[1,9,10], 52[10,47],
    58[9,61], 59[67], 60[1,67],
    61[73], 63[1], 66[73], 81, 83, 84,
    256[184,185,187], 265
Horibata, K. , 225[14], 235[14], 259
Horn, V. , 230[39,40,41], 231[39,40],
    238[79], 239[79], 243[79], 244[127],
    257[79], 260, 261, 263
Huang, C. Y. , 52[49], 54[49,50],
    72[49,50], 83
Hubbard, J. S. , 44[11,18], 47[18],
    49[18], 52[11], 54[18], 81, 82
Hudgin, R. L. , 171[103,110],
    174[125], 178, 179
Hudson, B. G. , 150[72], 152[72],
    177
Hulett, F. M. , 248[155], 257[155],
    264
Hunt, J. B. , 47[28], 49[28,43,44],
    50, 51[28,44], 52[43], 53[43],
    54[43], 55[28,43], 56[43], 66[43],
    70[28], 82, 83
Hutchinson, M. A. , 241[110],
    243[110,124], 255[110], 262, 263
Hütter, R. , 247, 248[151], 264
Hwang, L. H. , 243[123], 263

I

Imai, M. , 240[97], 262
Imamoto, F. , 224, 236[3], 237[3],
    238[3,80,84,85], 239[3], [90],
    259, 261, 262
Ingram, L. O. , 248[160], 249[160],
    264
Inoue, H. , 194[122], 196[122],
    197[122], 201, 204, 220
Irias, J. J. , 6[10], 40
Iritani, N. , 188[107,108,109],
    197[108], 219
Isaakidou, I. , 99[32], 100[32], 132
Isano, H. , 188[109], 219
Ito, J. , 227[22], 228[22], 230[40],
    231[40], 238[86,87], 239[92],
    243[119], 260, 262, 263
Ito, K. , 236[65], 261

Ivatt, R. J. , 167, 178
Iyengar, C. W. , 292[53], 305
Iyer, K. S. , 147, 176

J

Jabbal, I. , 171[110], 178
Jackson, E. N. , 237[70], 238[76],
    258[198], 261, 265
Jackson, L. E. , 225[9], 259
Jackson, R. L. , 287[45,46],
    289[45,46], 304
Jackson, R. W. , 240[105],
    248[105], 257[105], 262
Jackson, W. T. H. , 31[38], 41
Jacob, F. , 3[4], 4[4], 7[4], 13[4],
    39, 237[67], 261
Jacob, M. I. , 182[8], 215
Jacobs, R. A. , 197[129,133],
    220
Jacobson, B. E. , 186[82], 190[82],
    218
James, S. A. , [37], 132
Janosko, N. , 255 [181] , 265
Jenness, R. , 138[1], 171[107],
    172[107], 174, 178
Jensen, R. A. , 240[98,107],
    244[107], 245[98], 246[98],
    247[98], 248[160], 249[160],
    254[107], 262, 264
Johansen, J. T. , 94, 99[29],
    102[29], 103[29], 104[29],
    111[29], 127[29], 132
John, K. M. , 287[47], 289[47],
    304
Johnson, R. C. , 273[27], 287[27],
    296[65], 304, 305
Jolles, P. , 147[51], 176
Jordan, E. , 96[22], 106[22], 132
Jourdian, G. W. , 141, 156[23],
    175
Jütting, G. , 186[86,88], 218

K

Kalckar, H. M. , 90[3,5], 91[5,11],
    92[5], 94[5], 96[22], 99[5,11,25,

Kalckar, H. M. (cont'd)
    26,27,28,29,30], 100, 102[29],
    103[29], 104[29,30], 105[5],
    106[22,38], 108[46,50],
    109[46], 111[26,29,46,50],
    112[55], 113[55,57], 114[5,57,
    60], 115[63], 116[63], 119[46,
    55], 120[63,75], 127[29,46,63],
    129[4'], 130[8',10'), 131,132,
    133, 134
Kallen, R. G., 183[41], 216
Kaminogawa, S., 153[77], 177
Kan, K. W., 290[49], 296[49,50],
    301[91], 304, 306
Kanarek, L., 147[50], 176
Kandutsch, A. A., 270, 303
Kane, J. F., 240[98], 245[98],
    246[98], 247[98], 248[160],
    249[160], 262, 264
Kannangara, C. G., 214[146],
    215[146], 221
Kano, Y., 238[84], 262
Kaplan, N. O., 115[61], 133
Kaplan, S., 255[179], 265
Katsunuma, T., 256[184,185,187],
    265
Katunuma, N., 256[184], 265
Kawachi, T., 300[83], 306
Kawai, H., 121, 134
Kaziro, Y., 183[63], 186[85],
    190[85], 217, 218
Keenan, T. W., 149[68], 171[68],
    177
Kenyon, G. L., 229[37], 260
Keresztesy, J. C., 184[65], 217
Ketley, J. N., 111[54], 113, 133
Khatra, B. S., 173, 178
Kida, S., 226[16], 232[49], 259,
    261
Kilburn, E., 197[130], 220
King, N. L. R., 143[40], 176
Kingdon, H. S., 44[2,11,12,18,20,
    25], 45[34,36], 47[18,25,34,
    36], 48[25], 49[18], 52[2,11,
    12,25], 54[18], 58[2], 60[2,
    12], 63, 70[20,25], 71[20], 81,82

Kirkwood, S., 108[47], 109[47,
    51,52], 111,114, 119[47],
    127[47], 129[6'], 133, 134
Kirschbaum, B. B., 174, 178
Kirschner, K., 19[57,58,59],
    38[57,58,59], 41, 224, 225[4],
    226[4], 234[4], 255[4], 259
Kitchen, B. J., 157[82], 163[88],
    169[98], 177, 178
Kiyosawa, I., 149[60], 158[60],
    164[60], 176
Kiyosawa, K., 149[65], 154[65],
    177
Klapper, M. H., 31[39], 41
Klee, C. B., 149[62], 150[62],
    158, 163[83], 167, 177
Klee, W. A., 147, 149[62],
    150[62], 158, 163[83], 167,
    176, 177
Klein, H. P., 214, 220
Kleinschmidt, A. K., 183[50,51,
    52,55,57], 186[51], 188[51,
    52], 190[113], 193[50,51,55,
    57,113], 194[50,51,52,55,113,
    123], 195[50,55,113,123,124,
    125], 196[50,51,52,55,113],
    197[50,51,52,55,113], 199[50,
    123,124,125],200[125],
    201[51,52,55,113,125],
    202[51,52,125], 204[51,125],
    205[50,52], 206[50], 207[50,
    51,52,55,113], 209[50,51,52,
    113], 210[51,52], 212[52,113],
    214[113], 217, 219, 220
Kloos, W. E., 246[146], 247[146,
    147], 264
Klotz, I. M., 6[11], 8[11],
    31[11,39], 40, 41, 75[88], 84
Klug, A., 6[13], 40
Knappe, J., 184[72], 186[72,76,
    80,86,87,88], 190[76,80],
    193[76,80], 218
Koehler, D. F., 274[28],
    276[28], 280[28], 281[28],
    287[40,41], 296[28], 297[28],
    298[28], 304

Koga, S., 288[43], 304
Kohn, L. D., 249[194], 265
Kohno, T., 257[191], 265
Kominami, E., 256[184], 265
Koshland, D. E., Jr., 5[8], 11,
   13,19,24[30], 25[30], 26[30,31],
   30[34,37], 31[40], 33[42],
   34[24,46], 35[34], 36[37,40,
   42,48], 37[37,40,49], 38[49,
   60,61,62], 40, 41, 67[77,80],
   84, 108[49], 111[49], 133
Kosow, D. P., 186[75], 190[75],
   218
Kostner, G., 288[44], 304
Kowalsky, A., 108[49], 111[49], 133
Kraml, M., 281[35], 304
Krigbaum, W. R., 143, 176
Krishnapillai, V., 244[132], 263
Krogh, A., 8[16], 40
Kronman, M. J., 143[42], 176
Kugler, F. R., 143, 176
Kuhn, N. J., 139[18], 171, 175
Kula, M. R., 256[186], 265
Kumagai, H., 227[23], 228[23],
   234[52], 260, 261
Kurahashi, K., 96[22], 106[22],
   132
Kuwano, M., 238[82], 261

L

LaChance, J.-P., 186[88], 218
Lacy, A. M., 255[180], 265
Lahiri, A. K., 45[34], 47[34], 82
Landel, A. M., 108[48], 111[48],
   133
Landrey, J. R., 292[53], 305
Lane, M. D., 182[1,2,3,4,5,6],
   183[1,2,50,52,55,57,61,62],
   184[1,2], 185, 186[1,75,77],
   187, 188[52,92,93,94,95,99,
   100,101,103,104], 190[1,75,77,94,
   113,116], 191[61,62,99,100],
   192[61,100,104], 193[1,50,51,
   55,57,61,77,92,99,101,104,

113,116,119,120], 194[50,51,
   52,55,99,113,119,123], 195[50,
   55,113,119,123,124,125],
   196[50,51,52,55,113,147],
   197[50,51,52,55,94,113,119,
   134],199[50,123,124,125],
   200[125], 201[51,52,55,113,
   119,125], 202[5,51,52,94,119,
   125,135], 204[51,125,135],
   205,206[50], 207[3,4,50,51,
   52,55,94,113,119], 209[50,51,
   52,113,119,138], 210[51,52,
   119], 211, 212[2,52,93,94,
   95,113], 213[3,5,93,120,141],
   214[1,3,4,5,94,113,120], 215,
   217, 218, 219, 220, 221,
   299[71,72], 305
Langdon, R. G., 182[34], 216
Langer, R., 131[12'], 135
Langerman, N. R., 6[11], 8[11],
   31[11], 40
Lara, J. C., 249[161], 255[161],
   264
Lardy, H. A., 189[111], 219
Largen, M. T., 241[109,112],
   242[109], 243[109], 257[109],
   262
Lea, D. J., 227[21], 260
Lebowitz, J., 229[27], 260
Lee, W. E., 225[9], 259
Leef, J. L., 248[154], 264
Leelavathi, D. E., 149[70],
   171[70], 177
Lehman, E. D., 150[72], 152[72],
   177
Lehninger, A. L., 197[126], 220
Leloir, L. F., 87, 131
Lentz, P. I., Jr., 20[27], 40
Leonard, K. R., 195[124,125],
   199[124,125], 200, 201[125],
   202, 220
Levi, A. J., 296[59], 305
Levin, A. P., 229[36], 260
Levintow, L., 44[3], 81

Levitzki, A., 24[30], 25[30],
  26[30], 30[37], 31[40,41],
  33[42,43,44], 36[37,40,42,48],
  37[40], 38[53,55], 40, 41
Levy, H. R., 183[56], 194[56], 217
Levy, R. I., 288[42], 304
Lewis, P. N., 143[35], 176
Ley, J. M., 171[107], 172[107],
  178
Li, S. L., 224[6,7], 229[6,7],
  243[126], 251[6,7,169,170,171],
  252[169,170,173], 253[7], 259,
  263, 264
Lienhard, G. E., 118[67], 133
Liess, K., 44[10], 52[10], 81
Liggins, G. L., 245, 263
Lin, J. Y., 119[72], 122[84],
  123[84,86], 133, 134
Lin, T. Y., 148[58], 168[58], 176
Lindquist, L. C., 118[71], 124[71],
  127[71] , 133
Lineweaver, H., 23, 40
Linzell, J. L., 171[106], 178
Lipkin, E., 127[88], 134
Lochmüller, H., 186[79,81], 190[79,
  81], 193[81], 218
Lombardi, B., 149[70], 171[70],
  177
Lorch, E., 183[40], 186[86,88],
  216, 218
Lorusso, D. J., 296[60], 305
Lovett, M. A., 259[199], 265
Lowenstein, J. M., 183[41], 194
  [122], 196[122], 197[122], 201,
  204, 216, 220
Luterman, D. L., 66[76], 84
Lux, S. E., 287[47], 288[42,48],
  289[47,48], 304
Lynen, F., 182[21,24], 183[21,44,
  47], 186[77,79,81,83,84,87,88],
  188[47,83,96], 190[77,79,81,
  83], 193[77,81], 205[84], 210
  [84,140], 212[47], 214[96],
  216, 217, 218, 219, 220

M

McClure, W. R., 189[111], 219
McConnell, H., 34, 41
McCoy, K. E., 274[28,29],
  276[28,29], 278[29,31],
  279[31], 280[28,29], 281[28,
  29], 282[31], 291[52], 292[52],
  293[52], 296[28,29], 297[28,
  29], 298[28,29], 299[31],
  300[31], 301[31,91], 304, 305,
  306
McGarrahan, K., 270[9], 303
McGeoch, D., 236[64], 237[64],
  261
McGeoch, J., 236[64], 237[64],
  261
McGuire, E. J., 141, 156[23],
  171[110], 173[114], 175, 178
McKenzie, L. M., 149[59], 170[59],
  171[105], 176, 178
McNamara, D. J., 299[73], 305
McPherson, A., Jr., 20[27], 40
McPhie, P., 235, 261
McQuade, J. F., 242[118],
  257[118], 263
MacQuarrie, R. A., 38[52], 41
Magee, S. C., 150[71], 151[73,74],
  152[74,75], 153[74,75], 177
Magni, G., 44[24], 59[24,66],
  60[24], 61[24], 62[66], 80[66],
  81[66], 82, 84
Magnuson, J. A., 147[47], 176
Magnuson, N. S., 147[47], 176
Maitra, U. S., 109[44], 112, 118,
  132, 133
Maizels, N. M., 258[197], 265
Majerus, P. W., 197[129,130,
  133], 220
Mallory, F. B., 292, 305
Mamoon, A.-M., 182[30], 216
Mangum, J. H., 44[24], 59[24,66],
  60[24], 61[24], 62[66], 80[66],
  81[66], 82, 84

Mannering, G. J., 296[60], 305
Manney, T. R., 255[181], 256, 265
Manning, J. A., 296[58], 305
Mantel, M., 58[61], 83
Margolin, P., 224, 238[78], 259, 261
Margolis, S., 299[77], 305
Maron, E., 147[52,53], 176
Martell, A. E., 55[52], 56[52], 83
Martin, D. B., 182[39], 183[39, 42,45,48,49], 204[49], 216, 217
Martin, R. G., 239[89], 262
Martinez, L., 232[47], 261
Masui, H., 301[89], 306
Matchett, W. H., 235[58], 256, 261, 265
Matsuhashi, M., 183[44,47], 188[47, 96], 212[47], 214[96,144], 216, 217, 219, 220
Matsuhashi, S., 183[44,47], 188[47, 96], 212[47], 214[96], 216, 217, 219
Matsumoto, H., 241[114], 263
Matsuo, H., 105, 132
Matthes, K. J., 182[35,37], 216
Matthews, B. W., 8[15], 40
Mattock, P., 139[16], 142[27], 147[27], 148[16], 156[16], 170[16], 175
Maurer, R., 244[133], 246[133], 253[175], 254[176], 255[176], 256[176], 263, 264
Mawal, R., 139[14], 149[60], 150[71], 151[71,74], 152[74,75], 153[74,75], 158[60], 164[60,89], 175, 176, 177
Maxwell, E. S., 87[2], 90[3,4], 91[4], 95[4], 108[50], 109[2], 111[50], 120[74], 121[74], 131, 133
Mecke, D., 44[1,10], 52[10,47], 60[1], 63[1], 81, 83
Meduski, J. W., 246[144], 255[144], 263

Meers, J. L., 44[6], 81
Meighen, E. A., 75[86,87], 84
Meister, A., 44[3,5], 81
Melhorn, D. K., 241[111], 242[111], 257[111], 262
Melville, D. B., 184[64,65], 217
Mercer, W. D., 38[56], 41
Merriman, C. R., 169[99], 170[99], 178
Meyer, R. G., 255[177], 256 [177], 264
Miles, E. W., 227[23,24], 228[23, 24], 229[33], 234[52,53], 235[24], 249[194], 258[30], 260, 261, 265
Miller, A. L., 183[56], 194[56], 217
Miller, R. E., 44[7], 48, 49[41], 50, 55[41], 66[75], 81, 83, 84
Mills, S. E., 229[28,29,34], 242[28,29], 249[161,162, 163, 164, 165], 254[28,29,164], 255[161,164], 256[190], 260, 264, 265
Mizobuchi, H., 153[77], 177
Mobberley, M. L., 270[10], 303
Mockrin, S. C., 38[61], 41
Monod, J., 3[4], 4, 5, 7[4], 13, 14, 19[5], 34[5], 39, 67[79], 84, 237[67], 261
Moore, T. C., 249[167], 255[167], 264
Morikawa, N., 239[90,91], 262
Morino, Y., 6[9], 40
Morre, D. J., 149[68], 171[68], 177
Morris, H. P., 197[129], 220
Morrison, J. F., 139[14], 157[81], 158[81,86,87], 159[81,86,87], 160[81,86,87], 161[81,86,87], 162[81], 163[81,86,87], 164[81,86,87,89], 175, 177
Morse, D. E., 236[64], 237[64, 66,69,71], 238[81,82], [91], 261

Moss, J., 182[1,2,3,4,5,6], 183[1,
    2,55,57], 184[1,2], 185, 186[1],
    188[99,100,101], 190[1,99,100,
    113], 191[99,100], 192[100],
    193[1,55,57,99,101,113,119],
    194[55,99,113,119], 195[55,
    113,119], 196[55,113], 197[55,
    113,119,134], 201[5,55,113,119],
    202[5,119], 206[5], 207[3,4,55,
    113,119], 209[113,119], 210[119],
    211, 212[2,113], 213[3,5,141],
    214[1,3,4,5,113], 215, 217,
    219, 220, 299[71,72], 305
Mosteller, R. D., [91], 236[60],
    261, 262
Moyer, A. W., 184[64], 217
Mozingo, O. R., 184[65,66,67,
    69,70], 217, 218

Muirhead, H., 13[21], 40
Murgola, E. J., 239[93], 262
Murphy, T. M., 229[28], 242[28],
    254[28], 260
Musso, R. E., 129[3'], 134,
    258[196], 265
Myers, J. S., 130[9',11'], 134,
    234, 261

                    N

Nagao, R. T., 249[167], 255[167],
    264
Nagasawa, T., 149[65], 154[65],
    177
Nakanishi, S., 188[107,108],
    197[108,131,132], 219, 220
Nakao, E., 183[63], 217
Nelsestuen, G. L., 108[47],
    109[47,52], 111, 114, 119[47],
    127[47], 129[6'], 133, 134
Nelson, D. O., 296[60], 305
Némethy, G., 11[24], 13, 19,
    34[24], 40, 67[80], 84
Nervi, A. M., 188[106], 190[106,
    114,117], 191[106,114],
    192[106], 193[106,114], 219

Nester, E. W., 246, [141], 263
Neufeld, E. F., 121[79], 134
Neurath, H., 148[55], 176
Niall, H. D., 129, 134
Nichol, L. W., 31[38], 41
Nihoul-Deconinck, C., 147[50],
    176
Noel, J. K. F., 38[51], 41
Nolan, L. D., 129[5'], 134
North, A. C. T., 142[31], 147[31],
    168[31], 175
Novick, A., 3[1], 39
Novikova, M. A., 121[78], 134
Noyes, C., 45[34], 47[34], 82
Numa, S., 183[44,53,54,59],
    186[83,84], 188[83,96],
    190[83], 194[59,121], 196[128],
    197[108,121,128,131,132],
    205, 210[84,140], 212[53],
    214[96], 216, 217, 218, 219,
    220

                    O

Ochoa, S., 186[85], 190[85], 218
Ockner, R. K., 296[58], 305
Ogata, Y., 241[116], 263
Ogawa, S., 13[22], 34[22], 40,
    41
Okamoto, T., 240[97], 262
Okazaki, T., 188[108], 197[108],
    219
Oliver, E. J., 61[72], 84
Oliver, R. M., 6[14], 40
Olsen, K. W., 149[63], 150[63],
    151[63], 164[63], 177
Osborne, J. C., 174, 178
Ottesen, M., 94, 99[29], 102[29],
    103[29], 104[29], 111[29],
    127[29], 132
Ozawa, H., 149[69], 171[69], 177

                    P

Packer, E. L., 229[37], 260
Palmiter, R. D., 153, 177

Pardee, A. B., 3[3], 4[6], 13, 39, 40

Parker, R. A., 281[33], 304

Partridge, C. W. H., 240[102], 262

Pascal, M.-C., 244[129], 263

Pastan, I., 129[3'], 134, 258[196], 265

Pauling, L., 13, 40

Peaker, M., 171[106], 178

Pedersen, K. O., 97[23], 132

Peisach, J., 13[22], 34[22], 40

Pendergast, M., 195[124], 199[124], 220

Perham, R. N., 37[50], 41

Perrin, D. D., 66[74], 84

Perutz, M. F., 13[21], 30[35,36], 35, 40, 41

Peters, B. P., 174[124], 179

Petersen, W. E., 139[19,20], 175

Phillips, D. C., 142[31], 143[36, 37], 147[31], 168[31], 175, 176

Piersen, D. L., 240[107], 244[107], 254[107], 262

Pierson, D., 248[160], 249[160], 264

Pigiet, V., 75[87], 84

Pinteric, L., 171[110], 178

Polakis, S. E., 182[2,6], 183[2, 62], 184[2], 187[91,92], 188[91,92,100,101,102,103, 104], 190[100], 191[62,100], 192[100,104], 193[91,92,101, 104], 212[2], 215, 217, 218, 219

Pollard, H. M., 197[127], 220

Popják, G., 182[29,36,38], 216, 299, 305

Poppenhausen, R. B., 296[58], 305

Porter, J. W., 182[8,9,32], 215, 216, 299[80], 305

Porter, R. W., 55[53], 56[53], 83

Powell, J. T., 174, 178

Prager, E. M., 147[54], 176

Proctor, A. R., 240[108], 246[146], 247[146], 262, 264

Provost, P., 256[183], 265

Purich, D., 59[71], 60[71], 61[71], 84

Q

Queener, S. F., 244[135], 245[135], 263

Qureshi, A. A., 299[80], 305

R

Rachmeler, M., 226[18], 259

Randerath, E., 99[28], 132

Rasmussen, R. K., 214, 220

Rawitch, A. B., 143[43], 169[43], 176

Ray, M., 131[11'], 135

Reed, L. J., 6[14], 40

Reed, W. D., 299[72], 305

Reiser, R., 300, 306

Reithel, F. J., 149[66,67], 177

Reyes, H., 296[59], 305

Rhee, S. G., 54[51], 66[76], 83, 84

Riedel, B., 183[53,54,59], 194[59], 212[53], 217

Riepertinger, C., 186[81], 190[81], 193[81], 214[142], 218, 220

Rilling, H. C., 273[26], 286[26], 296, 304

Ringelmann, E., 183[53,54,59], 186[83,86,87,88], 188[83], 194[59], 210[140], 212[53], 217, 218, 220

Ritter, M. C., 268[1,2], 270[15], 272[1,20], 273[1,2,15,20,21], 274[1,2], 275[2], 276[20], 278[1,2,20], 279[2,20], 280[1, 2,20], 281[20,33], 283[32], 286[1,2,20], 287[1,2,20], 288[20,48], 289[48], 290[49], 296[49], 300[87], 302[86], 303, 304, 306

Robb, F., 243[124,125], 263
Robbins, F. M., 147[46], 176
Roberts, C. F., 248[156], 264
Robinson, E. A., 120[75], 134
Rocha, V., 229[29,34], 242[29], 254[29], 260
Roden, L., 165[91], 167[93,94], 177, 178
Rodstrom, R., 90[6,7], 91[6,7, 13], 92[6,13], 94[6,13], 95[6,7,13], 96[6,7], 97, 99[13], 105[6,7], 131
Rodwell, V. W., 299[73], 305
Ronan, R., 287[47], 289[47], 304
Rose, I. A., 111[53], 133
Rose, J. K., 236[61,62], 237[61, 62], 238, 261
Rose, N. E., 247[147], 264
Roseman, S., 141, 156[23], 171[110], 173[113,114], 175, 178
Rosemeyer, M. A., 167, 178
Ross, P. D., 49[44], 50[44], 51[44], 55[56], 57[56], 83
Rossmann, M. G., 20[27], 40
Roth, C. W., 246, [141], 263
Roth, J. R., 257[191], 265
Roth, S., 173[114], 178
Roth, T. F., 240[96], 262
Rothblat, G. H., 296[68,69], 305
Rubin, M. M., 19[26], 40, 67[81], 84
Rudney, H., 299[76], 300[83], 305, 306
Ryden, H. M., 227[21], 260
Ryder, E., 182[3], 183[50,51,52, 93], 186[51], 188[51,52,94], 190[94], 193[50,51,120], 194[50,51,52], 195[50], 196[50, 51,52], 197[50,51,52,94], 199[50], 201[51,52], 202[51, 52,94,135], 204[51,135], 205[50, 52], 206[50], 207[3,50,51,52, 94], 209[50,51,52], 210[51,52],

212[52,93,94], 213[3,93,120], 214[3,94,120], 215, 217, 219, 220
Rystedt, L., 147[48], 176

S

Sakaguchi, K., 248[159], 249[159], 254[159], 255[159], 264
Salazar, J., 255[181], 265
Sanderson, K. E., 244[128], 263
Sanford, K., 120[75], 134
Sawula, R., 240[100,101], 244[100], 245[100,101], 262
Scallen, T. J., 273[24,25], 274, 277,287[25], 303
Scanu, A., 288[43], 304
Scatchard, G., 23, 40
Schachman, H. K., 19[25], 40, 75[86,87], 84, 91, 94[19], 131, 132
Schachter, H., 171[103,110], 178
Schanbacher, F. L., 156, 158[85], 167[85], 177
Schellenberg, K. A., 111[54], 113, 133
Scheraga, H. A., 143[35], 176
Schevitz, R. W., 20[27], 40
Schlesinger, D. H., 129, 134
Schlesinger, M. J., 229[35], 260
Schlessinger, D., 238[82], 261
Schlessinger, J., 31[41], 38[55], 41
Schmid, C., 182[23], 216
Schmidt, D. V., 171[108], 178
Schmidt, E., 197[127], 220
Schmidt, F. W., 197[127], 220
Schött, E., 256[185], 265
Schroepfer, G. J., Jr., 270[17], 276[17], 303
Schulman, R. G., 13[22], 34, 40
Schultz, G. E., 229[31], 258[31], 260
Schumaker, V. N., 38[51], 41

Schuster, I., 19[58,59], 38[58, 59], 41
Schuster, M. W., 273[24], 277[24], 303
Schutt, H., 61[73], 66[73], 84
Schutzbach, J. S., 121[81], 134
Schwarz, A. K., 246[143], 255[143], 263
Schwarz, H. B., 167[94], 178
Scrutton, M. C., 189[112], 219
Seaton, J. D., 270[14,17], 271[14], 272[14], 276[17], 303
Segal, A., 44[13,21,22,23], 49[21], 50, 55[21], 58[22,23], 59[23], 60[23], 61[13,23], 62[23], 63[21,22], 64[22], 65[22], 66[21,22], 67[22], 68[22], 69[22], 70[22], 79[13], 80[13], 81, 82
Seidman, I., 209[138], 220
Semmett, W. F., 138, 174
Serutton, M. C., 6[10], 40
Seyama, Y., 99[25,27], 113[57], 114[57], 129[4'], 132, 133, 134
Shah, S. N., 273[27], 287[27], 296[64,65], 304, 305
Shaper, J. H., 149[63], 150[63], 151[63], 164[63], 177
Shapiro, B. M., 6[12], 40, 44[2,8, 11,12,15,19,25], 45[19,35], 46[32], 47[15,19,25,29,30,32, 35], 48[25,29,32], 49[19,30], 50, 52[2,8,11,12,15,25,48], 54[19], 56[29,30], 57[8,30], 58[2,8], 59[70], 60[2,12], 63[2,48], 66[15,29], 70[25], 71[8], 73[15,19,32,48], 74[29, 48], 75[48], 78[19,32], 81, 82, 83, 84
Shapiro, D. J., 299[73], 305
Sharma, V. S., 66[74], 84
Shaw, W. N., 182[33], 216
Shelton, E., 48[41], 49[41], 50[41], 55[41], 58[58], 83

Shibaev, V. N., 121[78], 134
Shifrin, S., 72, 73[84], 84
Shimizu, N., 236[63], 237[63], 261
Shimizu, Y., 236[63], 237[63], 261
Shrago, E., 183[58], 217
Siegel, M. I., 187[91], 188[91], 193[91], 218
Sikes, S., 232[47,48], 241[111], 242[111], 244[131], 257[111], 261, 262, 263
Sipirstein, M. D., 300[84], 306
Sklar, J., 258[196], 265
Skralant, H. B., 273[25], 274[25], 277[25], 287[25], 303
Skylar, J., 129[3'], 134
Slakey, L. L., 299[80], 305
Sly, W. S., 197[133], 220
Smiley, I. E., 20[27], 40
Smith, F., 109[51], 133
Smith, M. B., 197[129], 220
Smith, O. H., 226, 227[19,20], 260
Smyrniotis, P. Z., 49[43], 50[43], 52[43], 53[43], 54[43], 55[43], 56[43], 66[43], 83
Smyth, D. G., 107, 132
Snell, E. E., 6[9], 40
Sokawa, Y., 183[63], 217
Sokoloff, L., 696[69], 305
Sokolski, W., 188[99, 101], 190[99], 191[99], 193[99, 101], 194[99], 219
Somerville, R. L., 240, 262
Sommers, P. B., 143, 176
Spennetta, T., 183[58], 217
Spiro, M. J., 156[80], 177
Spiro, R. G., 156[80], 177
Spizizen, J., 245[137], 263
Spudich, J. A., 244[127], 263
Squires, C., 258[195], 265
Squires, C. L., 182[23], 216, 236[61,62], 237[61,62], 261
Srikantaiah, M. V., 273[25], 274[25], 277[25], 287[25], 303

Stadtman, E. R., 6[12], <u>40</u>, 44[2,7,8, 11,12,13,14,15,16,<u>17</u>,18,20,21, 22,23,24,25], 45[31,35,39], 46[32], 47[15,18,25,27,29,30, 31,32,35], 48[25,29,32,41], 49[18,21,30,31,41,43], 50[41, 43], 51[14], 52[2,8,11,12,15, 25,43,48,49], 53[43], 54[18, 43,49,50], 55[21,41,43], 56[17, 29,30,43], 57[8,17,30,57], 58[2,8,22,23,57,58,59,60,63], 59[23,24,59,60,66,71], 60[2,12, 14,23,24,60,71], 61[13,14,23, 24,71], 62[23,66], 63[2,21,22, 48], 64[22], 65[22], 66[15,21, 22,29,43,75], 67[22], 68[22], 69[22], 70[20,22,25], 71[8,20], 72[49,50], 73[15,27,31,32,48], 74[27,29], 75[27,48], 76[27], 77[27], 78[27,31,32], 79[13], 80[13,66,89], 81[66,89], <u>81</u>, <u>82</u> <u>83</u>, <u>84</u>, 182[30], <u>216</u>

Stallcup, W. B., 38[42,62], <u>41</u>

Stanisich, V., 244[132], <u>263</u>

Stark, G. R., 107, <u>132</u>, 202[136], 203, <u>220</u>

Stasiowski, S., 238[79], 239[79], 243[79], 257[79], <u>261</u>

Stathakos, D., 92[17], 99[31,32], 100[31,32], 102[31], 112[55], 113[55], 115[63], 116[63], 119[55], 120[63], 127[63], 130[8'], <u>132</u>, <u>133</u>, <u>134</u>

Steinberg, D., 270[12], <u>303</u>

Steiner, R. F., 174, <u>178</u>

Steinman, H. M., 142[34], 146[34], <u>176</u>

Steinrauf, L. K., 184[72], 186[72], <u>218</u>

Sternlicht, H., 229[37], <u>260</u>

Stetson, K., 240, <u>262</u>

Stiller, E., 288[43], <u>304</u>

Stjernholm, R., 186[82], 190[82], <u>218</u>

Stoll, E., 182[3,5], 183[61], 188[95,100], 190[61,100],

191[61,100], 192[61,100], 193[61,120], 201[5], 202[5], 206[5], 207[3], 212[95], 213[3,5,120], 214[3,5,120], <u>215</u>, <u>217</u>, <u>219</u>, <u>220</u>

Stone, S. E., 270[15], 273[15], <u>303</u>

Stoolmiller, A. C., 167[93], <u>178</u>

Stouthamer, A. H., 241[115], <u>263</u>

Strosberg, A. D., 147[50], <u>176</u>

Stumpf, P. K., 182[23], 188[97, 98], 214[97,98,145,146], 215[98,146], <u>216</u>, <u>220</u>, <u>221</u>

Sturtevant, J. M., 51[46], <u>83</u>

Subbarayan, C., 299[80], <u>305</u>

Sugiyama, T., 299[71,72], <u>305</u>

Sumper, M., 214[142], <u>220</u>

Suskind, S. R., 255[177], 256[177, 183], <u>264</u>, <u>265</u>

Suyama, Y., 255[179], <u>265</u>

Swaisgood, H., 33[45], <u>41</u>

Swan, I. D. A., 143[39], <u>176</u>

Swyrd, E. A., 299[79], <u>305</u>

Szilard, L., 3[1], <u>39</u>

T

Tabocik, C., 296[63], <u>305</u>

Tai, H., 296[62], <u>305</u>

Tamaki, H., 238[88], <u>262</u>

Tan, L. Y., 174[124], <u>179</u>

Tanahashi, N., 138[6], 140[6], 141[6,22], 147[22], 149[65], 154[65], 170[6], <u>175</u>, <u>177</u>

Tani, S., 238[84,85], <u>262</u>

Tatsuno, T., 105[35,36], <u>132</u>

Tatum, E. L., 247, <u>264</u>

Teebor, G., 209[138], <u>220</u>

Tempest, D. W., 44[6], <u>81</u>

Tenenbaum, H., 33[44], <u>41</u>

Thomou, H., 99[32], 100[32], <u>132</u>

Thorpe, D., 230[41], <u>260</u>

Tietz, A., 182[8,9,29,31], <u>215</u>, <u>216</u>

Timmons, R. B., 49[45], 52[49], 54[49,50], 66[76], 72[49,50], <u>83</u>, <u>84</u>

Titchener, E. B. , 182[7,10,12, 13,14,15], 183[12], 215, 216
Tochikura, T. , 121, 134
Toth, J. , 288[43], 304
Traub, W. , 184[71], 186[74], 218
Trayer, I. P. , 139[16], 142[27], 147[27], 148 [16,61] , 149[61], 150[61], 156[16], 158[61], 164[61], 170[16], 175, 177
Trockman, R. W. , 270[14,17], 271[14], 272[14], 276[17], 303
Troedsson, H. , 120[75], 134
Tronick, S. R. , 57, 58[57], 83
Trotter, J. , 184[73], 218
Trzeciak, W. H. , 301[90], 306
Tsai, C. M. , 120, 121, 134
Tsai, H. , 256[186], 265
Turneer, M. , 147[44], 176
Twarog, R. , 245, 247, 263, 264

U

Ullrey, D. , 108[46], 109[46], 111[46], 112[55], 113[55], 119[46,55], 127[46], 133
Umbarger, H. E. , 3[2], 39
Ungar, F. , 290[49], 296[49,50], 301[91], 304, 306
Utter, M. F. , 6[10], 40, 189[112], 219

V

Vagelos, P. R. , 182[22,39], 183 [22,39,42,45,60], 188[105,106], 190[106,114,115,117], 191[105, 106,114,115], 192[106], 193[105, 106,114], 204, 216, 217, 219
Valentine, R. C. , 6[10,12], 40, 46, 47[32], 48, 73[32], 78[32], 82
Vanaman, T. C. , 138[10], 139[16], 141[10], 142[26,27,28,30,31], 146[28], 147[27,30,31], 148[16], 156[10,16], 158[26], 163[26], 168[31], 170[16], 175

van Baalen, C. , 248[160], 249[160], 264
Vogel, H. J. , 248, 264
Voight, B. , 19[57], 38[57], 41

W

Wagner, M. , 189[111], 219
Wagner, R. R. , 171[104], 178
Waite, M. , 183[43,46], 186[78, 89], 190[78], 216, 217, 218
Wakil, S. J. , 182[8,10,11,12,13, 14,15,17,18,19,26,32], 183[11, 12,14,19,46], 186[78,89], 190[78], 215, 216, 217, 218
Walker, L. , 170[101], 178
Walsh, K. A. , 148[55], 176
Walton, G. M. , 301[89], 306
Wang, S. F. , 90[9], 124[9,89], 126[89], 127[9], 131, 134
Ward, D. N. , [37], 132
Ward, L. , 117[66], 133
Warms, J. V. B. , 182[33], 216
Warner, R. C. , 183[50,51,52], 186[51], 188[51,52], 193[50, 51], 194[50,51,52], 195[50], 196[50,51,52], 197[50,51,52], 199[50], 201[51,52], 202[51, 52], 204[51], 205[50,52], 206[50], 207[50,51,52], 209[50,51,52], 210[51,52], 212[52], 217
Watanabe, T. , 241[116], 263
Watkins, W. M. , 138, 139, 175
Watson, H. C. , 38[56], 41
Watt, G. D. , 51[46], 83
Weber, A. , 238[88], 262
Weber, B. H. , 47[33], 82
Wedler, F. , 47[33], 82
Wee, T. G. , 117, 118, 130[7'], 133, 134
Wegman, J. , 247[148], 264
Weissman, S. , 129[3'], 134, 258[196], 265
Wenger, B. , 186[80], 190[80], 193[80], 218
Westhoff, G. W. , 237[72], 261

Wheelock, J. V., 173[112], 178
White, E. A.. 225[14], 235[14], 259
Whitfield, H. J., Jr., 230[45], 260
Whitt, D. D., 245[140], 263
Widholm, J. M., 249[166], 264
Wiegand, U., 186[80], 190[80], 193[80], 218
Wiesmeyer, H., 106[40], 108[40], 109[40], 132
Wilcox, P. E., 142[24], 175
Wilson, A. C., 147[54], 176
Wilson, A. N., 184[67,68,69], 217, 218
Wilson, D. A., 224[8], 229[8], 259
Wilson, D. B., 90[8], 105[8], 106, 107[8], 108[41], 109[41], 119[8, 41], 127[41], 129[2'], 131, 132, 134
Wilson, D. M., 229[37], 260
Wilson, J. D., 296[67], 305
Winter, W. P., 148[55], 176
Winzor, O. J., 31[38], 41
Wiskocil, R., 224, 225[4], 226[4], 234[4], 255[4], 259
Witholt, B., 122[85], 134
Witiak, D. T., 281[33], 304
Wohlheiter, J. A., 241[113], 262
Wohlhueter, R. M., 58[64], 61[73], 66[73], 84
Wolf, D. E., 184[65,66,67,69,70], 217, 218
Wolf, D. H., 45[38], 59[38,67], 60[67], 82, 84
Wood, H. G., 186[79,81], 190[79, 81], 193[81], 218
Woodwin, C. D., 299[77], 305
Woolfolk, C. A., 44[11,15,16,17], 45[31], 47[15,31], 49[31], 52[11, 15], 56[17], 57[17], 66[15], 73[15,31], 78[31], 81, 82
Wrigley, N. G., 6[10], 40
Wu, C.-W., 55[53], 56[53], 83

Wulff, K., 44[1,10], 52[10,47], 60[1], 63, 81, 83
Wütrich, K., 13[22], 34[22], 40
Wyman, J., 4[5], 5, 13[5], 14, 19[5], 28, 30, 34[5], 39, 40, 67[78,79,82], 70, 71, 84

Y

Yamagishi, M., 183[57], 190[113], 193[57,113], 194[113], 195[113], 196[113], 197[113], 201[113], 207[113], 209[113], 212[113], 214[113], 217, 219
Yamamoto, S., 296[61], 305
Yamane, T., 13[22], 34[22], 40
Yamauchi, K., 153[77], 177
Yang, H. L., 236[61,62], 237[61,62], 261
Yaniv, M., 237[68], 261
Yanofsky, C., 107, 132, 224[1,5,6, 7], 225[1], 226[18], 227[20, 26], 229[6,7], 230[1,38,39, 40,41], 231[38,39,40,42,43,44], 232[1,43], 233[1], 234[1], 235[59], 236[59,60,61,62], 237[61,62,66,68,69,70,71], 238[75,76,79,80,81,87], 239[79,92,93], 243[79,119, 126], 244[127], 251[7,169,170, 171], 252[169,170,172], 253[7, 174], 255[1,6], 257[79], 258[195,198], 259[199], [91], 259, 260, 261, 262, 263, 264, 265
Yarmolinsky, M. B., 106[40], 109[40], 132
Yasunobu, K. T., 142[24], 175
Yates, R. A., 3[3], 39
Yeh, J., 48[40], 52[48], 58[58], 63[40,48], 64[40], 66[40], 67[40], 73[48], 74[40,48], 75[48], 83
York, S. S., 234, 235[54], 261

Yoshida, A., 90, 128[10], 131
Yot, P., 129[3'], 134, 258[196], 265
Yourno, J., 257[191], 265
Yphantis, D. A., 94, 132
Yu, P. H., 256[186], 265
Yura, T., 240[97], 262

Z

Zalkin, H., 243[120,123], 257[120], 263

Zamenhof, S., 246[144], 255[144], 263
Zarkowsky, H., 127[88], 134
Zilversmit, D. B., 294, 305
Zubay, G., 236[61,62], 237[61, 62], 261
Zwergel, E., 183[62], 188[100, 104], 190[100], 191[62,100], 192[100,104], 193[104], 202[135], 204[135], 217, 219, 220

SUBJECT INDEX

A

Acetyl Coenzyme A carboxylase,
  181-215
  animals from, 193-214
    citrate activation, 212-214
    filamentous forms, 193-199
    molecular properties, 196
    paracrystalline form, 199-201
    protomer-polymer equilibrium,
      204-212
    protomeric form, 201-204
  E. coli from, 189-193
    intersubunit translocation, 194
    protein components, 190-193
  reaction catalyzed, 182-183
    biotinyl group, 184-187
    partial reactions, 187-189
  yeast and plants from, 214-215
N-acetyllactosamine, 141
Adair equation, 9
Adenyltransferase, 45, 58-60
Anthranilate synthetase, 240, 243
ApoLP-Gln-I, 287-289
ApoLP-Gln-II, 287-289

B

Biotin, 186-189
Biotin carboxylase, 189-191

C

Carboxyl carrier protein, 190-192
Carboxyltransferase, 192-193
Cholesterol Synthesis, 269-283

Chromatography, affinity, 163-165,
  167
Citrate, 212-214
Cooperativity in proteins, 8-12
  Adair equation, 9
  allosteric dimer, 10
  Hill coefficient, 8
  models, 13-27
    comparison of, 26
    diagnostic tests for, 23-26
    equations of state, 27-28
    Koshland-Nemethy-Filmer,
      sequential, 19-23
    Monod-Wyman-Changeux,
      concerted, 14-19
  negative, 12, 37-38
  positive, 12
CTP synthetase, 36-37

F

Fatty acid synthetase, 182

G

β-(1-4) Galactosyltransferase,
  149-155
  catalytic properties, 155-158
    inhibitors, 157-158
    metals, 157
    substrate specificity, 155-157
  complexes, 163-165
  distribution, 149
  multiple forms, 150-155
    effect of proteases, 152-153

β-(1-4)Galactosyltransferase (cont'd)
  physical and chemical properties,
    149-150
  steady-state kinetics, 158-163
Galactosyltransferases, other,
    165-167
Gene duplication, 148
Glutamine Synthetase, E. coli,
    43-81
  activation by divalent cations, 48-52
    kinetics, 49-50
    thermodynamic parameters, 51
  binding effectors in, 70-72
  binding of effectors and substrates,
    54-57
    ADP, 56
    AMP, 55
    manganeous, 55
    tryptophan, 55
  determinants, antigenic, 57-58
  formation of hybrids, 75-78
    succinylated derivatives, 75
  γ-glutamyl transfer, 52-54
  interactions, subunit, 62-78
    biosynthetic catalysis in, 63-70
    effect of adenylation, 64-67
  metabolic control, 78-81
    cascade system, 81
    multiple binding sites, 80
  physical and chemical properties,
    45-58
    structure, 45-58
  regulation by adenylylation and
    deadenylylation, 58-62
  adenylation-deadenylylation,
    58-59
  adenyltransferase, 59-60
  uridylytransferase, 61-62
  stability in, 72-75
Glyceraldehyde-3-phosphate dehy-
    drogenase,
  rabbit, 37-38
  yeast, 38

Glyco-α-lactalbumin, 173
Glycoprotein biosynthesis, 142,
    171-173

H

Hemoglobin, 8, 30, 34-36
HMG-CoA reductase, 299-301

I

Inhibition, allosteric, 17-18, 22-23
Inhibition, reductive, 112-116,
    129-130
Interactions, subunit in proteins,
    1-39
  allosteric control, 3-4
  structure of multisubunit proteins,
    4-8
  CTP-synthetase, 36-37
  energetics of subunit interac-
    tions, 28-31
  half-of-sites reactivity, 31-34,
    38
  hemoglobin, 34-36
  rabbit muscle glyceraldehyde-
    phosphate dehydrogenase,
    37-38
  yeast glyceraldehyde-3-phos-
    phate dehydrogenase, 38-39

L

α-Lactalbumin, 138, 141-149,
    168-170
  chemical modification, 168-170
  comparison to lysozyme, 142-149
    amino acid sequence by, 142-143
    conformational studies by,
    143-147

α-Lactalbumin (cont'd)
  evolutional relationships,
    148-149
  immunological techniques by, 147
  distribution, 138
Lactose, 138, 171
Lactose synthetase, 137-174
  biological significance, 170-173
    lactose and glycoprotein bio-
      synthesis, 171-173
  biosynthesis of lactose, 138-139
  interactions between α-lactalbumin
    and galactosyltransferase,
      167-168
  requirements for two proteins,
    140-142
    α-lactalbumin, 141
    galactosyltransferase, 141-142
Lipid carrier protein, 281-282
Lysosyme, 142-149

                  N

NAD$^+$, enzyme-bound, 88-90, 99,
    106, 110
NAD$^+$-glycohydrolase, 100

                  P

P$_{II}$ proteins, 59-62
Proteins
  cation activation, 48-52
  complex, 158, 225-227
  conformation, 27-28, 225
  heterologous, 6-7
  hybrid, 75-78
  interactions, 1-39
  isologous association, 5-6
  isologous dimer, 6
  modifier, 163
  multiple forms, 150-154
  oligomeric, 5

paracrystalline, 199-201
polymerization, 183, 193-199
reactivity, half-of-sites, 31-34,
    38
regulator, 61
subunits, 4-8, 92-98, 119, 224-
    225, 255
symmetrical dimer, 106

                  R

Reactions, partial, 187-189

                  S

Squalene and Sterol Carrier Pro-
    tein, 267-302
  biological role, 268-270
  discovery, 270-274
  distribution, 296-297
  functions, 282-296
    bile acid synthesis, role in,
      291-292
    cholesterol synthesis, role in,
      286-287
    membrane formation, role in,
      292-294
    other proteins, relationships to,
      294-296
    plasma lipoproteins, relation-
      ship to, 287-290
    steroid synthesis, role in,
      290-291
  oligomer form, 279-282
  lipid binding, 280-282
  protomer form, 277-279
  purification, 274-277
  regulatory role, 297-301
Δ^7-Sterol-Δ^5 dehydrogenase,
    271-273, 292
  requirements for soluble frac-
    tion, 273

T

TDP-glucose, 108
TDPG-oxidoreductase, 124-126
Tryptophan Synthetase, 223-259
  amino acid sequence, 249-256
    α chains in, 249-253
    β chains in, 253-254
  chemical modification, 233-234
  enteric bacteria from, 227-234
  subunit reactions, 228-229
  enzymatic reactions, 225-226
  enzyme synthesis, regulation of,
    235-240
  evolution, 256-258
  evolutionary relationships, 241-258
    chromosomal organization, 241-249
      enteric bacterial in, 241-244
      other bacteria in, 244-249
  fungi and yeasts from, 255-256
  minor reactions, 226-227
  mutational modification, 230-233
  reaction mechanism, 234-235
  subunit structure, 224-225

U

UDP-D-xylo-4-hexosulose, 88
UDP-galactose-4-epimerase, 85-131
  E. coli enzyme, 105-120
    reaction mechanism, 108-112
      substrate specificity, 109
    reactions, other, 116-119
      autoxidation, 116-118
    reductive inhibition, 112-116
      galactose or arabinose with
        5'UMP by, 113-114

$NaBH_4$ by, 114
  nucleotides by, 114-116
  structural properties, 106-108
  subunits, 119-120
epimerase and sugar transforma-
    tion, 122-128
  enzyme-NAD^+ transfer reac-
    tions, 126-128
    common properties, 127
  function of enzyme-NADH,
    122-124
  TDPG-oxidoreductase, 124-126
properties, general, 87-90
reaction mechanism, 88-89
sources, other, 120-121
  mammalian, 120-121
  plants, 121
  yeast, torula, 121
  yeast enzyme (Candida), 90-105
    asymmetry of enzyme, 105
    heterogeneity, 91-92
    isoelectric point, 102-104
    NAD^+ exchange, 100-102
    peptide hydrogen exchange,
      102
    properties, general, 90-91
    prosthetic group, 99
    subunits, 92-98
      p-CMB or guanidine hydro-
        chloride by, 94-95
      dimers and tetramers, 95-96
      low ionic strength at, 97-98
Uridyltransferase, 61

X

Xylosyltransferase, 167